NETWORK CONTROL AND ENGINEERING FOR QOS, SECURITY AND MOBILITY, III

IFIP – The International Federation for Information Processing

IFIP was founded in 1960 under the auspices of UNESCO, following the First World Computer Congress held in Paris the previous year. An umbrella organization for societies working in information processing, IFIP's aim is two-fold: to support information processing within its member countries and to encourage technology transfer to developing nations. As its mission statement clearly states,

> *IFIP's mission is to be the leading, truly international, apolitical organization which encourages and assists in the development, exploitation and application of information technology for the benefit of all people.*

IFIP is a non-profitmaking organization, run almost solely by 2500 volunteers. It operates through a number of technical committees, which organize events and publications. IFIP's events range from an international congress to local seminars, but the most important are:

- The IFIP World Computer Congress, held every second year;
- Open conferences;
- Working conferences.

The flagship event is the IFIP World Computer Congress, at which both invited and contributed papers are presented. Contributed papers are rigorously refereed and the rejection rate is high.

As with the Congress, participation in the open conferences is open to all and papers may be invited or submitted. Again, submitted papers are stringently refereed.

The working conferences are structured differently. They are usually run by a working group and attendance is small and by invitation only. Their purpose is to create an atmosphere conducive to innovation and development. Refereeing is less rigorous and papers are subjected to extensive group discussion.

Publications arising from IFIP events vary. The papers presented at the IFIP World Computer Congress and at open conferences are published as conference proceedings, while the results of the working conferences are often published as collections of selected and edited papers.

Any national society whose primary activity is in information may apply to become a full member of IFIP, although full membership is restricted to one society per country. Full members are entitled to vote at the annual General Assembly, National societies preferring a less committed involvement may apply for associate or corresponding membership. Associate members enjoy the same benefits as full members, but without voting rights. Corresponding members are not represented in IFIP bodies. Affiliated membership is open to non-national societies, and individual and honorary membership schemes are also offered.

NETWORK CONTROL AND ENGINEERING FOR QOS, SECURITY AND MOBILITY, III

IFIP TC6 / WG6.2, 6.6, 6.7 and 6.8 Third International Conference on Network Control and Engineering for QoS, Security and Mobility, NetCon 2004 on November 2-5, 2004, Palma de Mallorca, Spain.

Edited by

Dominique Gaïti
Université Technique de Troyes, France.

Sebastià Galmés
Universitat de les Illes Balears, Spain.

Ramon Puigjaner
Universitat de les Illes Balears, Spain.

Dominique Gaïti
Université Technique de Troyes,
France.

Sebastià Galmés
Universitat de les Illes Balears,
Spain.

Ramon Puigjaner
Universitat de les Illes Balears,
Spain.

A C.I.P. Catalogue record for this book is available from the Library of Congress.

IFIP TC6/WG6.2, 6.6, 6.7, and 6.8 International Conference on Network Control and Engineering for QOS, Security, and Mobility (3rd : 2004 : Palma de Mallorca, Spain)
Network Control and Engineering for QOS, Security and Mobility, III / IFIP TC6/WG 6.2, 6.6, 6.7, and 6.8 Third International Conference on Network Control and Engineering for QOS, Security, and Mobility, NetCon 2004 on November 2-5, 2004, Palma de Mallorca, Spain; edited by Dominique Gaïti, Sebastià Galmés, Ramon Puigjaner.
p.cm. (The International Federation for Information Processing; 165)
Includes bibliographical references and index.
ISBN 1-4899-8624-3 Printed on alk. paper / 0-387-23198-6 (eBook)
1. Computer networks—Management—Congresses. 2. Computer networks--Quality control—Congresses. 3. Computer networks—Security measures—Congresses. 4. Mobile computing—Congresses. I. Gaïti, Dominique. II. Galmés, Sebastià. III. Ramon Puigjaner. IV. International Federation for Information Processing. V. Title. VI. International Federation for Information Processing (Series); 165.

TK51-5.5I342418 2004
004.6—dc22

2004058944

ISBN: 1-4899-8624-3 / (eB00K) 0-387-23198-6 Printed on acid-free paper.

Softcover re-print of the Hardcover 1st edition 2005

9 8 7 6 5 4 3 2 1 SPIN 11325277 (HC) / 11325574 (eBook)

springeronline.com

Contents

Preface

This volume contains the proceedings of the Third International Conference on Network Control and Engineering for Quality of Service, Security and Mobility (Net-Con'2004), celebrated in Palma de Mallorca (Illes Balears, Spain) during November 2-5, 2004. This IFIP TC6 Conference was organized by the Universitat de les Illes Balears and sponsored by the following Working Groups: WG6.2 (Network and Internetwork Architectures), WG6.6 (Management of Networks and Distributed Systems), WG6.7 (Smart Networks) and WG6.8 (Mobile and Wireless Communications).

The rapid evolution of the networking industry introduces new exciting challenges that need to be explored by the research community. The adoption of Internet as the global network infrastructure places the issue of quality of service among one of the hot topics nowadays: a huge diversity of applications with quite different service requirements must be supported over a basic core of protocols. Also, the open and uncontrolled nature of Internet enforces the need to guarantee secure transactions among users, thus placing security as another hot topic. Finally, the explosion of mobility and its integration as part of the global infrastructure are probably now the most challenging issues in the networking field.

According to these trends, the intention of the conference was to provide a forum for the exchange of ideas and findings in a wide range of areas related to network control and network engineering with a focus on quality of service, security and mobility control. The main program covered three days and included six sequential sessions and a poster session. Also, the program was enriched by a keynote speech and two invited talks offered by prestigious and world-renowned researchers in the networking field: Guy Pujolle from the University of Paris 6 (France), who imparted the keynote speech, Harry Perros from the North Carolina State University (USA) and Özgur B. Akan, from the Georgia Institute of Technology (USA). The main conference program was complemented by a variety of stimulating and high-quality tutorials.

Dominique GaÏti
Sebastià Galmés
Ramon Puigjaner

Committees

GENERAL CHAIR
R. Puigjaner (Universitat Illes Balears, Spain)

PROGRAM CO-CHAIRS
D. Gaïti (University of Troyes, FR)
S. Galmés (Universitat Illes Balears, ES)

STEERING COMMITTEE
A. Casaca (INESC, PT)
A.A. Lazar (Columbia University, US)
Al-Naamany (Sultan Qaboos University, OM)
O. Martikainen (Micsom, SF)
G. Pujolle (LIP6, FR)
J. Slavik (Testcom, CZ)
O. Spaniol (RWT Aachen, DE)

TUTORIAL CHAIR
J.-L. Ferrer (Universitat Illes Balears, ES)

FINANCIAL CHAIR
B. Serra (Universitat Illes Balears, ES)

PROGRAM COMMITTEE

A. Al-Naamany (Sultan Qabous University, OM)
F. Arve Aagesen (Norwegian University, NO)
G. Bianchi (Universita di Palermo, IT)
A. Benzekri (Université Paul Sabatier, FR)
C. Blondia (Univestiy of Antwerpen, BE)
R. Boutaba (University of Waterloo, CA)
A. Casaca (INESC, PT)
O. Cherkaoui (UQAM, CA)
W. Dabbous (INRIA, FR)
F. Davoli (Univesita di Genova, IT)
J. Domingo (Universitat Politècnica Catalunya, ES)
O. Duarte (Universidade Federal de Rio de Janeiro, BR)
A. El Sherbini (National Telecommunication Inst. EG)
J. Escobar (Centauritech, PA)
L. Fratta (Politecnico de Milano, IT)
G. Haring (Wien Univeristät, AT)
D-Y. Hu (Inst. of Network Technology, CN)
L. Huguet (Universitat Illes Balears, ES)
V. B. Iversen (Technical University of Denmark, DK)
F. Kamoun (Université La Manouba, TU)
U. Korner (Lund University, SE)
G. Leduc (Université de Liège, BE)
G. Omidyar (Institute for Communications Research, SG)
G. Pacifici (IBM - US)
H. Perros (North Carolina State University, US)
G. Pujolle (LIP6, FR)
F.J. Quiles (Universidad de Castilla La Mancha, ES)
T. Saito (Toyota, JP)
B. Serra (Universitat Illes Balears, ES)
J. Slavik (Testcom, CZ)
O. Spaniol (RWT Aachen, DE)
Y. Stavrakakis (Universtiy of Athens, GR
Y. Takahashi (Kyoto University, JP))
F. Tobagi (Stanford University, US)

ORGANISING COMMITTEE

L. Carrasco (Universitat Illes Balears, ES)
I. Furió (Universitat Illes Balears, ES)
M. Payeras (Universitat Illes Balears, ES)

Reviewers

A. Al-Naamany
F. A. Aagessen
G. Bianchi
V. Benetis
A. Benzekri
A. Bermúdez
R. Boutaba
S. Calomme
B. Caminero
A. Casaca
O. Cherkaoui
P. Cuenca
F. Davoli
R. Deca
O. Duarte
A. El Sherbini
J.-M. François
L. Fratta
M. Ghaderi
D. Gaïti
S. Galmés
M. Gambardella
A. Garrido
S. Hallé
G. Haring
B. Helvik
L. Huguet
V. B. Iversen
J. Janecek
F. Kamoun
K-T Ko
O. Kure
N. Laoutaris
G. Leduc
F. Lo Piccolo
C. Matteo
G. Omidyar
A. Panagakis
H. Perros
R. Puigjaner
G. Pujolle
F. J. Quiles
R. Reda
F. Ricciato
N. Rico
B. Serra
M. Sichitiu
B. Simak
F. Skivée
J. Slavik
O. Spaniol
Y. Stavrakakis

Y. Takahashi
D. Tianshu Li
L. Truong
L. Tzevelekas
A. Vaios
S. Waharte
J. Xiao
Y. Xin
L. Xu

Part One: Network Policy

CONFIGURATION MODEL FOR NETWORK MANAGEMENT

Rudy Deca[1], Omar Cherkaoui[2] and Daniel Puche[3]
[1,2]University of Quebec at Montreal; [3]Cisco Systems, Inc.

Abstract: As today's networks increase in size and complexity and new network services are deployed, the management becomes more complex and error-prone and the configurations can become inconsistent. To enforce the configuration consistence and integrity, it is necessary to enhance the validation capabilities of the management tools. The Meta-CLI Model presented in this paper captures the dependences among the configuration components and the network service properties and translates them into validation rules. It also translates the device configuration information into tree-like models and checks their integrity and consistence using theses rules.

Key words: network management; network services; integrity and consistence validation; configuration rules; configuration constraints; configuration model.

1. INTRODUCTION

The constant growth of the Internet implies the creation and deployment of an ever increasing number of network services, each of which is becoming more complex in its turn. In this context, the network configuration becomes more difficult and error prone. Some of the causes are the diversity of configuration approaches and information repositories used by the configuration tools.

In plus, the main network configuration approaches, such as the command line interfaces (CLIs),[1,2] the Simple Network Management Protocol (SNMP),[3] the policy-based management (PBM),[4] the NetConf protocol,[5] etc., lack configuration rules that capture the dependences and constraints that characterise the network service configuration and lack adequate logical formalisms that could be applied to the information repositories to validate the configuration integrity and consistence.

These approaches therefore do not take into account the dependences and hierarchy that exist among the device parameters that express the service at device-level, the heterogeneity of the configuration means and the interactions between heterogeneous management and configuration modes.

The Meta-CLI Model proposes a solution for these configuration inconsistencies. Our model captures the features of the CLI, such as the context and parameter dependences of the commands, as well as the service properties into validation rules. It translates the configuration information into trees and validates them using the appropriate rules. Based on the Meta-CLI Model, we have implemented the *ValidMaker* module, and incorporated it in a policy provisioning VPN tool.

This section presents the dependences between the commands and/or parameters that are translated into Meta-CLI concepts and discusses the properties of the network service configurations. Section 2 presents the concepts, operations, functions and properties of the Meta-CLI Model tree structures and the validation rules and procedures that translate and validate the service properties. Section 3 presents the *ValidMaker* tool and Section 4 draws conclusions.

1.1 Configuration Dependences

Several types of dependences exist in the network device configurations.

A. ***Syntactic parameter constraints*** The CLI commands' syntax enforces the constraints regarding the order and number of parameters.

1. *Fixed number of parameters* The number of parameters (which can be mandatory or optional) must be correct, otherwise the command fails.

EXAMPLE The command to configure a primary IP address on an interface requires 2 parameters: the interface *IP address* and the *mask.*

2. *Fixed sequential order of parameters.* In a configuration command or statement, the parameters are ordered.

EXAMPLE In an extended access list, the order of the *IP address* and the *mask* parameters in the above-mentioned command is fixed.

B. ***Parameter and command existence dependences*** Some parameters and commands can only exist in specific circumstances.

1. *The existence of a parameter depends on another*

EXAMPLE In an access list, the *timeout* parameter, which specifies the duration of a temporary access list entry, can exist only if the access list entry is *dynamic* (i.e. has the *dynamic* parameter).

2. *Context dependences* The existence of a parameter or a command depends on the context (mode). Thus, a parameter can only be configured, modified or created in an equipment using a command in a specific configuration context (mode).

EXAMPLE When specifying the *IP address* of an interface, the name of the interface must be specified by a prior command that "opens" it to the user, who can then access and manipulate its resources or parameters (e.g.: IP address, MTU, bandwidth, etc.).

3. ***Result dependence*** The order of some commands can be fixed. In this case, the success of a command depends on the successful completion of a previous one.

EXAMPLE The configuration of a link bundle consists of bundling several similar physical point-to-point links between 2 routers into 1 logical link. By default, at each change in bandwidth in a link bundle, the combined amount of bandwidth used on all active member links is propagated. Optional features are available, by configuring the following parameters.

I. *Automatic propagation*, which sends the bandwidth changes to upper layer protocols for the bandwidth threshold.

II. *Bandwidth threshold*, which sets the value of the threshold. If the actual bandwidth is smaller or equal, it is propagated, otherwise the nominal bandwidth is transmitted.

If we invert the configuration order of these 2 parameters, the threshold will not be set, since it requires the existence automatic propagation.

C. *Value dependences among parameters*

1. *Parameters of the same command are dependent*

EXAMPLE When specifying a classful *IP address*, the net mask must be consistent with the first two bits of the IP address. For instance, the B class IP address starts with the bits *10* and has the mask *255.255.0.0.*

2. *Parameters of different commands on the same device are dependent.*

EXAMPLE An access list is identified by its *number* parameter, which is referenced when the access list is attached to an interface. If we change this ID number in one place, we should change it likewise in the other, lest the functionality is lost. Parameters that reference each other may have the same or different names.

3. *Parameters on different devices are dependent*

EXAMPLE The connectivity between 2 devices requires the IP addresses of 2 interfaces directly connected to be on the same subnet.

D. ***Parameter Hierarchy and Service Composition*** In a configuration, there is a hierarchy of elements from simple to complex, namely from the parameters, going through several levels of aggregation, up to the network services. This grouping expresses the common goal and of the components and the dependences that exist among them.

1. *Grouping parameters under commands.* At the bottom, several parameters can be configured simultaneously using commands or statements, which group them based on logical links.

EXAMPLE An access list entry command specifies several packet parameters to be matched, such as: *source* and *destination addresses*, *direction* of the packet, *protocol*, *port*, whether it is *dynamic* or not, *timeout*, etc. Various dependences among some of these parameters have already been highlighted in the examples in the previous paragraphs § A and § B.1.

2. *Grouping commands under services.* The commands can be grouped as well, if they serve a common goal, i.e. a feature or a service. For instance, an access list is composed one or more access list entries, which are bound by the common access list number.
3. *Network service composition.* A network service can rely on simpler network services, according to a recursive, hierarchical model.

EXAMPLE. A Virtual Private Network (VPN) service requires: a network tunneling, e.g. through LSPs provided by an MPLS protocol, BGP connectivity (e.g. by means of neighbors), for the provider's backbone network, and IGP connectivity between the customer sites and the backbone, e.g. by means of RIP or OSPF protocols.

In complex services, such as BGP MPLS-based VPNs, VLANs, etc., there are multiple dependences at different hierarchical levels.

1.2 Dependences among network service components

Due to their complexity, the dependences can exist at different levels within network services, from parameters to sub-services.

1. *Parameter and command dependences* As already shown, the parameters and commands that compose the services can have their intrinsic, lower-level, dependences, which can be either independent or dependent on the device environment (software and hardware).

EXAMPLE The dependence C.1 is generic, whereas D.2 is command implementation-specific.

2. *Sub-service dependences.* At the top of the service hierarchical structure, there are higher-level dependences among the component sub-services. These dependences are dependent on the service- and network-level information, e.g. network topology, technology, protocols, etc., rather than on the devices, and span multiple equipments.

EXAMPLE If several customer sites use overlapping address spaces and are connected to more than one VPN, the corresponding PEs must ensure traffic isolation among the various VPNs by enforcing specific constraints on their forwarding tables.[6]

These properties are high-level and generic, and need to be transposed into concrete, lower-level properties, by adapting to concrete network and equipment environments, in order to be applicable to the configuration.

For instance, a VPN service can be materialized by a provider tunneling technology such as the Multi-protocol Label Switching (MPLS). The MPLS may run on an IP network and use the multi-protocol Border Gateway Protocol (MP-BGP) for VPN routing advertising among the edge routers and the MP-BGP may use the direct neighbors for configuration on the edge routers. We will explain these concepts and features in the following example.

1.2.1 MPLS VPN Service Example

A VPN is a private network constructed within the network of the service provider, which ensures the connectivity and privacy of customer's communications traffic.[7] For this purpose, the sites (which are contiguous parts of networks) of the customer network are connected to the provider's network through direct links between the interfaces of the devices that are at the edges of these networks, i.e. the Customer Edge device (CE) and the Provider Edge device (PE), as shown in Figure 1 (site 3 has been omitted from this representation, to save space).

The VPN service in the example is configured on the PE routers using the IOS[1] commands, without loss of solution's generality and validity since, as mentioned before, the generic service properties must be transposed into concrete properties by mapping the configuration components to specific CLIs, such as IOS, JUNOS,[2] etc. The provider specifies VPN routing information (by means of the VPN Routing and Forwarding tables – the *VRFs*) on the PE routers' interfaces that are connected to the corresponding CE routers of the customer's sites. (There are many implementations of the VPN service. Our example uses the BGP MPLS-based VPN.) Figure 2(a) presents some highlights from the configuration file of the *PE-1* router, which are explained in the following paragraphs. (A configuration file contains lists of commands, grouped by contexts, which customize the parameters of various configuration components, e.g. interfaces, protocols, security, etc. of network equipments, such as routers and switches.)

Commands 1-4 create and define the VRF *Blue*. This VRF is then assigned by command 8 to the appropriate interface that connects the PE router to the customer's network. The PE router must be directly connected to the CE router of the customer site and learn local VPN routes. Command 5 configures the interface to be used for this router's connectivity inside the provider network and command 6 assigns it an IP address.

Command 10 illustrates the configuration of the OSPF process in the VRF *Blue* and command 11 specifies the network directly connected to the router (and containing the interface IP address configured by command 9). Command 12 ensures that the OSPF process advertises the VPN routing information learned from across the provider's network by means of the BGP protocol, into the CE router.

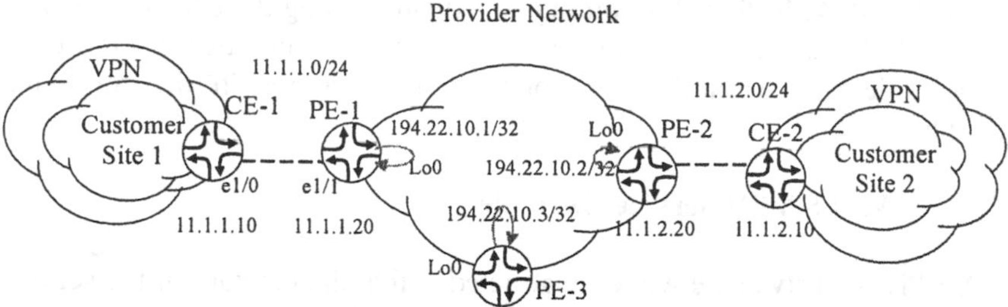

Figure 1. VPN example. Customer sites 1 and 2 are linked through CE routers to PE routers, which communicate over the service provider's network

The PE router exchanges customer and provider VPN routing information with other PE routers from the provider's network using the MP-BGP. Command 13 configures MP-BGP routing process by specifying the local autonomous system and commands 14-19 add the routing information necessary for the connection to other PE routers across the provider's network, i.e. autonomous system, interface type, number and IP address used by the remote PEs. Notice that we use the simplest method to configure the BGP process between PEs, namely the *direct neighbor configuration*.

The BGP needs also its own address family, *vpnv4*, which allows it to carry VPN-IPv4 in order to uniquely identify the addresses of different customers (commands 23, 25) and to advertise the extended community attribute (commands 24, 26).

1.2.2 VPN Configuration Properties

In this example, we can describe many properties, but we will restrict ourselves here to only two properties that the configuration must have with respect to the *neighbor* command and its relationships with other commands (from the same PE router and from other PE routers).

PROPERTY 1 The *address* of the *interface* of a PE router that is used by the BGP process for *PE connectivity* must be defined as BGP process *neighbor* in all of the other PE routers of the provider.

EXAMPLE In Figure 2(a), a *neighbor* with the address *194.22.10.2* is configured by commands 14-16. This corresponds to the IP address of the Loopback0 interface of router PE-2. Conversely, the *PE-1* router's *Loopback0* IP address, i.e. *194.22.10.1*, must be defined as a *neighbor* of the BGP processes of the other PE routers (PE-2, PE-3, etc.).

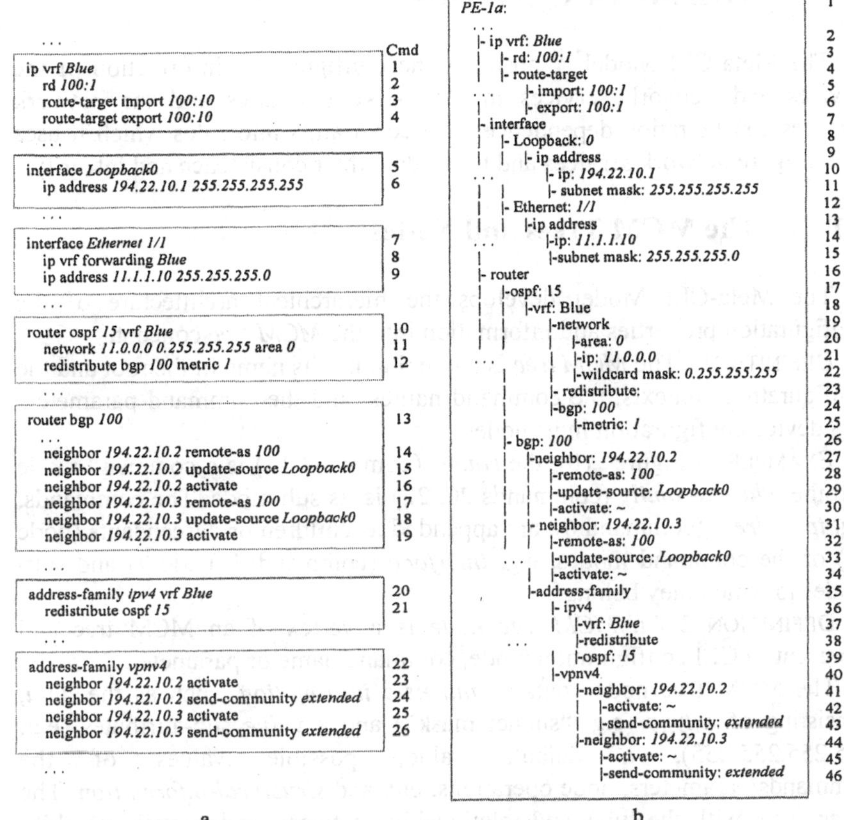

...
Cmd
ip vrf Blue 1
rd 100:1 2
route-target import 100:10 3
route-target export 100:10 4
...
interface Loopback0 5
ip address 194.22.10.1 255.255.255.255 6
...
interface Ethernet 1/1 7
ip vrf forwarding Blue 8
ip address 11.1.1.10 255.255.255.0 9
...
router ospf 15 vrf Blue 10
network 11.0.0.0 0.255.255.255 area 0 11
redistribute bgp 100 metric 1 12
...
router bgp 100 13
...
neighbor 194.22.10.2 remote-as 100 14
neighbor 194.22.10.2 update-source Loopback0 15
neighbor 194.22.10.2 activate 16
neighbor 194.22.10.3 remote-as 100 17
neighbor 194.22.10.3 update-source Loopback0 18
neighbor 194.22.10.3 activate 19
...
address-family ipv4 vrf Blue 20
redistribute ospf 15 21
...
address-family vpnv4 22
neighbor 194.22.10.2 activate 23
neighbor 194.22.10.2 send-community extended 24
neighbor 194.22.10.3 activate 25
neighbor 194.22.10.3 send-community extended 26
...
a
PE-1a: 1
|- ip vrf: Blue 2
|- rd: 100:1 3
|- route-target 4
|- import: 100:1 5
|- export: 100:1 6
|- interface 7
|- Loopback: 0 8
|- ip address 9
|- ip: 194.22.10.1 10
|- subnet mask: 255.255.255.255 11
|- Ethernet: 1/1 12
|-ip address 13
|-ip: 11.1.1.10 14
|-subnet mask: 255.255.255.0 15
|- router 16
|-ospf: 15 17
|-vrf: Blue 18
|-network 19
|-area: 0 20
|-ip: 11.0.0.0 21
|-wildcard mask: 0.255.255.255 22
|- redistribute: 23
|-bgp: 100 24
|-metric: 1 25
|- bgp: 100 26
|-neighbor: 194.22.10.2 27
|-remote-as: 100 28
|-update-source: Loopback0 29
|-activate: ~ 30
|- neighbor: 194.22.10.3 31
|-remote-as: 100 32
|-update-source: Loopback0 33
|-activate: ~ 34
|-address-family 35
|- ipv4 36
|-vrf: Blue 37
|-redistribute 38
|-ospf: 15 39
|-vpnv4 40
|-neighbor: 194.22.10.2 41
|-activate: ~ 42
|-send-community: extended 43
|-neighbor: 194.22.10.3 44
|-activate: ~ 45
|-send-community: extended 46
b

Figure 2. Selected commands that configure a VPN service in the configuration file of the provider edge router *PE-1* (a), and the corresponding MCM tree *PE-1a* that models them (b).

PROPERTY 2 The address family *vpnv4* must *activate* and *configure* all of the BGP *neighbors* for carrying only VPN IPv4 prefixes and advertising the extended community attribute.

EXAMPLE In Figure 2(a), the commands 23, 24 activate and advertise the extended community attribute of the *neighbor 194.22.10.2*, configured by the commands 14-16 under the BGP process. We have here an instance of a command whose various parameters are configured in two different contexts.

2. THE META-CLI MODEL

The Meta-CLI Model[8–10] abstracts the configuration information of the devices and network services into tree-like structures and the network services configuration dependences and constraints into rules which it uses to configure network services and to validate their consistence and integrity.

2.1 The MCM Trees and Nodes

The Meta-CLI Model develops the hierarchical architecture of the configuration properties and information into the *MCM tree* concept.

DEFINITION 1 The *MCM tree* is a tree that has its name in the root and the configuration contexts, the command names, and the command parameters of a device configuration in its nodes.

EXAMPLE In Figure 2(b), the *router* (command 13) is a command mode and the *address-family* (commands 20, 22) is its sub-mode. The commands, e.g. *ip address* (command 6), are appended as children or descendants (node 10) of the command modes, e.g. *interface* (command 5, node 7) and sub-modes to which they belong.

DEFINITION 2 An *MCM tree node* is a vertex of an MCM tree and represents a CLI configuration mode, command name or parameter.

The MCM tree node contains *intrinsic information,* such as the *data*, consisting of a *type* (e.g. "subnet mask") and a *value* sub-attributes (e.g. 255.255.255.255), a default value, possible values of the commands/parameters, node operations, etc. and *structural information*. The latter deals with the links and relationships between nodes, such as: child nodes and the *path*, which consists of the *data* of the ancestor nodes starting from the root.

DEFINITION 3 A *node type* represents a class or category of configuration modes, commands or parameters. The type of a node *N* is denoted by *N.type.*

DEFINITION 4 A *node value* represents the value of the command or parameter modeled by the node. The value of a node *N* is denoted by *N.value.* The type of a MCM tree node is unchangeable whereas the value may be changed.

DEFINITION 5 The *node data* represent the *type* and *value* of the node. The data of a node *N* are denoted by *N.data.*

EXAMPLE The type, value and data of node 41 in Figure 2(b) are: *N.type = neighbor, N.value = 194.22.10.2* and *N.data = neighbor:194.22.10.2,* respectively.

2.2 The Validation Rules

The Meta-CLI Model translates the CLI properties and combines them with intrinsic tree properties into validation rules.

DEFINITION 6 A *validation rule* is a condition (constraint) or combination of conditions that one or several MCM trees (and their components) must satisfy.

The validation rules abstract the CLI properties of the commands and configurations.

A few rule examples follow. A first group of simple rules indicate that a node must have a specific attribute.

- A node *N* has a given value *V*, i.e.: N.value = V (1a)
- A node *N* has a given data *D*, i.e.: N.data = D (1b)

EXAMPLE (1a) is *false*, if *N* is node 27 in Figure 2(b) and *V* = *194.22.10.1*.

DEFINITION 7 A *node reference* is the purposeful equality of the value or data of two nodes. It represents the conceptual and functional *identity* of some information located in two (or more) nodes.

We may thus have the value reference or data reference:

- A node *N* references the value of node *P*, i.e.: N.value = P.value (2a)
- A node *N* references the type of node *P*, i.e.: N.data = P.data (2b)

The following group of such rules checks whether a sub-tree or a tree *S* contains a node *N* that has a given type *T*, value *V* or data *D*, respectively.

$\exists$N $\in$ S, N.type = T (3a)

$\exists$N $\in$ S, N.value = V (3b)

$\exists$N $\in$ S, N.data = D (3c)

EXAMPLE (3a) is *true,* for *S* being the sub-tree of node 8 and *T* = *ip*.

Other simple rules are the following:

- A node has an ancestor with a given type, value or data. (4)
- Two (sub)trees S_1 and S_2 contain (at least) a node each, N_1 and N_2, respectively, having the same given value *V*, i.e.:
  $\exists$N1, N2 , N1 $\in$ S1, N2 $\in$ S2, N1.value = N2.value (5a)
- Two (sub)trees S_1 and S_2 contain (at least) a node each, N_1 and N_2, respectively, having the same given data *D*, i.e.:
  $\exists$N1 N2 , N1 $\in$ S1, N2 $\in$ S2, N1.data = N2.data (5b)

There are also more complex rules. For instance, the following rule corresponds to the configuration properties 1 and 2, stated in section 2.

- Let *D* be some node data, *T* a node type and $\{T_i \mid i = 1,\ldots, n, (n > 1)\}$, a set of MCM (sub)trees. If a tree T_i has a node N_D of the given data *D*, and N_D has a descendant N_T of the given type *T*, then all of the other trees T_j, contain two nodes N_x and N_y of identical data such that their values are equal to the value of N_T. (6)

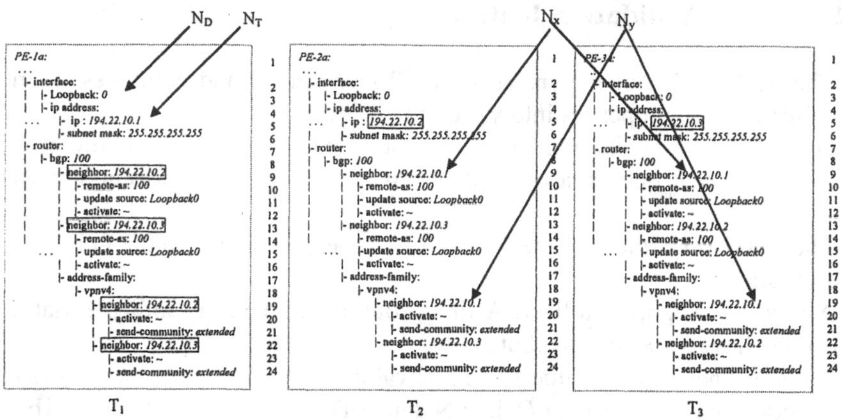

Figure 3. Three MCM trees, *PE-1a*, *PE-2a* and *PE-3a*. The arrows indicate the nodes of the first instance of rule (6).

We may apply this rule to the MCM trees *PE-1a, PE-2a* and *PE-3a*, shown in Figure 3, for *D* = *Loopback:0* and *T* = *ip*. The arrows depict the rule instance that has *i* = *1* and *j* = *2* and *3*. We see in the tree *PE-1a* that node 5 (corresponding to N_T), of type *ip* is a descendent of command 3 (corresponding to N_D) having the data *Loopback:0* and that the nodes 9 (corresponding to N_x) and 19 (corresponding to N_y), of *PE-2a* and *PE-3a*, respectively, reference the node 5 of *PE-1a*.

A validation algorithm compares the set of value of nodes 5 of *PE-2a* and *PE-3a* with the sets of values of the nodes 9, 13 of *PE-1a*. It also compares the set of data of nodes 9, 13 with the set of data of the nodes 19, 22. This is shown in Figure 3 by the boxes surrounding the respective leaves. Since these set equalities are true, the MCM trees are valid with respect to the first instance of the rule.

3. META-CLI *VALIDMAKER* MODULE FOR CONFIGURATION MANAGEMENT

The Meta-CLI *ValidMaker* module implements the Meta-CLI Model solution for the validation of the network configurations. It provides consistence and synchronization of the configurations achieved with network configuration management tools.

The configurations of the network equipments such as the routers, switches, etc., are provided by configuration management tools, based on the centralized information stored in a network management information base.[11] The configurations can be also changed in a CLI-based mode. The interactions between these heterogeneous modes of operation are managed by the *ValidMaker* module, based on the configuration information from the network equipments, the configuration tools and the network information model.The role of this module is to validate the device configurations in a dynamic and heterogeneous context. Since the changes were neither predicted by, nor included in, the configuration solution provided by the tools that enabled the deployment of the services, the tools cannot intervene in a coordinated fashion. The *ValidMaker* module intervenes to validate, maintain, restore and control the consistence of the network. It can be embedded in each service provisioning tool and ensure thus an error-free, consistent service provisioning.

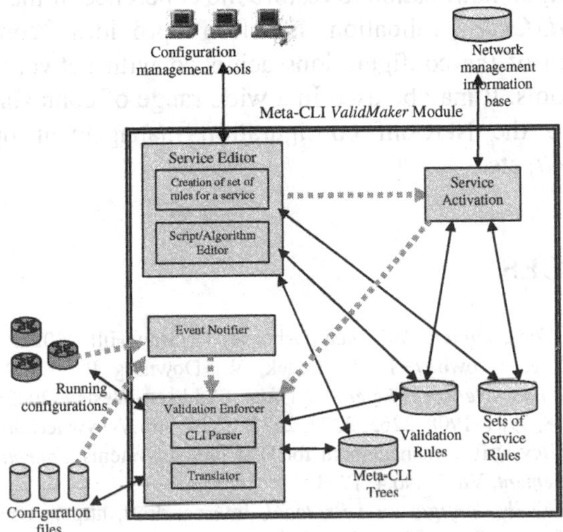

Figure 4. Conceptual View of the Meta-CLI ValidMaker.

The components of the *ValidMaker* module are presented in Figure 4. The *Service Editor* allows the service creator to generate a set of rules, algorithms and scripts, for each service, the *Service Activation* allows the network administrator to activate network services for the specific network equipments, the *Event Notification* monitors the status of the network and the *Validation Enforcer* enforces the rules on the equipment configurations.

4. CONCLUSIONS

The Meta-CLI Model offers multiple possibilities of utilization in various contexts of the configuration management process: service configuration and consistence validation, control of the consistence of data stored in management information bases, network device testing, etc.

This paper presents the application of the Meta-CLI Model to the validation of network service configuration consistence and integrity. The *ValidMaker* module creates validation rules, algorithms and strategies for all of the intermediate steps of the configuration process performed by the service provisioning tools. The validation rules are grouped in sets representing snapshots of the service configuration process and are therefore assigned to validate each configuration step performed by the provisioning tools. When the configuration is complete, the groups of validation rules, algorithms and scripts that are associated with the service remain on standby and intervene upon notification to restore the coherence of the system.

The *ValidMaker* validation solution provides consistence and synchronization of the configurations achieved with network configuration management tools. It may be used in a wide range of contexts, e.g. it can be embedded into the NetConf configuration management operations like *validate*, *commit*, etc.

REFERENCES

1. B. Hill, *Cisco: The Complete Reference* (Osborne, McGraw Hill, 2002).
2. T. Thomas II, R. Chowbay, D. Pavlichek, W. Downing III, L. Dwyer III, and J. Sonderegger, *Juniper Networks Reference Guide* (Addison-Wesley, 2002).
3. RFC 1155–1158, 1901–1908, 2263, 2273, 2573, 2574, http://www.ietf.org/rfc/rfcxxxx.txt.
4. M. Sloman, Policy Driven management for Distributed Systems, *Journal of Network and Systems Management,* Vol.2, No.4, 1994.
5. R. Enns, *NETCONF Configuration Protocol,* Internet-draft, http://www.ietf.org/internet-drafts/draft-ietf-netconf-prot-03.txt, June 2004.
6. R. Bush and T. G. Griffin, Integrity for Virtual Private Routed Networks *22nd Annual Joint Conf. of IEEE Computer & Comm. Society* (INFOCOM 2003), San Francisco, USA, 2003.
7. I. Pepelnjak and J. Guichard, *MPLS and VPN Architectures* (Cisco Press, 2001).
8. O. Cherkaoui and R. Deca, *Method and Apparatus for Using a Sequence of Commands Respecting at Least One Rule*, US Patent Application N° 60/458,364, March, 2003.
9. R. Deca, O. Cherkaoui, and D. Puche, A Validation Solution for Network Configuration, *Proceedings of the 2nd Annual Communications Networks and Services Research Conference* (CNSR 2004), Fredericton, NB, Canada, May, 2004.
10. S. Hallé, R. Deca, O. Cherkaoui, and R. Villemaire, Automated Verification of Service Configuration on Network Devices, Article accepted at the *Conference on Management of Multimedia Networks and Services,* (MMNS 2004), San Diego, USA, Oct. 2004.
11. O. Cherkaoui and M. Boukadoum, Managing Network Based VPN with Policies, Paper submitted to *The Telecommunications Journal*, 2003.

ON-LINE CONTROL OF SERVICE LEVEL AGREEMENTS

Manoel Camillo Penna and Rafael Romualdo Wandresen
Pontifical Catholic University of Parana, PPGIA, Rua Imaculada Conceicao, 1155 80215-901 Curitiba PR Brazil {penna, wandresen}@ppgia.pucpr.br

Abstract: Service Level Agreement (SLA) is used as the corner stone for building service quality management (SQM) systems. SLA and the processes associated with them, establish a two-way accountability for service, which is negotiated and mutually agreed upon, by customer and service provider. It defines a set of service level indicators and their corresponding Service Level Objectives (SLO), which defines a threshold for the indicator value. Service quality assessment can be accomplished in two modes, on-line and offline. Off-line service level evaluation is performed only at the end of the period of service delivery, whereas on-line service evaluation supports continuous supervision of service quality. This paper presents a method for on-line control of SLA that evaluates indicator value each time an event that changes its value occurs. The method also computes the deadline to reach the corresponding SLO, what is important for pro-active control.

Key words: Service Quality Management, Service Level Agreement

1. INTRODUCTION

In recent years there were many research efforts on service quality management (SQM). Many authors explored service management under infrastructure viewpoint, for example, quality of service in active networks[1,2] and IP networks[3,4,5,6]. Many experiments were also target to the construction of generic SQM platforms[7,8,9]. From those, we borrow some of the architectural concepts presented in section 2 where we introduce a three layer functional architecture. This architecture is presented in order to provide a framework for conceptual understanding of involved concepts. Particularly we simplify the management layer and just consider inference and control functions.

Inference and control functions can be implemented in two modes, on-line and off-line (see Figure 1). In on-line mode, the indicators are evaluated in the very moment an event that changes an indicator value arrives to the system. In off-line mode, service level indicators are evaluated at pre-defined periods, for example, daily, weekly or monthly. Both modes are necessary: on-line mode takes only into account data available on arriving of incoming events, in contrast with off-line mode, which takes into account all information collect during the assessment period. The last can consider data that is not available when events arrive to SQM system, for example events related to faults that can not be imputed to service provider, allowing more precision on indicator calculus. See Lewis[10] and Wandresen[11] for a more detailed discussion on two modes.

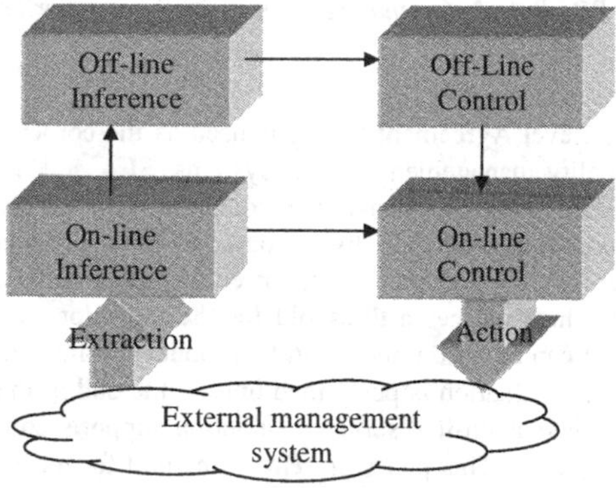

Figure 1. Inference and control functions for service quality management

The main concern of this work is with the questions related to on-line inference and control functions in SQM systems. In section 2 we present a functional architecture for SQM systems. The goal is to provide a conceptual basis for sections 3, where we discuss a method and depicts an algorithm used for on-line inference and control. Finally in section 4 we present some conclusions and future work.

2. SQM FUNCTIONAL ARCHITECTURE

In this section we present a functional architecture for SQM systems, which is organized on three logical layers: data collection layer, inference and control layer and presentation layer. At the first layer, extraction and mediation functions interact with external management systems to obtain data for SQM. Typically, extraction function gets data to calculate indicators and mediation function gets information to populate SQM database.

Extraction functions get external data and interact with SQM database to obtain the necessary information to construct service fault events (SF events) as described in section 3. Inference algorithms collect service fault events, calculate new indicators' values and deadlines, and deliver them by means of service quality events (SQ event), which are used at presentation layer to construct service level reports and supervision panels. The whole architecture is illustrated on Figure 2.

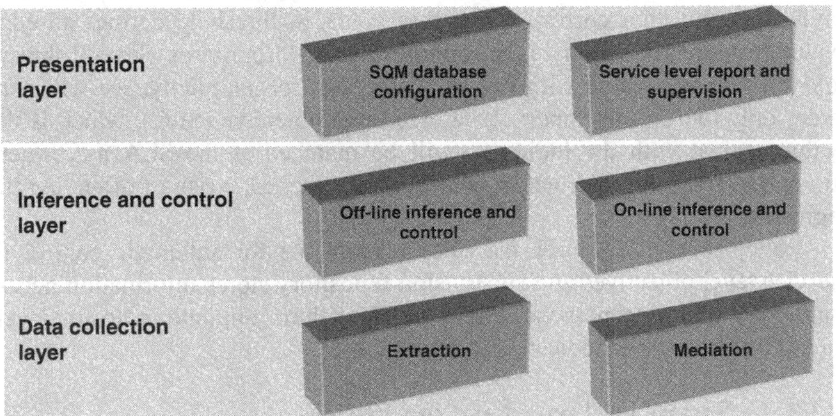

Figure 2. SQM functional architecture

2.1 Presentation layer

Presentation layer includes graphical user interface for SQM database configuration and result presentation functions. SQM database configuration includes service, customer and SLA registration, whereas service level reports and supervision panels present results.

2.1.1 SQM database configuration

SQM system collects and handles a large set of information to achieve its goal, and organize them in five repositories: service inventory, contract repository, policy repository, supervision repository and indicator repository. Service inventory contains the service list and respective attributes. It must be flexible allowing modeling of service information according to business needs. Service inventory also stores references to components of the infrastructure that supports the service (service elements). This is necessary when algorithms that compute indicator values rely on information originated at service elements, as their availability or other technical parameters.

Contract repository relates service instances to customers. In many cases this information already exists in other corporate information systems and should be obtained from them by integration. Mediation function is the architectural component for integration. Contract repository also stores SLA information, which relates service instances to a set of indicators and thresholds. SLA provides the basis for SQM by establishing a two-way accountability for service, which is negotiated and mutually agreed upon by customer and service provider. An SLA is determined by a set of service level indicators and their corresponding thresholds. A threshold defines an edge value for the indicator that is meaningful for SQM purposes. Several thresholds can be associated with one indicator, but for simplicity, we will consider only two in this paper: (i) service level objective (SLO), which is the value against with the indicator, will be matched at the SLA assessment time; (ii) alert threshold, which points, when reached, a risk of offending the agreement.

Policy repository stores the rules responsible for automatic control of service level objectives and supervision repository stores information necessary for monitoring purpose. Inference algorithms compute indicators' values and store them at indicator repository.

2.1.2 Service level report and supervision

Effective service level reporting is the medium of communication that demonstrates the value of service and can serve as an excellent management tool. Reporting can be broadly divided into tow categories: service level reporting and service supervision.

A service level report presents the performance obtained during service deployment within a pre-defined period. It presents, in a structured way, the measured indicators' values and compares them with the quality thresholds established in the agreements. They show the values stored in the indicator repository by off-line inference functions. When the service quality goals are not attained, the service level report includes the failure causes and shows the corrective actions that had been taken. However, quality goals may not be attained due to circumstances out of service provider control. In this case they should not be considered by service level evaluation algorithms, that is, service fault events which can not be imputed to the service provider must be excluded from calculus.

Service supervision is accomplished by presentation of two panels: indicator and alarm panel. Indicator panel presents all managed indicators in an organized way, by clients and service. It allows a managerial valuation of the service offer through the identification of those who have offended SLA or with offense risk. Alarms panel presents in a framed way the services alarms. They present events that cause modifications on service state and show deadlines for service degradation. As discussed before, an SLA defines several indicators for one service. The service state is defined as being the worst state among them. Service state evaluation depends on SLM events produced by on-line inference algorithm, and keeps the worst indicator state as the current service state.

In this paper service supervision uses two thresholds for each indicator, alert threshold and service level objective (SLO). They determine three states for each indicator: normal state means indicator current value is better than alert threshold; warning state means its value is between alert threshold and SLO; and violated state means that the value is worst than SLO.

2.2 Inference and control layer

Indicator evaluation and automatic triggering of management actions happen at the inference and control layer. Inference functions compute indicators' values; compare them with service level thresholds and fires control actions in order to pursue management objectives. Typical control actions are updating supervision panels, sending alert messages and starting management interactions with external systems. Data for indicator evaluation is obtained from external management systems by extraction function at the data collect layer.

2.2.1 Off-line inference and control

Off-line inference performs indicator evaluation by means of a scheduling mechanism that drives periodically corresponding computing algorithms. It reads service fault events prepared by extraction function and stores the calculated values in indicator repository. Computed values must remain stored, at least, up to the end of the assessment cycle. For example, all service fault events occurred last month must remain stored until this monthly assessment is performed, that is, service level reports are delivery and accept by customers. This is necessary because some events can be considered non-pertinent at assessment time even if they were considered at collection time.

Off-line control actions are fired through rules evaluation stored in policy repository. A rule contains a condition and an action. Conditions are a logical expressions made by variables (indicators) and by logical and arithmetical operators. When the condition is evaluated to TRUE, the corresponding action is fired. Off-line control actions unlink computing procedures, for example, reconfiguration of the network that supports service delivery or a routine for penalties and bonus account if the agreement is violated. Off-line control actions to update the indicator panel are also fired by changes in service state.

2.2.2 On-line inference and control

On-line inference performs indicator evaluation for alarm and alert generation. It is based on an event's handling mechanism: for each incoming event, indicator's value is updated, considering the information existing in service fault event. Also, a new deadline for indicator state change is computed. An SQ event is raised for each state change.

2.3 Data collection layer

Data collection layer gets the external systems data and can be divided into tow categories: extraction and mediation. Extraction functions get data for indicator account whereas mediation functions get information to fill service and SLA repositories.

2.3.1 Extraction function

Extraction functions interact with outside management systems to collect information used for indicator account. Extraction algorithms depend on data to be collected, including its origin, shape, access mode and previous treatment that it should receive. Data can be stored in log files, databases, spreadsheets or general files. It is also possible that data is not available at collection time, and a graphical interface should be provided.

As data origin can be multiple and heterogeneous, extraction function must perform the following tasks: convert multiple unities from origin data to a unified unity; synchronize data production periodicity making them available at appropriate frequency to the off-line account algorithms; summarize collected data and exclude the ones that are not necessary, ensuring that those data will get to the inference and control layer, according to their needs; and build, send and store SF events to account algorithms.

2.3.2 Mediation function

Mediation function interacts with outside systems to collect and store information at the SQM system repositories. They import customer, service, and SLA data to SQM databases when necessary.

3. ON-LINE CONTROL OF SLA

3.1 Method for indicator and deadline evaluation

This section presents a method and for on-line control of service level agreements. They can be applied to SLA established on indicators whose values depend on service fault events (SF events), which indicate the beginning and the end of a service unavailable interval. It also outlines the algorithm to implement the method.

Extraction functions build SF events by handling data reported by external management system. When building SF events, extraction functions relate infrastructure alarms with a customer-service pair. Moreover they assure events are delivered according to the following rules: don't send duplicated events; don't send an event with UP notification type before the corresponding event with DOWN notification type; provide a growing identification for events related to the same customer-service pair.

The method is based in mathematical functions that describe indicator's behavior based on occurrence of events throughout time. Figure 3 illustrates the idea: the horizontal axis represents SF events occurrences (e1 and e2) on time (t0 and t3), and the vertical axis represents the indicator value. It has a starting value (Q0) and two known thresholds: alert threshold (Q1) and service level objective (Q2). The occurrence of e1 event (at t0) characterizes the beginning of an unavailability interval for the service, and affects the indicator value according to its formula. The occurrence of e2 event characterizes the end of the unavailability interval, from when the indicator will have its value unchanged until the beginning of the next unavailability interval. With this information it is possible to compute t1 and t2, which are the moments when indicator value will reach thresholds Q1 and Q2, respectively.

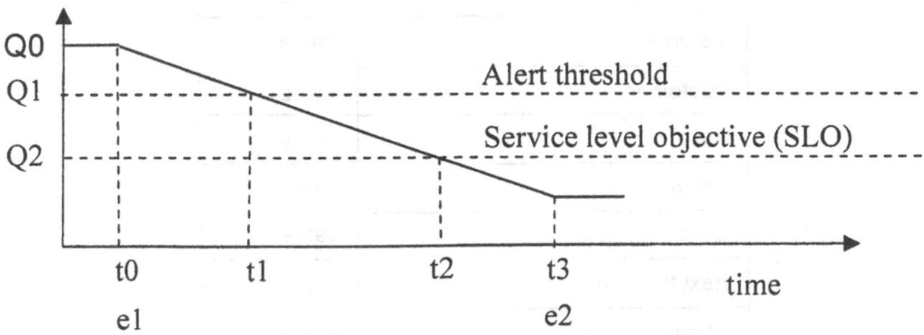

Figure 3. Indicator value and deadline to reach thresholds

3.2 Information flow

The structure of SF event is shown in figure 4. Event identifier is a grown number that uniquely identifies the event for a customer-service pair. Notification type says if the event corresponds to the beginning or the end of an unavailability interval, carrying respectively DOWN or UP value. The same event identifier is used in two related DOWN and UP notifications. The event also carries service and customer identifiers and a timestamp informs the time and date of event occurrence.

event identifier	value
notification type	value
service	value
customer	value
timestamp	value

Figure 4. Service fault event

SQ events are sent when service quality condition is changed (see figure 5). Event identifier is a grown number that uniquely identifies the event for a customer-service pair. The same identifier is used for all SQ events generated when handling SF events related to a DOWN-UP pair, that is, having the same identifier. The SQ event also carries service and customer identifiers, informs what indicator is being reported, the indicator value and state, the next threshold to be reached (Q1 or Q2), and the deadline to reach it.

event identifier	value
service	value
customer	value
indicator	value
value	value
state	value
next threshold	value
deadline	value

Figure 5. Service quality event

Some intermediary data is stored in a control register, which structure is presented in figure 6. It contains the event identifier, service and customer identifiers, a timestamp for last DOWN event occurred for this service-customer pair (DT), the number of SF events for this service-customer pair (NSF), and the previous value calculated for each indicator.

event identifier	value
service	value
customer	value
DT	value
NSF	value
previous value [i]	value

Figure 6. Control register

3.3 Indicators and formulas

To demonstrate the method we will consider mean time to restore service (MTRS) indicator, which is one of the most important and used service level indicators. Initially we present its definition (equations 1) and graphic (figures 7); then we deduce the formula that calculate its current value (equations 2) and the formula that calculate the deadline to reach the corresponding threshold (equations 3). Mean time to restore service (MTRS) is the mean of all unavailability intervals (TRS) observed during the evaluation period.

$$MTRS = \frac{\sum_{n} TRS_n}{n} \qquad (1)$$

The value of MTRS indicator depends on the occurrence of SF events. A SF event with DOWN notification type starts an unavailability interval when MTRS value starts increasing as shown in figure 10. SF events *e1* and e3 have notification type DOWN, and start two unavailability periods, whereas *e2* and *e4*, with UP notification type, end the corresponding periods. *PMTRS* is the indicator value at the end of the previous unavailability period and *dl_1* is the deadline when indicator value will reach the alert threshold (*Q1*).

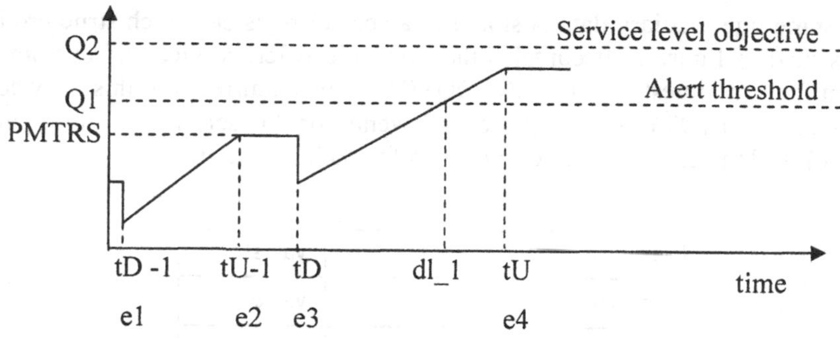

Figure 7. MTRS indicator value and deadline to reach thresholds

Equation 2 calculates MTRS value at the end of an unavailability interval (*tU*), started at *tD*. The value is calculated from MTRS previous value (PMTRS) at the end of the last unavailability interval.

$$MTRS_{(n)} = \frac{\sum_n TRS_{(n)}}{n} \quad \text{or} \quad \sum_n TRS_{(n)} = MTRS_{(n)} \times n$$

Considering event (n+1):

$$MTRS_{(n+1)} = \frac{\sum_n TRS_{(n)} + TRS_{(n+1)}}{n+1}$$

Substituting $\sum_n TRS_{(n)}$ by $MTRS_{(n)} \times n$

$$MTRS_{(n+1)} = \frac{MTRS_{(n)} \times n + TRS_{(n+1)}}{n+1} \quad \text{or}$$

$$MTRS = \frac{PMTRS \times n + (tU - tD)}{n+1} \tag{2}$$

Equation 3 computes the deadline to reach threshold *Qi* at the beginning of an unavailability interval, that is, at the reception of a SF event with DOWN notification type and timestamp *tD*. The deadline to reach *Qi* is deduced from (2) by substituting SA by *Qi* and *tU* by *dl_i*.

$$dl\_i = tD + (n+1) \times Qi - n \times PMTRS \tag{3}$$

3.4 Algorithm

The algorithm to implement the method is quite simple, and is outlined bellow. It is, however, necessary to implement a very sophisticated mechanism to handle incoming and outgoing events. A mechanism, named event dispatcher, receives and processes incoming SF events, then builds and sends corresponding SQ events. It calls an evaluation component to calculate the values according to equations 2 and 3.

Event dispatcher reads incoming SF events and verifies its notification type. For SF events with DOWN notification type it calls the evaluation component to compute the deadline to reach next threshold by applying equation 3. Then it prepares an SQ event and sets a timer that expires at computed deadline. When deadline expires, the associated SQ event is sent to be presented at service alarm panel. When the notification type is UP, the corresponding timer is unset. A new SQ event is built to clear previous alarm condition, and also to inform new indicator's value, computed according to equation 2.

4. CONCLUSION AND FUTURE WORK

This work presented a method for on-line service level management. It is the base for an algorithm that computes indicator's value for each incoming event and also, that computes the deadline for reaching the next service level threshold. The information rendered on-line by the algorithm in SQ events is important because it allows pro-active management. Operational management actions, as reconfiguring the service, can be started based on it. An SQM system with the outlined architecture was fully implemented and is in use in two telecommunication operators in Brazil. The algorithm has been implemented and a benchmark against off-line methods is available in Wandresen[11].

The service level indicators considered in this paper are all based on SF events, which are related intervals where service is unavailable. Such intervals are characterized by two timestamps, one at the beginning and the other at the end of the unavailability period. The extension of the algorithm for other kind of service level indicators can be considered. The only output of the algorithm is SQ events. Another important extension is to consider control actions based on policies.

REFERENCES

1. R. Boutaba; A. Polyrakis. Projecting Advanced Enterprise and Service Management to Active Networks. IEEE Network – February 2002.
2. M. Brunner; B. Platner; R. Stadler. Service Creation and Management in Active Networks. Communication of the ACM – April 2001.
3. G. Cortese; P Cremonese; A. Diaconescu; S. D'Antonio; M. Espósito; R. Fiutem; S. Romano. Cadenus: Creation and Deployment of End-User Services in Premium IP Networks. IEEE Communications Magazine, January 2003 – Vol. 41 n.1.
4. R. Maresca; M. D'Arienzo; M. Espósito; S. Romano; G.Ventre. An Active Network Approach to Virtual Private Networks. In Proceedings of ISC2002, July 2002.
5. A. Kittel; R. Bhatnagar; K. Hum; M. Weintraub. Service Creation and Service Management for Advanced IP Network and Services – An Experience Paper. In Proceedings of IM2001
6. Dinesh Verma. Supporting Service Level Agreements on IP Networks. New Ridres, Indianapolis, 1999.
7. A. Keller; G. Kar; H. Ludwig; A. Dan; J. Hellerstein. Managing Dynamic Services: A Contract Based Approach to a Conceptual Architecture. Journal of Network Operations and Management, 2002
8. A. Keller; H. Ludwig. The WSLA Framework: Specifying and Monitoring Service Level Agreement for Webservices. Journal of Network and Systems Management, 11(1), 2003.
9. P. Trimintzios; I. Andrikopoulos; D. Goderis; Y. T'Joens. An Architectural Framework for Providing QoS in IP Differentiated Services Networks. In Proceedings of IM2001.
10. L. Lewis. Service Level Management for Enterprise Networks. Artech House, London, 1999.
11. R. Wandresen. Proposta de um Modelo Computacional Baseado em Eventos para Gerência de Níveis de Serviço em Telecomunicações. Dissertação de Mestrado, PPGIA. PUCPR, 2003.

REVENUE-AWARE RESOURCE ALLOCATION IN THE FUTURE MULTI-SERVICE IP NETWORKS

Jian Zhang, Timo Hamalainen, Jyrki Joutsensalo
Dept. of Mathematical Information Technology, University of Jyvaskyla, FIN-40014 Jyvaskyla, Finland, Email: zhang@cc.jyu.fi, timoh@mit.jyu.fi, jyrkij@mit.jyu.fi

Abstract: In the future IP networks, a wide range of different service classes must be supported in a network node and different classes of customers will pay different prices for their used node resources based on their Service-Level-Agreements. In this paper, we link the resource allocation issue with pricing strategies and explore the problem of maximizing the revenue of service providers in a network node by optimally allocating a given amount of node resources among multiple service classes. Under the linear pricing strategy, the optimal resource allocation scheme is derived for the case that no firm Quality-of-Service (QoS) guarantees are required for all service classes, which can achieve the maximum revenue in a network node; moreover, the suboptimal allocation scheme is proposed for the case that all classes have their firm QoS (mean delay) requirements, which can satisfy those required QoS guarantees while still being able to achieve very high revenue close to the analytic maximum one.

1. INTRODUCTION

Resource allocation in the multi-service communication networks presents a very important problem in the design of the future multi-class Internet. The main motivation for the research in this field lies in the necessity for structural changes in the way the Internet is designed. The current Internet offers a single class of 'best-effort' service, however, the Internet is changing. New sophisticated real-time applications (video conferencing, video on demand, distance learning, etc) require a better and more reliable network performance. Moreover, these applications require firm performance guarantees from the network where certain resources should be reserved for them. The problem of optimal resource allocation for satisfying an end-to-end Quality-of-Service (QoS) requirement in a network of schedulers is usually addressed by partitioning the end-to-end QoS requirement into local requirements and then solving them in each individual node[5,7]. Hence, the problem of further mapping traffics' local QoS requirements (delay in particular) to their resource allocations in a single network node is of practical importance. On the other hand, in the future multi-class Internet, users will have to pay the network service providers for their used resources based on pricing strategies agreed upon in their Service-Level-Agreements (SLA). From the service providers' point of view, the optimal resource allocation scheme for their revenue maximization is very desirable. In this paper, we addressed the problem of optimizing the resource allocation in a network node for satisfying both the local QoS requirements of multi-class traffics and the revenue maximization of service providers.

Pricing research in the network has been quite intensive during the past few years[2,9,10]. Also a lot of work[12,8,4] has been done concerning the issues of resource allocation and fairness in a single-service environment. The combination of pricing strategies and resource allocation among multiple service classes have not been analyzed widely. A number of works[1,11,6] recently use end-users' utility as the maximizing objective for resource allocation schemes. All of these approaches have a common objective in maximizing the network performance in terms of the users' utility. Our research differs from these studies by linking the resource allocation scheme together with certain pricing strategies to maximize the revenue of a service provider under a given amount of resources.

This paper extends our previous QoS and pricing research[3] and addresses the problem of revenue maximization in a network node by novel revenue-aware resource allocation schemes. Specifically, in a network node supporting multiple service classes, packets are queued in a multi-queue system, where each queue corresponds to one service class. The service provider will receive certain revenues or suffer certain penalties based on given pricing strategies whenever serving a packet. For the case that the service classes supported in a network node are all delay-insensitive, i.e., all service classes have no firm QoS (mean delay, in this paper) requirements, we derive the optimal resource allocation scheme by which the maximum revenue can be obtained. Moreover, when the service classes supported in a network node are all delay-sensitive, i.e., firm QoS (mean delay) guarantees are required for all classes, a suboptimal resource allocation scheme is proposed, which can satisfy those local QoS guarantees while still achieving very high revenue. The simulation results demonstrated the performances of our proposed revenue-aware resource allocation schemes.

The rest of the paper is organized as follows. In Section 2, the linear pricing strategy, which is used in this paper, is generally defined. Revenue-aware resource allocation is investigated in Section 3, where optimal/suboptimal allocation schemes are derived. Section 4 contains the simulation part evaluating the performances of our proposed resource allocation schemes. Finally, in Section 5, we present our concluding remarks.

2. PRICING STRATEGY

As we know, linear, flat and piecewise linear strategies are believed to the most used one in practice. In this paper, our study concentrates on the revenue-maximizing issue under the linear pricing strategy and the analysis under the flat pricing strategy is postponed to its sequel. The solution to the piecewise linear pricing strategy is a straightforward extension to the above two cases. First some parameters and notions are defined. We consider a network node which supports multiple service classes. Here, incoming packets are queued in a multi-queue system (each queue corresponds to one service class) and the resources in the network node (e.g. processor capacity and bandwidth) are shared amongst those service classes. The number of classes is denoted by m. Literature usually refers to the gold, silver and bronze classes; in this case, $m = 3$. The metric of QoS considered in this paper focuses on *packet delay*. We use d_i to denote class i packet delay in the network node. For each service class, a pricing function $r_i(d_i)$ is defined to rule the relationship between the QoS (packet delay here) offered to class i customers and the price which class i customers should pay for that QoS. Obviously, it is non-increasing with respect to the delay d_i. Specifically, the linear pricing strategy for class i is characterized by the following function.

Definition 1: *The function*

$$r_i(d_i) = b_i - k_i d_i, \qquad i = 1, 2, ..., m, b_i > 0, k_i > 0 \tag{1}$$

is called *linear pricing function*, where b_i and k_i are positive constants and normally $b_i \geq b_j$ and $k_i \geq k_j$ hold to ensure differentiated pricing if class i has a higher priority than class j (in this paper, we assume that class 1 is the highest priority and class m is the lowest one).

From Eq. (1), it is observed that the constant shift b_i determines the maximum price paid by class i customers and the growing rate of penalty paid to class i with delay depends on the slope k_i. Clearly, the set of eligible pricing functions should show that the constant shift b_1 and the slope k_1 from the highest priority class are both maximal. This is actually what we expect based on the requirements of the Service-Level-Agreement.

3. REVENUE-AWARE RESOURCE ALLOCATION

In this section, we consider a network node with capacity C bits/s, supporting m service classes totally. In the node, each class has its own queue. Assume that the queues corresponding to different classes are infinite in length and the packets in the same queue are served in the order they arrive. As most traffic arrival processes have been proven to be Poisson process, the used source traffic model for each class consists of Poisson arrivals and an exponential packet length distribution in this paper. The arrival rate for the m classes is λ_1,λ_2,...,λ_m (packets/s), respectively. We use $\bar{L_i}$ to denote the mean packet length (in bits) of class i. As mentioned above, d_i is used to denote class i packet delay in the node, which consists of the waiting time in queue i and the service time. The share of node capacity allotted to class i is specified by parameter w_i, which is called the weight of class i. Obviously, the constraints for w_i, $1 \leq i \leq m$ are $\sum_{i=1}^{m} w_i = 1$ and $w_i \in (0, 1]$. As a necessary stability condition, $\sum_{i=1}^{m} \lambda_i \bar{L_i} < C$ is required.

As class i packets arrive at queue i with rate λ_i and they are guaranteed to receive a portion of node capacity $w_i C$, the analytic mean packet delay $\hat{\bar{d_i}}$ of class i in the node can be estimated as $\hat{\bar{d_i}} = \frac{1}{\frac{w_i C}{\bar{L_i}} - \lambda_i} = \frac{\bar{L_i}}{w_i C - \lambda_i \bar{L_i}}$ based on the queuing theory. Its natural constraint is $w_i C > \lambda_i \bar{L_i}$ due to the fact that delay can not be negative.

Note that a service provider will obtain a revenue or penalty whenever serving one packet. Hence, the metric of revenue used in this paper is the revenue gained per time unit by a service provider. Unless stated otherwise, we shall hereafter refer to the revenue per time unit as revenue. We use the above analytic mean packet delay $\hat{\bar{d_i}}$ to estimate class i packet delay d_i. Then the revenue F gained by a service provider in the node may be defined as follows when the linear pricing function in Eq. (1) is deployed:

$$F = \sum_{i=1}^{m} \lambda_i r_i(d_i) = \sum_{i=1}^{m} \lambda_i (b_i - \frac{k_i \bar{L_i}}{w_i C - \lambda_i \bar{L_i}}) \quad (2)$$

3.1 Case 1: the resource allocation scheme when no firm QoS guarantees are required

In this subsection, we derive the optimal resource allocation scheme for the case that no firm QoS (delay) guarantees are required for all classes. In other words, packet delay d_i of class i can be any positive value in this case. Then, the issue of revenue maximization in a network node can be formulated as follows based on the revenue definition in Eq. (2):

$$max \quad F = \sum_{i=1}^{m} \lambda_i (b_i - \frac{k_i \bar{L_i}}{w_i C - \lambda_i \bar{L_i}}) \tag{3}$$

$$s.t. \quad \sum_{i=1}^{m} w_i = 1, \ \ 0 < w_i \leq 1 \tag{4}$$

$$w_i C > \lambda_i \bar{L_i} \tag{5}$$

Theorem 1. *When no firm delay guarantees are required for all classes, the globally maximum revenue F obtained in a network node is achieved by using the following optimal resource allocation scheme:*

$$w_i = \frac{\sqrt{\lambda_i k_i \bar{L_i}}(C + \frac{\sum_{j=1}^{m} \sqrt{\lambda_j k_j \bar{L_j}}}{\sqrt{\lambda_i k_i \bar{L_i}}} \lambda_i \bar{L_i} - \sum_{j=1}^{m} \lambda_j \bar{L_j})}{C \sum_{j=1}^{m} \sqrt{\lambda_j k_j \bar{L_j}}} \tag{6}$$

i = 1, 2, ..., m and it is unique when $w_i \in (0, 1]$.

Proof: Based on Eqs. (3) and (4), we can construct the following Lagrangian equation.

$$P = P(w_1, w_2, ..., w_m) = \sum_{i=1}^{m} \lambda_i (b_i - \frac{k_i \bar{L_i} d_0}{w_i - \lambda_i \bar{L_i} d_0}) + \sigma(1 - \sum_{i=1}^{m} w_i) \tag{7}$$

Set partial derivatives of *P* in Eq. (7) to zero, i.e., $\frac{\partial P}{\partial w_i} = \frac{\lambda_i k_i \bar{L_i} C}{(w_i C - \lambda_i \bar{L_i})^2} - \sigma = 0$. It follows that $\sigma = \frac{\lambda_i k_i \bar{L_i} C}{(w_i C - \lambda_i \bar{L_i})^2}$, leading to the solution:

$$w_i = \sqrt{\frac{\lambda_i k_i \bar{L_i}}{C\sigma}} + \frac{\lambda_i \bar{L_i}}{C}, \ \ i = 1, 2, ..., m. \tag{8}$$

Substituting Eq. (8) to Eq. (4), we get $\sqrt{\sigma} = \frac{\sum_{i=1}^{m} \sqrt{\lambda_i k_i \bar{L_i} C}}{C - \sum_{i=1}^{m} \lambda_i \bar{L_i}}$ and when this $\sqrt{\sigma}$ is substituted back to Eq. (8), the closed-form solution in Eq. (6) is obtained.

Due to the constraint $w_i C > \lambda_i \bar{L_i}$ in (5), obviously, $\sum_{j=1}^{m} w_j C = C > \sum_{j=1}^{m} \lambda_j \bar{L_j}$. Hence, the closed-form solution in Eq. (6) $w_i > 0$. Moreover, this inequality holds: $\lambda_i \bar{L_i} - \frac{\sqrt{\lambda_i k_i \bar{L_i}} \sum_{j \neq i, j=1}^{m} \lambda_j \bar{L_j}}{\sum_{j \neq i, j=1}^{m} \sqrt{\lambda_j k_j \bar{L_j}}} \leq C$, leading to in Eq. (6) the numerator less than the denominator and thus $w_i < 1$. Hence, we can conclude that the closed-form solution in Eq. (6) $w_i \in (0, 1]$ and it is an eligible weight.

To prove that the closed-form solution in Eq. (6) is the only optimal one in the interval (0, 1], we consider the second order derivative of *P*: $\frac{\partial^2 P}{\partial w_i^2} =$

$-\frac{2\lambda_i k_i \bar{L_i} c^2}{(w_i C - \lambda_i \bar{L_i})^3} < 0$ due to the constraint $w_i C > \lambda_i \bar{L_i}$ in (5). Therefore, the revenue F is strictly convex with the allotted set of weights $\{w_1, ..., w_i, ..., w_m\}$ for the interval $0 < w_i \leq 1$, having one and only one maximum. Hence, the closed-form solution w_i in Eq. (6) is the optimal weight of class *i*, *i*=1,2,...,m. This completes the proof. **Q.E.D.**

Furthermore, the maximum revenue obtained in a network node can be calculated as follows.

Theorem 2. *When the optimal resource allocation scheme in Theorem 1 is deployed, the maximum revenue obtained in a network node is*

$$F_{max} = \sum_{i=1}^{m}(\lambda_i b_i) - \frac{(\sum_{i=1}^{m} \sqrt{\lambda_i k_i \bar{L_i}})^2}{C - \sum_{i=1}^{m} \lambda_i \bar{L_i}} \tag{9}$$

Proof: Substituting the optimal weight in Eq. (6) to Eq. (2), the maximum revenue F_{max} which can be obtained in a network node is

$$\begin{aligned} F_{max} &= \sum_{i=1}^{m}(\lambda_i b_i - \frac{\lambda_i k_i \bar{L_i} \sum_{i=1}^{m} \sqrt{\lambda_i k_i \bar{L_i}}}{\sqrt{\lambda_i k_i \bar{L_i}}(C - \sum_{i=1}^{m} \lambda_i \bar{L_i})}) = \\ &= \sum_{i=1}^{m}(\lambda_i b_i) - \frac{(\sum_{i=1}^{m} \sqrt{\lambda_i k_i \bar{L_i}})^2}{C - \sum_{i=1}^{m} \lambda_i \bar{L_i}} \end{aligned} \tag{10}$$

Q.E.D.

3.2 Case 2: the resource allocation scheme when the firm QoS guarantees are required

In this subsection, we derive the resource allocation scheme in a network node which should satisfy the required QoS guarantees of all classes while still achieving as higher revenue as possible. In this paper, the firm QoS guarantee of class *i* means that the mean packet delay $\bar{d_i}$ of class *i* must be less than the given value D_i, i.e., $\bar{d_i} \leq D_i$. We use the analytic mean packet delay $\hat{\bar{d_i}}$ to estimate $\bar{d_i}$, i.e., $\hat{\bar{d_i}} = \frac{\bar{L_i}}{w_i C - \lambda_i \bar{L_i}} \leq D_i$, leading to $w_i \geq \frac{\bar{L_i}}{C}(\lambda_i + \frac{1}{D_i})$. Thus, the required minimum weight of class *i* for satisfying its firm mean delay guarantee is acquired, which we use $w_{i,minimum}$ to denote: $w_{i,minimum} = \frac{\bar{L_i}}{C}(\lambda_i + \frac{1}{D_i})$. Then, the issue of revenue maximization in Case 2 can be formulated as below:

$$max \quad F = \sum_{i=1}^{m} \lambda_i (b_i - \frac{k_i \bar{L_i}}{w_i C - \lambda_i \bar{L_i}}) \tag{11}$$

$$s.t. \quad \sum_{i=1}^{m} w_i = 1, \quad 0 < w_i \leq 1 \tag{12}$$

$$w_i \geq w_{i,minimum} \tag{13}$$

Note that the constraint in (5) has been contained in (12).

To address the above issue, we first calculate the optimal solution by Eq. (6) and it is referred to as $w_{i,optimal}, i = 1, 2, ..., m$ hereafter. Then, if the inequality $w_{i,optimal} > w_{i,minimum}$ holds for all classes, then the optimal solution $w_{i,optimal}, i = 1, 2, ..., m$ by Eq. (6) is the optimal resource allocation scheme in this case because not only can it achieve the maximum revenue but it can also satisfy the firm mean delay guarantees.

If the inequality ($w_{i,optimal} > w_{i,minimum}$) can not hold for all classes, the optimal solution $w_{i,optimal}$ is not eligible to be the resource allocation scheme in this scenario as it cannot satisfy all the firm mean delay guarantees. Additionally, as the constraint in (12) is derived based on the analytic mean delay $\hat{\bar{d}}_i$, the weight allotted to class *i* in this scenario should actually satisfy the following inequality in (13) to ensure the fulfilment of the firm QoS guarantee of class *i*: $\bar{d}_i \leq D_i$.

$$w_i \geq w_{i,minimum} + \epsilon_i \qquad (14)$$

where ϵ_i is a small positive constant and may be set at different values depending on the accuracy of the implemented firm QoS guarantee. In other words, there is no optimal resource allocation scheme in this scenario. However, as Theorem 1 shows that the revenue *F* is strictly convex to the allotted set of weights and only has one maximum, the suboptimal resource allocation scheme can be derived by having the suboptimal weight ($w_{i,suboptimal}$) as near $w_{i,optimal}$ as possible and meanwhile satisfying the constraints in Eqs. (11) and (13). We propose a feasible approach to derive the suboptimal allocation scheme for this scenario below.

The value of ($w_{i,optimal} - w_{i,minimum}$) is defined as the weight distance of class *i*. First, all $w_{i,minimum}$ (*i*=1,2,...,m, *i* is class index) are sorted by the defined weight distance in descending order so that we acquire a new series of weights denoted by $w_{p,minimum}$ (*p*=1,2,...,m, *p* is position index). Obviously, there is one mapping between class index *i* and position index *p*. Next, all $w_{i,minimum}$ are sorted by size also in descending order to achieve another series of weights denoted by $w'_{p,minimum}$ (*p*=1,2,...,m). Then, the suboptimal weight is derived as follows:

$$w_{p,suboptimal} = w_{p,minimum} + \frac{w'_{p,minimum}}{\sum_{p=1}^{m} w_{p,minimum}} (1 - \sum_{p=1}^{m} w_{p,minimum}) \qquad (15)$$

for $p = 1, 2, ..., m$. As mentioned above, there is the mapping between class index *i* and position index *p*, hence, the suboptimal weight $w_{i,suboptimal}$ of class *i* can be acquired by $w_{p,suboptimal}$.

4. SIMULATIONS AND RESULTS

In this section we present some simulation results to illustrate the effectiveness of our revenue-aware resource allocation scheme for maximizing the revenue of service providers while also satisfying the firm mean delay guarantees if needed. A number of simulations have been conducted under different parameter settings. A representative set of these simulations are presented herein. In accord with the presentation format in Section 3, this section also consists of two subsections, one for the case that no mean delay guarantees are required in the simulations and the other one for the case that the supported service classes all have the settings of their firm mean delay guarantees in the simulations. The configuration of all simulations in this section is described below.

Throughout this section, we shall focus on a network node with capacity $C = 10^6$ bits/s and the number of service classes supported $m = 3$ (namely, gold, silver and bronze classes). The base arrival rates and the mean packet lengths of the above three classes are provided as follows: for the gold class, λ_1=10 packet/s, for the silver class, λ_2=15 packet/s, for the bronze class, λ_3=20 packet/s, and $\bar{L}_i$=3360 bits, i=1,2,3. A multiplicative *load factor* $\rho > 0$ is used to scale these base arrival rates to consider different traffic intensities; i.e., $\lambda_j \rho$ will be used in the simulations as the class-j arrival rate. The set of linear pricing functions deployed is $\{r_1(d_1) = 200 - 10d_1, r_2(d_2) = 150 - 5d_2, r_3(d_3) = 80 - 2d_3\}$ (the time unit of delay is *ms* here).

4.1 Case 1 simulations

In this case, the simulations were made for evaluating the performance of the optimal resource allocation scheme derived by Theorem 1. The simulation-generated revenue value by the optimal allocation scheme will be compared with the maximum revenue value calculated by Theorem 2, which is called the analytic maximum revenue value hereafter. In the simulations, the proportional resource allocation scheme, which proportionally allocates the resource amongst all service classes, is also employed for comparison. Specifically, the *proportional* scheme allots the weight of a class i as follows: $w_i = \frac{\lambda_i \bar{L}_i}{\sum_{j=1}^{m}(\lambda_j \bar{L}_j)}$, i = *1, 2, ..., m*. Note that this proportional scheme is a natural way to allocate network resources.

Furthermore, we first investigate the evolution of the simulation-generated revenue by the optimal scheme with the time. In this scenario, only the above base arrival rates were used, i.e., load factor ρ=1. Additionally, a set of given weights (w_1=0.60, w_2=0.25, w_3=0.15) was also employed for further illustrating the performance of the optimal allocation scheme. Figure 1(a) presents the simulation results when the above set of linear pricing functions is used, where the x-axis represents the time (the measurement period is 100 seconds here) and the y-axis represents the revenue.

It is observed from Figure 1(a) that the largest simulation-generated revenue value is always achieved by the optimal allocation scheme compared to those by the proportional scheme and the given set of weights. Moreover, the simulation-generated revenue value by the optimal scheme is always quite close to the analytic maximum revenue value, which demonstrates the effectiveness of Theorem 1 and 2. As the parameters in Eq. (9) for calculating the analytic maximum revenue are invariable with time in this scenario, the analytic maximum revenue value does not change with time in Figure 1(a); whereas, as the simulation-generated packet delays are variable, the simulation-generated revenue values vary with time in the figure.

Next we evaluate the performance of the optimal allocation scheme when different traffic intensities are fed into the network node. Figure 1(b) shows the simulation results, where the x-axis represents the load factor and the y-axis represents the revenue. It is seen in Figure 1(b) that the optimal allocation scheme achieves the largest revenue in the simulations under all traffic intensities, which is also very close to the analytic maximum revenue. Moreover, both revenue curve grow almost linearly under light and medium loads. This is as expected because few penalties will be incurred under such loads. Under heavy loads, both curves start to level off as the penalties start to grow faster than the revenues. Although the revenue curve of the proportional scheme also grows under light loads, it starts to decrease much earlier as the penalties incurred by the proportional scheme are much larger than the ones by the optimal scheme under the same traffic load.

Based on the above simulation results, we can conclude that the Theorem 1 does derive the optimal resource allocation scheme for the case that no firm QoS guarantees are required, which can achieve the maximum revenue under different traffic intensities.

4.2 Case 2 simulations

In this subsection, we evaluate the performances of our derived revenue-aware resource allocation schemes for the case that all classes require their firm QoS (mean delay) guarantees. Throughout this subsection, the firm mean delay guarantees for the gold, silver and bronze classes are set as follows: $\bar{d_1} \leq 10ms, \bar{d_2} \leq 15ms, \bar{d_3} \leq 30ms$. According to the analysis in Section 3, the optimal weight $w_{i,optimal}$ and the required minimum weight $w_{i,minimum}$ should first be calculated, respectively, then the optimal or suboptimal resource allocation scheme in this case can be derived based on the comparison of them. Obviously, if the sum of the calculated $w_{i,minimum}$ exceeds 1 for any load

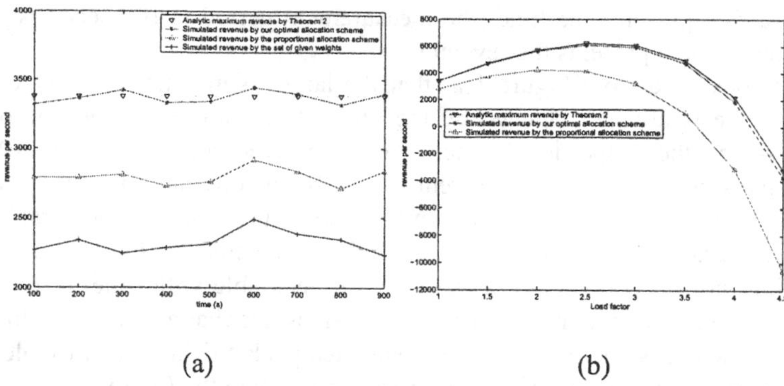

Figure 1. Case 1 simulations: (a) Revenue comparison as function of time (load factor $\rho = 1$); (b) Revenue comparison as function of load factor ρ.

Table 1. The calculated $w_{i,optimal}$ and $w_{i,minimal}$, i=1, 2, 3, for Case 2 simulations

	$w_{1,opt}$	$w_{2,opt}$	$w_{3,opt}$	$w_{1,min}$	$w_{2,min}$	$w_{3,min}$
$\rho = 1$	0.3733	0.3446	0.2821	0.3696	0.2744	0.1792
$\rho = 1.5$	0.3599	0.3436	0.2965	0.3864	0.2996	0.2128
$\rho = 2$	0.3464	0.3426	0.3110	0.4032	0.3248	0.2464

factor, it means that the node capacity is not enough to satisfy those firm QoS guarantees under such load intensity.

Specifically, in this case, $\sum_{i=1}^{3} w_{i,minimum} > 1$ holds for any load factor ρ which is more than 2.5. Hence, the load factors ρ=1, ρ=1.5 and ρ=2 were used to generate the traffic loads in the following simulations. The above set of linear pricing functions is employed in the simulations, then the calculated $w_{i,optimal}$ and $w_{i,minimum}$ for load factor ρ=1, ρ=1.5 and ρ=2 are summarized in Table 1. Based on the analysis in Section 3, we know that, if $w_{i,optimal} > w_{i,minimum}$ holds for all classes under a load intensity, then the optimal resource allocation scheme ($w_{i,optimal}$, i=1,2,3) exists for that load intensity. Specifically, Table 1 shows that the optimal allocation scheme exists for the traffic intensity: ρ=1, and thus the optimal weights ($w_{i,optimal}$ shown in Table 1 were also used in this case simulations for this load factor ρ=1.

For the scenarios that $w_{i,optimal} > w_{i,minimum}$ does not hold for all classes, the suboptimal resource allocation scheme has to be derived, which should satisfy the required mean delay guarantees while still being able to achieve high revenue. One feasible approach was proposed in Section 3 to derive such suboptimal allocation scheme. Below we illustrate how to calculate the suboptimal weight by that approach.

From Table 1, we observe that the suboptimal weight has to be derived for these two load intensities: ρ=1.5, ρ=2. Furthermore, under the above load intensities, it is observed from Table 1 that the weight distance of the gold class ($w_{1,optimal} - w_{1,minimum}$) is always the largest and the one of the bronze class ($w_{3,optimal} - w_{3,minimum}$) the smallest among the weight distances of the three classes; moreover, the required minimum weight of the gold class ($w_{1,minimum}$) is also always the largest and the one of the bronze class the smallest among all $w_{i,minimum}$, i=1,2,3. Hence, for this special scenario, the formula in Eq. (14) for deriving the suboptimal weight by our proposed approach can be expressed as follows:

$$w_{i,suboptimal} = w_{i,minimum} + \frac{w_{m+1-i,minimum}}{\sum_{i=1}^{m} w_{i,minimum}}(1 - \sum_{i=1}^{m} w_{i,minimum}) \quad (16)$$

for i=1, 2, 3, which exactly shows that the left portion of nodal capacity is distributed among all service classes based on their weight distances. Then, the suboptimal weights ($w_{i,suboptimal}$, i=1,2,3) for the above load intensities may be calculated by Eq. (15) and summarized as follows: for $\rho = 1.5$, $w_{1,suboptimal} = 0.4104$, $w_{2,suboptimal} = 0.3333$ and $w_{3,suboptimal} = 0.2563$; for $\rho = 2$, $w_{1,suboptimal} = 0.4097$, $w_{2,suboptimal} = 0.3333$ and $w_{3,suboptimal} = 0.2570$.

Additionally, another set of weights denoted by $w_{i,comparison}$, i=1,2,3 are also employed in the simulations for the comparison and calculated by equation: $w_{i,comparison} = w_{i,minimum} + \frac{w_{i,minimum}}{\sum_{i=1}^{m} w_{i,minimum}}(1 - \sum_{i=1}^{m} w_{i,minimum})$, i=1,2,3, which means that the left portion of nodal capacity is allotted to class i only based on the size of its $w_{i,minimum}$. They are summarized below: for $\rho = 1$, $w_{1,comparison} = 0.4490$, $w_{2,comparison} = 0.3333$ and $w_{3,comparison} = 0.2177$; for $\rho = 1.5$, $w_{1,comparison} = 0.4299$, $w_{2,comparison} = 0.3333$ and $w_{3,comparison} = 0.2368$; for $\rho = 2$, $w_{1,comparison} = 0.4138$, $w_{2,comparison} = 0.3333$ and $w_{3,comparison} = 0.2529$.

Then, the above derived optimal/suboptimal weights and the above comparison weights were used in the simulations, respectively. Figure 2 presents the simulation results, which shows that the simulation-generated revenue by the optimal/suboptimal weights is always higher than the one by the comparison weights and it is also pretty close to the analytic maximum revenue.

Moreover, the simulation-generated mean packet delay by the optimal/suboptimal allocation schemes are: 9.9882 *ms*, 11.5756 *ms* and 16.0049 *ms* for ρ=1; 9.4695 *ms*, 13.3077 ms and 22.2440 *ms* for ρ=1.5; 9.9636 *ms*, 14.8071 *ms* and 28.0530 *ms* for ρ=2, which shows that the simulation-generated mean packet delays by the derived optimal/suboptimal scheme satisfy the required firm mean delay guarantees ($\bar{d_1} \leq 10ms, \bar{d_2} \leq 15ms, \bar{d_3} \leq 30ms$). Therefore, it is demonstrated that our derived revenue-aware resource allocation scheme (optimal/suboptimal scheme) in Case 2 is an eligible one, which satisfies all

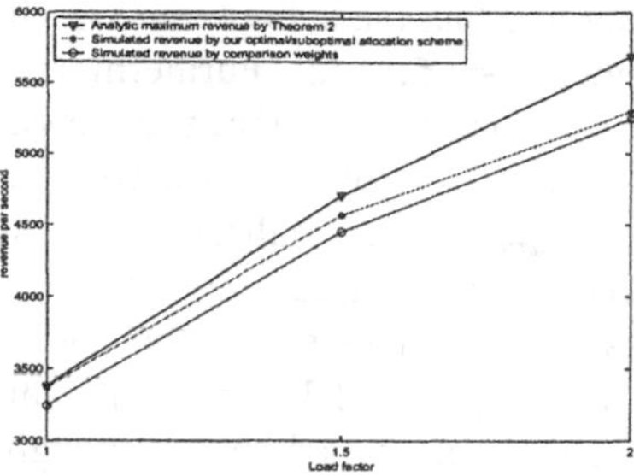

Figure 2. Revenue comparison as function of load factor ρ in Case 2 simulations

required firm QoS (mean delay) guarantees while still achieves very high revenue (pretty close to the analytic maximum one).

5. CONCLUSIONS

In this paper, we link the resource allocation issue with the pricing strategies and explore the problem of maximizing the revenue of service providers by optimally allocating a given amount of resource among multiple service classes. Under the linear pricing strategy, the optimal allocation scheme is derived for the case that no firm QoS guarantees are required for all classes, which can achieve the maximum revenue in a network node; moreover, the suboptimal allocation scheme is proposed for the case that all classes have their firm QoS (mean delay) requirements, which can satisfy those required QoS guarantees while still being able to achieve very high revenue close to the analytic maximum one. The simulation results demonstrated the effectiveness of our proposed optimal/suboptimal resource allocation schemes.

In future work, the issue of revenue maximization under a flat pricing strategy will be investigated. Moreover, a revenue criterion as the admission control mechanism will be studied.

REFERENCES

1. Z. Cao, E. W. Zegura, "Utility Max-Min: An Application-Oriented Bandwidth Allocation Scheme," IEEE INFOCOM99, New York, USA, 1999.

2. C. Courcoubetis, F. P. Kelly, and R. Weber, "Measurement-based usage charges in communication networks," Oper. Res., Vol.48, no.4, 2000, pp. 535-548.

3. J. Joutsensalo, T. Hamalainen, M. Paakkonen, and A. Sayenko, "Revenue Aware Scheduling Algorithm in The Single Node Case," Journal of Communications and Networks, Vol.6, No.1 March 2004.
4. L. Massoulie, J. Roberts, "Bandwidth Sharing: Objectives and Algorithms," IEEE INFOCOM99, New York, USA.
5. R. Nagarajan, J. F. Kurose and D. Towsley, "Allocation of Local Quality of Service Constraints to Meet End-to-End Requirements," IFIP Workshop on the Performance Analysis of ATM Systems, Martinique, Jan. 1993.
6. C. Lee, J. Lehoczky, R. Rajkumar, D. Siewiorek, "On Quality of Service Optimization with Discrete QoS Options," IEEE Real-Time Technology and Application Symposium, June 1999.
7. D. H. Lorenz and A. Orda, "Optimal partition of QoS requirements on unicast paths and multicast trees," Proceedings of IEEE INFOCOM'99, pp. 246-253, March 1999.
8. S. H. Low, "Equilibrium Allocation of Variable Resources for Elastic Traffics," INFOCOM98, San Francisco, USA, 1998.
9. I. Ch. Paschalidis and J. N. Tsitsiklis, "Congestion-dependent pricing of network services," IEEE/ACM Transactions on Networking, vol.8, April 2000, pp. 171-184.
10. I. Ch. Paschalidis and Yong Liu, "Pricing in multiservice loss networks: static pricing, asymptotic optimality and demand subsitution effects," IEEE/ACM Transactions on Networking, vol.10, Issue: 3, June 2002, pp. 425-438.
11. S. Sarkar, L. Tassiulas, "Fair Allocation of Utilities in Multirate Multicast Networks," Proceedings of the 37th Annual Allerton Conference on Communication, Control and Computing, Urbana, Illinois, USA, 1999.
12. S. Shenker, "Fundamental Design Issues for the Future Internet," IEEE Journal on Selected Areas in Telecommunications, Vol.13, No.7, September 1995.

Part Two: Network Security

A KERBEROS-BASED AUTHENTICATION ARCHITECTURE FOR WLANS

Test beds and experiments

Mohamed Ali Kaafar,[1] Lamia Ben Azzouz[2] and Farouk Kamoun[2]
Laboratoire CRISTAL, Ecole Nationale des Sciences de l'Informatique. Université de la Manouba. Manouba, Tunisia. [1]*medali.kaafar@cristal.rnu.tn,*
[2]{*lamia.benzaaouz,farouk.kamoun*}*@ensi.rnu.tn*

Abstract: This work addresses the issues related to authentication in wireless LAN environments, with emphasis on the IEEE 802.11 standard. It proposes an authentication architecture for Wireless networks. This architecture called Wireless Kerberos (W-Kerberos), is based on the Kerberos authentication server and the IEEE 802.1X-EAP model, in order to satisfy both security and mobility needs. It then, provides a mean of protecting the network, assuring mutual authentication, thwarts cryptographic attack risks via a key refreshment mechanism and manages fast and secure Handovers between access points. In addition to authentication, Kerberos has also the advantage of secure communications via encryption.

Keywords: Wireless authentication, IEEE 802.1X, EAP, Kerberos, test beds.

1. INTRODUCTION

The convenience of IEEE 802.11-based wireless access networks has led to widespread deployment in many sectors. This use is predicated on an implicit assumption of access control, confidentiality and availability. However, this widespread deployment makes 802.11-based networks an attractive target for potential attackers. Indeed, many researches have demonstrated basic flaws in 802.11's encryption mechanisms[1,2] and authentication protocols[3].
Although the IEEE 802.11i framework is proposing solutions to deal with wireless networks security limitations, actually there is not a complete set of standards available that solves all the issues related to Wireless security. We

have then proposed a mobility-aware authentication architecture for 802.11 networks, based on the IEEE 802.11i works and exploiting the Kerberos protocol to overcome security limitations of Wi-Fi networks. In this paper, we propose a Kerberos-like authentication architecture, evaluate and experiment its security. We first begin by presenting concepts related to the IEEE 802.11i architecture such as the EAP-802.1X model and introduces the Kerberos protocol and related approaches in wireless networks. This is followed by a brief description of the proposed architecture (called W-Kerberos) and the authentication process. We then evaluate W-Kerberos on the basis of the IEEE 802.1X-EAP threat model. Next, we describe attacks experimented to test the architecture resistance to wireless vulnerabilities, and conclude with perspectives of this work.

2. THE IEEE 802.1X FRAMEWORK

The IEEE 802.1X standard[4] defines a port-based network access control using the physical characteristics of LAN infrastructures, to perform a data link layer access-control. This standard abstracts three entities [Figure1]:

- The supplicant, that wishes to access services, usually the client.
- The authenticator, which is the entity that wishes to enforce authentication before allowing access to its services, usually within the device the supplicant connects to.
- The authentication server, authenticating supplicants on behalf of the authenticator.

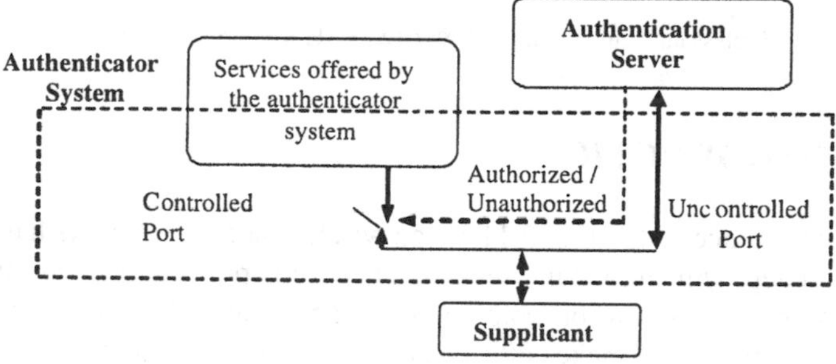

Figure 1. The IEEE 802.1X Set-up.

The IEEE 802.1X framework does not specify any particular authentication mechanism; it uses the Extensible Authentication Protocol (EAP)[5] as its

authentication framework. EAP is a protocol that supports exchange of information for multiple authentication mechanisms. The authenticator is responsible for relaying this information between the supplicant and the authentication server.
The authenticator's port-based access control defines two logical ports via a single physical LAN port. These are controlled and uncontrolled ports. The uncontrolled port allows uncontrolled exchange (typically information for the authentication mechanism) between the authenticator and other entities on the LAN, irrespective of the authentication state of the system. Any other exchange takes place via the controlled port.

3. THE KERBEROS PROTOCOL

Kerberos was developed as an open software at the Massachusetts Institute of Technology (MIT) as part of its Athena project[6]. The Kerberos architecture defines three entities: the client wanting to reach resources of a certain server, the service supplier or server, and the authentication Kerberos server, based on two distinct logical entities: An AS server (Authentication Server), responsible for the identification of clients, and a TGS server (Ticket Granting Service) which provides clients with access authorizations on the basis of an AS identification. These two entities are regrouped under the name of KDC to mean Key Distribution Center.

3.1 The Kerberos authentication process

The Kerberos authentication takes place in a set of steps as shown in [Fig.2] and described below:

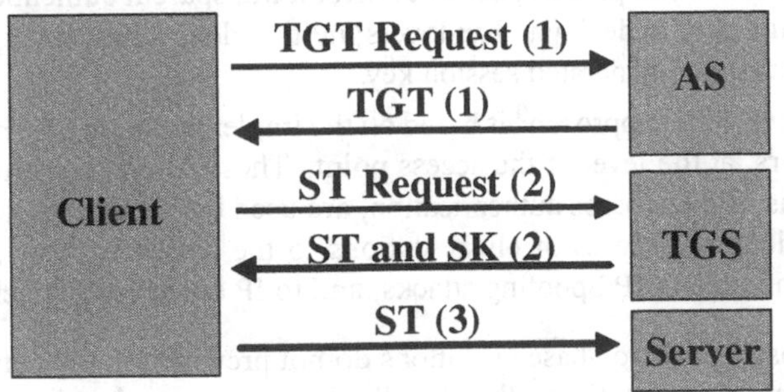

Figure 2. The Kerberos authentication process.

1 Before the client attempts to use any service of the network, a Kerberos Authentication Server AS must authenticate him. This authentication consists in obtaining an initial ticket request: Ticket Granting Ticket (TGT), which will be used subsequently to get credentials for several services.

2 When the client wants to communicate with a particular server, he sends a request to the TGS asking for credentials for this server. The TGS answers with these credentials encrypted by the user's key. The credentials consist of a temporary session key Sk and a ticket for the service supplier called Service Ticket ST, containing the client identity and the session key, all of them encoded with the server's key.

3 The client, wanting to reach a server's resources, transmits the ticket to this server. The session key, now shared by the client and the server, can be used to encrypt the next communications.

3.2 Kerberos in Wireless environments

The Kerberos use for authentication in WLAN environments has been considered several times. We discuss in the following some of these approaches.

- The "Symbol Technologies" approach is considering the SSID (Service Set Identity) as a service name shared between all the access points, to offer network access, and to make roaming easier. This approach does not offer a perfect forward secrecy. Thus, an attacker compromising an access point, could compromise the entire network.
- The IEEE 802.11e approach[7] is using EAP as an authentication method transporter, and the IAKerb protocol[8] for proxying the client messages to the authentication server. This approach, using a classic Kerberos authentication process, does not offer a transparent authentication and a secure way to deal with handovers; neither does it prevent cryptographic attacks on a generated session key.
- The IP Filter approach[9] is based on the implementation of some IP_tables filters, at the level of the access point. Those filters, constructed on the basis of a Kerberos authentication, are used by the access point to allow or deny the access of client stations to the network. This approach is vulnerable to IP Spoofing attacks, and to IP based DoS attacks[10].

The proposed Kerberos-based solutions do not prevent WLAN networks from cryptographic attacks nor do they handle fast and secure handovers. Thus, in our work, we have been interested by specifying a new Kerberos based architecture for WLAN. In the following subsections, we describe the proposed architecture called Wireless Kerberos: W-Kerberos.

4. W-KERBEROS OR KERBEROS FOR THE 802.11 NETWORKS

The proposed authentication process is based on tickets delivered by a W-Kerberos server. These tickets are going to direct the access points either to allow or not the traffic of a particular client. W-Kerberos system is composed of the client trying to have access to the network, the access points considered as the Kerberos service suppliers, offering the service of access to the network, and the W-Kerberos server as an authentication and roaming server. A W-Kerberos system is composed of the client trying to have access to the network, the access points considered as the Kerberos service suppliers, offering the service of access to the network, and the W-Kerberos server as an authentication and roaming server.

4.1 Initial authentication

This phase is typically initiated by the client terminal, which achieved a 802.11 association. In a first step, the client, receiving an EAP Request Identity from the access point, sends an EAP Response message, encapsulating an initial Service Ticket request (KRB-AS-REQ) [Fig.3]. The key used to encode the KRB messages is shared between the client and the Kerberos server and derived from the password provided by the client [1].

After receiving the EAP Response, a Kerberos authentication request is sent from the access point to the W-Kerberos authentication server on the non controlled IEEE 802.1X port. The authentication server consults then the basis of principals, fixes the session time (needed for key refreshment), and generates a session key. An answer message KRB-AS-REP containing the session key, the ticket encoded with the AP secret key, and some authentication information is sent to the client via the access point. This transmitted Data is encrypted with the client key. To have access to network resources, the client issues the ticket to the access point as a KRB-AP-REQ message encapsulated in an EAP Response packet. Thus, the client is now authenticated and authorised by the access point.

4.2 The key refreshment and Handover phases

W-Kerberos offers a secure channel for communications via encryption mechanisms where key exchange is dynamic. This avoids the possibility of passive attacks to retrieve encryption keys. Hence, in addition to the ticket validity time, a key refreshment mechanism based on a session time out, sent in the initial authentication ticket, is specified by our architecture[11]. Moreover,

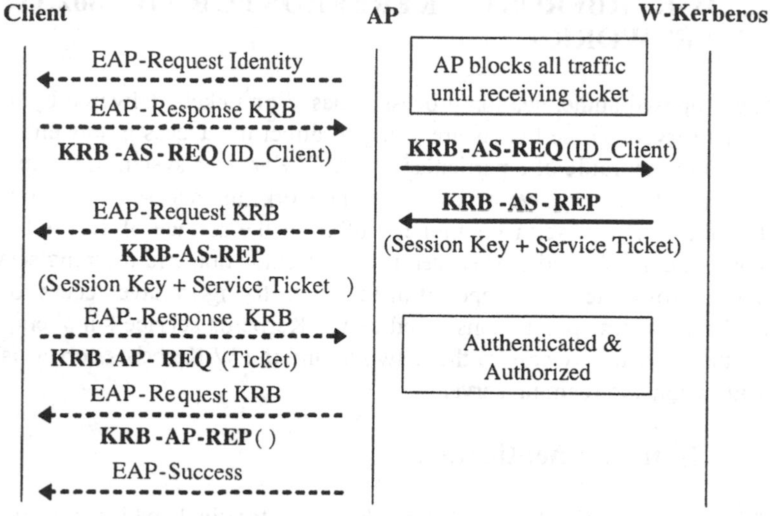

Figure 3. The Kerberos authentication process.

the WKerberos authentication is completely transparent to the client during a Handoff phase, in a way that no new authentication does take place. The client station only activates its context sending a ticket, which will be verified by the new access point[11].

5. W-KERBEROS EVALUATION

This section evaluates the W-Kerberos architecture on the basis of IEEE 802.11 networks threat model. The following points summarize security services offered by W-Kerberos to thwart different 802.11-specific attacks.

5.1 Confidentiality and Key derivation

Using the Kerberos key distribution to generate a MAC layer key, the W-Kereberos architecture allows the use of different available encryption mechanisms and thwarts different cryptographic attacks via the key refreshment phase.

5.2 Replay attacks and Integrity protection

W-Kerberos deals with replay attacks using EAP sequencing and Kerberos techniques in order to prevent an attacker from capturing a valid authentication message (or an entire authentication conversation) and replay it. Kerberos techniques are based on timestamps (and/or random numbers sequences)[12]. Further, caching authenticators within Kerberos V implementations denies any replays within the allowed time interval of Kerberos (5 minutes by default). Integrity protection provides data origin authentication and protection against unauthorized modification of information for authentication messages. W-Keberos uses an authenticator field to verify the authenticity of WKerb messages[6]. Moreover, the entities have to verify the EAP messages authenticity through an authenticator field added to the EAP messages.

5.3 Dictionary attacks resistance

Kerberos is known to be vulnerable to Dictionary attacks[13]. The W-Kerberos architecture, using a pass phrase to generate a master key could also be vulnerable to such type of attacks. In fact, where password authentication is used, passwords are commonly selected from a small set (as compared to a set of N-bit keys), which raises a concern about dictionary attacks. However, the possibility of using certificates and public key cryptography within a Kerberos environment has been studied[14]. While these techniques are certainly more secure, there is a compromise to consider: Using certificate based authentication means material authentication (in opposition to user one), public key infrastructure set up and less comfort of use for clients.

5.4 Mutual authentication and Protected results indications

This refers first to the ability for clients and access points to ascertain that they are communicating with authentic counterparts and to indicate whether they have successfully done it. Where EAP is tunnelled within another protocol that omits station authentication, there exists a potential vulnerability to man-in-the-middle attack[15,16]. W-Kerberos uses the Kerberos optional mutual authentication mechanisms, where both the access point and the client authenticate themselves. This authentication is mandatory within our proposal, so that rogue access point risks are prevented.

5.5 Denial of Service protection

Avoiding denial of Service (DoS) attacks within a wireless network is of a paramount importance for any security architecture. Performing a DoS attack is generally the first step that an attacker is taken to launch other, more clever attacks. Those attacks have been studied in many works[3,10], and some of them could be prevented by implementing some essential services. The per-packet authenticity used within a W-Kerberos authentication is one way to avoid some of these attacks. This prevent EAP-failure or EAPoL Logoff spoofing attacks[3] (trying to de-authenticate a legitimate client), and typically used to initiate a Man-in-the-Middle attack. Further, avoiding replay attack preserves 802.11 entities, especially access points, from flooding attacks.

6. TESTS AND EXPERIMENTS

In our experiments, we have tested three of the most known 802.11-specific attack implementations[12] (AirSnort, Monkey Jack, Void11) and observed the W-Kerberos behavior and attacks results.

6.1 AirSnort

AirSnort is a wireless LAN tool that recovers WEP encryption keys. It operates by passively monitoring transmissions, computing the encryption key when enough packets have been gathered. In the W-Kerberos implementation, based on the HostAP driver, dynamic WEP is used as the MAC layer protocol for frames encryption. Our test bed has consisted in multi sessions FTP transfers on five wireless stations running W-Kerberos. The table below presents the obtained results.

This test proves that key refreshment is a mean thwarting some cryptographic attacks, especially WEP crack tools requiring a minimal number of unique key-encrypted captured frames.

6.2 Monkey Jack

Monkey Jack is an implementation of a wireless Man-In-the-Middle attack. It is used within the AirJack toolbox, which is a free 802.11 device driver API, and an 802.11 development environment. AirJack offers many utilities as user space programs, such as wlan-Jack, essid-Jack, monkey-Jack, etc. The principle of the Monkey Jack attack is to take over connections at layer 1 and

Table 1. AirSnort results.

Key refreshment time(min)	***Key size (bits)***	***Average throughput***	***Average captures (frame/sec)***	***Observations and results***
0: Static key	128	2.8 Mbits/s	186	Key has been retrieved in 9095 seconds (about 2 hours 31min) capturing 1 681 138 ciphered frames.
0: static key.	64	3.6 Mbits/s	192	Key has been retrieved in 5390 seconds (about 1 hour 30 min) capturing 1 032 887 ciphered frames.
60	64	3.9 Mbits/s	194	Key has not been retrieved by AirSnort, collecting more than 5 200 000 ciphered frames.

2, and to insert attack machine between victim and access point. It consists of three main phases:

- *Phase 1:* De-authentication attack, sending de-authentication frames to the victim using the access point's MAC address as the source.
- *Phase 2*: Client capture, victim scans channels to search for a new access point, and then associates with fake access point on the attacker's machine. Fake access point is on a different channel than the real one, and is generally duplicating its ESSID.
- *Phase 3*: Connection to the access point, attacker's machine associates with the real access point and is now inserted between the two entities. It tries to pass frames through.

We have tested this attack on two different architectures. The first was based on EAP-MD5 as a non-mutual authentication method, and the second on a W-Kerberos scheme. Monkey Jack versus EAP-MD5 was a total success. The attacker has successfully de-authenticated the client station in about 2 seconds. The victim, then associates with the fake access point and the attacker passes through authentication frames. The attack lasts 5 seconds.

For the W-Kerberos architecture, the first phase of this attack was the same as the first test. The attack machine, via de-authentication frames (those frames being not authenticated), has been able to capture the victim station (phase 1 and 2). However, when trying to relay authentication frames, the access point must have knowledge of the access point secret key, to decrypt the session key in the service ticket. Moreover, any address modification is detected when verifying the authenticator field of the authentication messages, so the fake access point was unable to authenticate itself when asked to answer the authentication request of the client station. Thus, in our test the attack process failed and terminated with some errors on the attacker's machine interfaces.

6.3 Void11

Void11 is an open source implementation of some basic 802.11b DoS attacks. It mainly consists of:

- The *Deauth* tool (Network DoS): flooding wireless networks with de-authentication frames and spoofed ESSID, so that authenticated stations will drop their network connections.
- The *Auth* tool (Access point DoS): flooding access points with de-authentication frames and random stations addresses, so that access points will deny any service.

The W-Kerberos has been compared to material authentication test beds. The results for the "Deauth" tool are the following:

Table 2. Void11 Deauth tool results.

Clients	***Auth. Method***	***Auth. AP***	***Auth. Server***	***Test Time***	***Results***
Client1 MAC1	EAP-TLS	Cisco Aironet AP 350 Series	FreeRadius	12 sec	Deauthenticated
Client2 MAC2	EAP W-Kerb	HostAP W-Kerb	WKerberos Server	25 min	Associated, Authenticated

These results prove the resistance of the W-Kerberos to attacks based on forgery of spoofed EAP messages, and particularly de-authentication messages. The Man In the Middle attacks, being typically based, on such denial of service, are very reduced with an authenticator field in the EAP messages.

The second scenario has consisted in testing the hostAP soft access point, implementing the W-Kerberos authentication, and verifying its capacity of resistance to de-authentication frames flooding from random stations addresses. The results are the presented in [Table 3].

Table 3. Void11 Auth tool results.

AP	***Auth. Server***	***Observations & results***
Cisco AP 350	FreeRadius	The access point stops broadcasting beacons during 15 minutes. Authenticated clients are disassociated.
Cisco AP 1100	FreeRadius	Clients are disassociated, but re-associate periodically. The access point continues to broadcast beacons.
HostAP WKerb	W-Kerberos	Authenticated clients continue to be associated and authenticated, but the access point rejects any new authentication. Key refreshment messages are received at time.

The hostAP access point survived to this attack, and authenticated clients are not rejected. Further, their refreshment keys arrive at time, so there is no risk concerning any cryptographic attack. These results demonstrate the resistance of the W-Kerberos architecture based on the HostAP module to DoS attacks.

7. CONCLUSIONS AND FURTHER WORKS

In this paper, we have presented a Kerberos-based authentication architecture for Wi-Fi networks. Furthermore, we have evaluated this architecture on the basis of the IEEE 802.11 threat model. Experimentations have shown that key refreshment is a mean thwarting some cryptographic attacks, and that mutual authentication and protecting the authentication results guarantees a prevention against Man In the Middle attacks. Finally, W-Kerberos has been robust enough to resist to Denial of Service attacks.

The specified architecture provides then an effective means of protecting the network from unauthorized users and rogue access points, making then the possibility to steal valuable information ruled out, due to the fact that Kerberos provides mutual authentication. Moreover, this architecture is highly customizable, allowing the use of different encryption mechanisms and maintaining thus ability to plug-in different cryptographic algorithms.

The Ticket concept existing in the W-Kerberos protocol is well adapted to mobility needs within an IEEE 802.11 environment. Future works will expand this work considering the exploitation of proactive key distribution to specify a fast and secure handover, performance evaluation in different scenarios and especially handover overhead, and public key cryptography extension.

NOTES

1. For more details on key generation, see[6]

REFERENCES

1. Scott Fluhrer et al., "Weaknesses in the Key Scheduling Algorithm of RC4". In proceedings of the eighth Annual Workshop on Selected Areas in Cryptography, Toronto, August 2001.
2. Nikita Borisov et al., "Intercepting Mobile Communications: The Insecurity of 802.11". In Seventh Annual International Conference on Mobile Computing And Networking, Rome, Italy, July 2001.
3. M. Mishra, W.Arbaugh, "An initial Security Analysis of the IEEE 802.1X Standard". Technical report, University of Maryland. February 2002.
4. IEEE 8021X, "Port-based Network Access Control. IEEE Std 802.1x". IEEE Standard, June 2001.
5. L.Blunk, J. Vollbrecht, "PPP Extensible Authentication Protocol (EAP)". RFC 2284, March 1998.
6. J. Kohl, C. Neuman, "The Kerberos Network Authentication Service (V5)". RFC 1510, September 1993.
7. IEEE, "TGe Security Baseline Draft" Draft IEEE 802.11e, March 2001.
8. J. Trostle et al., "Initial and Pass Through Authentication Using Kerberos V5 and the GSSAPI (IAKERB)". Internet draft, October 2002.
9. A. Lakhiani, "A transparent authentication protocol for Wireless Networks", Master thesis, University of OHIO, March 2003.
10. D.B. Faria, D.R. Cheriton. "DoS and Authentication in Wireless Public Access Networks". In Proceedings of the First ACM Workshop on Wireless Security , Atlanta, September 2002.
11. M.A. Kaafar et al., "A Kerberos-based authentication architecture for Wireless Lans". In proceedings of the Third IFIP-TC6 International Conference on Networking (Networking 2004), Athens, Greece, May 2004.
12. C. Neuman et al. "The Kerberos Network Authentication Service (V5)", Internet Draft draftietf-krb-wg-kerberos-clarifications-05.txt, June 2003.
13. T. Wu, "A Real-World Analysis of Kerberos Password Security". In proceedings of the sixth Annual Symposium on Network and Distributed System Security, San Diego, February 1999.

14. B.Tung, et al., "Public Key Cryptography for initial authentication in Kerberos". Internet Draft draft-ietf-cat-kerberos-pk-init-18.txt, March 2001.
15. N.Asokan et al., "Man-in-the-middle in tunneled authentication protocols". Technical Report 2002/163, IACR ePrint archive, October 2002.
16. J. Puthenkulam et al., "The compound authentication binding problem". Internet draft draft-puthenkulam- eap-binding-01.txt, October 2002.

AN EFFICIENT MECHANISM TO ENSURE LOCATION PRIVACY IN TELECOM SERVICE APPLICATIONS

Oliver Jorns[1], Sandford Bessler[1] and Rudolf Pailer[2]
[1]*Telecommunications Research Center Vienna (ftw.), Donau-City-Strasse 1, A-1220 Vienna;*
[2]*mobilkom austria AG & Co KG, Obere Donaustrasse 29, A-1220 Vienna*

Abstract: Location and presence information will provide considerable value to information and communication services. Nevertheless, the users are still concerned about revealing their position data especially to un-trusted third party applications. Furthermore, legal restrictions are effective in most countries that regulate processing of personal data and the protection of privacy in electronic communications. In this paper we propose a novel privacy enhancement solution (PRIVES) which is targeted for location and presence services in the 3G service architecture and uses cryptographic techniques well suited to run in small devices with little computing and power resources. Once a user is granted the permission to localize another user, the location server generates a key used to create pseudonyms that are specific for the localized user. Passed from the watcher to the location server via the application, these pseudonyms identify both the watcher and the desired localized user at the location server, but are opaque to the application. The paper presents architecture and protocols of the proposed solution and discusses the performance increase in comparison with current implementations.

Key words: location privacy, pseudonyms, 3rd party applications, HMAC, Parlay-X web service, hash value, hash chain, one-time password, presence.

1. INTRODUCTION

The current next-generation service architecture enables the establishment of a new class of service providers and independent software vendors that design innovative services accessing the networks through open and standardized interfaces and protocols such as OSA/Parlay, or SIP/SIMPLE. A key factor for these services to be successful is their personalization, i.e.

the ability to take into account user's preferences, presence, location, etc. The existence of 3rd party service providers that are less trusted than the network operator on one side and the personalization need on the other side have lead to increasing privacy concerns and motivated us to perform research on efficient methods to protect user data.

Furthermore, the processing of personal data and the protection of privacy is regulated by law in most countries. The EU directive on privacy and electronic communications [1] mentions location data explicitly and establishes the following rules:

- Providers should minimize the processing of personal data and use anonymous or pseudonymous data where possible.
- Location data is listed as traffic data. Traffic data means any data processed for the purpose of the conveyance of a communication on an electronic communications network or for the billing thereof.
- Location data in the sense of traffic data means any data processed in an electronic communications network, indicating the geographic position of the terminal equipment of a user of a publicly available electronic communications service and may refer to the latitude, longitude and altitude of the user's terminal equipment, to the direction of travel, to the level of accuracy of the location information, to the identification of the network cell in which the terminal equipment is located at a certain point in time and to the time the location information was recorded. Location data other than traffic data is regulated to the same extent as traffic data.
- For the provision of value added services, the provider of a publicly available electronic communications service may process traffic data to the extent and for the duration necessary for such services, if the subscriber or user to whom the data relates has given his/her consent. Users or subscribers shall be given the possibility to withdraw their consent for the processing of traffic data at any time. The service provider must inform the users or subscribers, prior to obtaining their consent, of the type of location data which will be processed, of the purposes and duration of the processing and whether the data will be transmitted to a third party for the purpose of providing the value added service.
- Where consent of the users or subscribers has been obtained for the processing of location data, the user or subscriber must continue to have the possibility, using a simple means and free of charge, of temporarily refusing the processing of such data for each connection to the network or for each transmission of a communication.
- Location data may only be processed when it is made anonymous, or with the consent of the users or subscribers to the extent and for the duration necessary for the provision of a value added service.

Privacy requirements are also influenced by the type of the localizing application. We categorize localizing applications in 'Pull', 'Push' and 'Tracking', depending on the relationship between the localizing user, called 'watcher', and the user to be localized, called 'presentity'. These applications have in common, that the localization process is usually executed by some kind of middleware component, that may not be part of the trusted network provider's domain, but is operated by a 3rd party service or content provider. In general there is a relationship triangle between watcher, application and presentity (see Fig. 1).

Privacy enhancement technologies (PET) address four basic ISO requirements [2]: anonymity, pseudonymity, unlinkability and unobservability that are subject of a number of research projects dealing with address privacy, location privacy, service access privacy or authentication privacy [3].

In this paper we study the location privacy, an issue that arises today in all 2G and 3G mobile networks that start to offer location based services. Terminal (and user) localization is done either by the user himself, equipped with a GPS receiver or, in most cases by the network, using the signal strength of a few neighboring cells listening to the terminal. The location information (expressed for example in geographical coordinates) can be used by the localized user (push/pull applications) for example to direct a call to the nearest pharmacy or taxi, or by another user (tracking applications) who may run an application that schedules a service team. In either scenario there is an application or service that processes location data and that is owned by some third party commercial organization.

Any proposed schemes to improve privacy in mobile networks clearly have strong interoperability and standardization aspects, as they have to be approved by 3GPP or IETF. The IETF "Geographic Location/Privacy (geopriv)" Working Group has defined location privacy requirements [4] in which a Location Object (LO) plays the main role: it contains the location information to be transmitted from the Location Server entity (a part of the network operator) to the location requestor (watcher) as well as access rules for different users and is itself cryptographically protected. While this proposal is general and powerful, it implies that a large data amount has to be transmitted and requires from the watcher a lot of processing power.

As the first GSM and UMTS networks start to offer location based services, the 3GPP has tried to improve privacy by tightening the access control of users and applications on location information. Thus, in a privacy enhancement specification for UMTS Release 6 [5], a complete authorization relationship between three entities: requestor (watcher), application and presentity has to be defined in the Telecom Server in order to allow access to location information.

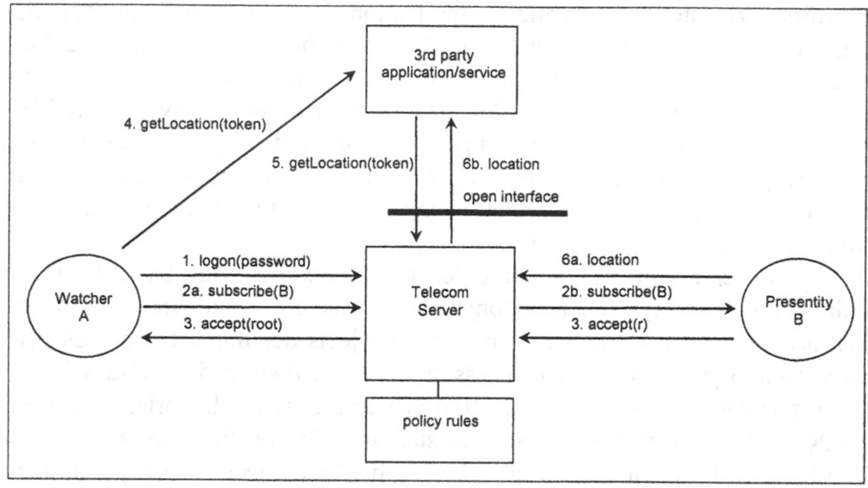

Figure 1. Architecture overview

The responsibility of the Telecom Server in Fig. 1 is to check the access rights of watcher and application (also for non-repudiation and monitoring reasons) and leads to complex processing, very large database tables and a tight coupling of all participating entities. The protection of the presentity identity is handled vaguely by proposing the use of aliases.

To overcome this situation, the approach described by Hauser et al.[6] can be used. It is based on pseudonyms which are exchanged between the watcher and the telecom service in order to make it impossible for 3rd party services to track the location of a certain presentity, store the location history or aggregate information from several services and create a profile. The difficulty in applying these schemes in practice arises from the use of public key infrastructures for signature, encryption and decryption processes which are computationally too expensive to be executed on today's mobile terminals. These considerations have motivated us to propose a more efficient Privacy Enhancement Scheme (PRIVES)[1] to be used with the architecture in Fig. 1 to protect the identity of localized users.

The rest of the paper is organized as follows: section 2 describes the architecture and the generic interactions between watcher, presentity, Telecom Server and application. Section 3 describes the mechanism used to generate pseudonyms at the watcher and at the Telecom Server. Section 4 gives preliminary implementation and performance results and section 5 concludes

[1] Verfahren zum Unwandeln von Target-Ortsinformation in Mehrwertinformation, pending Patent Nr.: A 363/2004

with future extensions and applicability of the proposed scheme in other scenarios.

2. SERVICE INTERACTIONS

In this section we describe shortly the service interactions between the system entities in Fig. 1: watcher, presentity, telecom server and 3rd party application. Basically, the watcher starts by establishing a trust relation with the presentity using a subscription/notify message pattern. We assume that the presentity accepts a subscription to his/her location information only, if the watcher is known and trusted. Alternatively, rules may be predefined and stored in form of policies in the Telecom Server. Both approaches can be combined with a group management system.

Subscription/acceptance messages precede authentication of the users to the Telecom Server. Other messages are needed to query the status of subscriptions, which are stored and forwarded when the users go online:

- `getBuddies()` returns the list of subscribed and accepted presentities
- `getPendingWatchers()` returns the list of watchers waiting an accept message for that presentity
- `getPendingSubscriptions()` returns the list of presentities that did not send an accept so far.

Returning to the general operation in Fig. 1, the accept message is mediated by the server which calculates a "root" value r and sends it to the watcher (step 3). The root is the initial value of a chain of one-time passwords (pseudonyms, tokens) that are subsequently sent in localization requests to the 3rd party application. These pseudonyms are used to identify the presentity in the (standardized) API methods between the 3rd party application and the Telecom Server (see Fig. 1, step 5). The pseudonyms sent by the application to the Telecom Server are used to authenticate and authorize the localization request and to retrieve the real user identity. The Telecom Server then retrieves presence or location information of the presentity and returns this data to the application.

3. USE OF HASH VALUES FOR AUTHENTICATION AND AUTHORIZATION

In order to reduce the calculation complexity on the watchers's and the presentity's device, we propose PRIVES, a scheme which authenticates and authorizes the watcher on the basis of hash values.

A hash value is the result of a hash function H which is a one-way function that it is easy to compute but computationally infeasible to invert. A hash function y = H(x) is defined as a function for which the effort of computing x, given the output y consisting of n bit, is 2^{n-1}. Hash functions are also required to be collision resistant, that means, finding two different inputs x and x' such that H(x) = H(x') requires an effort of $2^{n/2}$. The most commonly used hash functions are MD5 [7] by Ronald Rivest and SHA-1 [8] by NIST.

For repeated interactions such as requesting location information periodically, we use hash values that are calculated from the previous hash value. The idea to use hash values for authentication which are based on a chain of computations of hash values was first published by Lamport [9]. Starting with the shared secret the first computed hash value is used as input for the calculation of the next hash value and so on. After the client received n (number of hashes to compute) it calculates the n-1 hash value (h^{n-1}) on the basis of the shared secret and sends it back to the server. The server calculates the next hash value of this received one ($h(h_{n-1})$) and compares it with the nth computed hash value computed on the basis of the shared secret (h^n(shared_secret)). Equality of these values proves that the client owns the right shared secret. Now, the server decrements the value of n by one and stores this value. The next time the client authenticates, the server receives the hash value h^{n-2} (see Fig. 2).

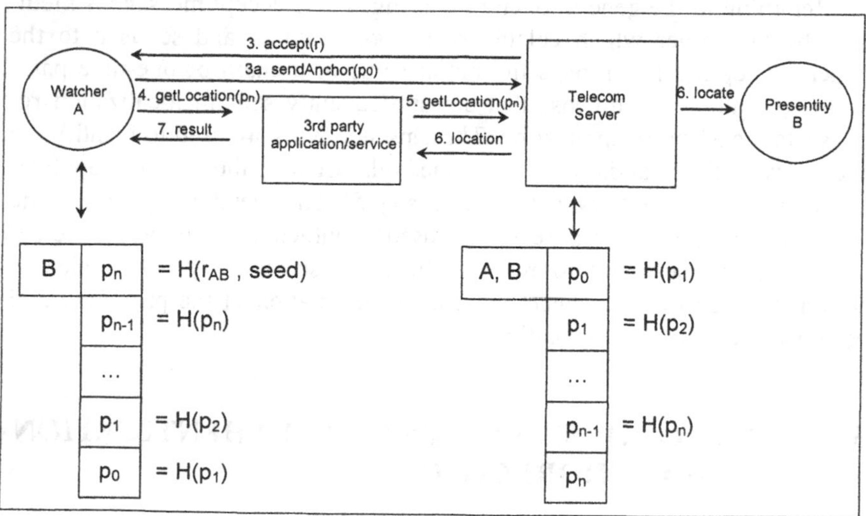

Figure 2: System Architecture based on Lamports Hash chains

The strength of security is reached because of the fundamental one-way property of hash values. The problem with a hash chain is that it has to be used in the reverse direction, i.e. first the nth value, then the n-1 value, etc., implying that n-values have to be generated first - a complex operation in small devices.

The One-Time Password (OTP) System [10] is similar to the solution of Lamport. Although both guarantee a high level of security, they fail in practical implementations because OTPs have to be used in reverse order of their creation, that is the ith OTP p_i is the $(n-i)^{th}$ value of the hash chain which is not feasible on mobile devices as our measurement results clearly show.

This means that all OTPs could be computed at once and stored in the user's terminal, which is probably infeasible in mobile devices that usually have only a very restricted amount of memory available.

Alternatively the OTP could be computed on demand from the root r, meaning that for every OTP the hash chain has to be partially rebuilt. On demand computation of the OTP p_i requires n-i applications of the hash function with the result that for using all OTPs of a hash chain of length n, a mobile device would have to compute hashes $n*(n+1)/2$ times. The computational effort of calculating one OTP has the order O(n) and the on demand calculation of all OTPs of a hash chain of length n increases with the order $O(n^2)$, which also restricts the applicability on mobile device platforms.

We use in PRIVES the keyed hash scheme called HMAC [11] that overcomes the computational problems of the former schemes and still guarantees high security without the need of computation of a large number of hash values in advance. HMAC is based on MD5 or SHA-1 and allows us to create a hash value from a previous one, using in addition the password as a secret key shared between the watcher and the Telecom Server.

3.1 Hash-chain generation and initialization

The only information watcher A needs is the root r_{AB}, which is a random number initially created by the Telecom Server after a successful subscription. All messages are synchronous (request - response), so that watchers have to refresh periodically the subscription status. As shown in Fig. 3, the hash values are created in parallel by the watcher and the Telecom Server from the previous value h_{n-1} and watcher's password by applying the HMAC function.

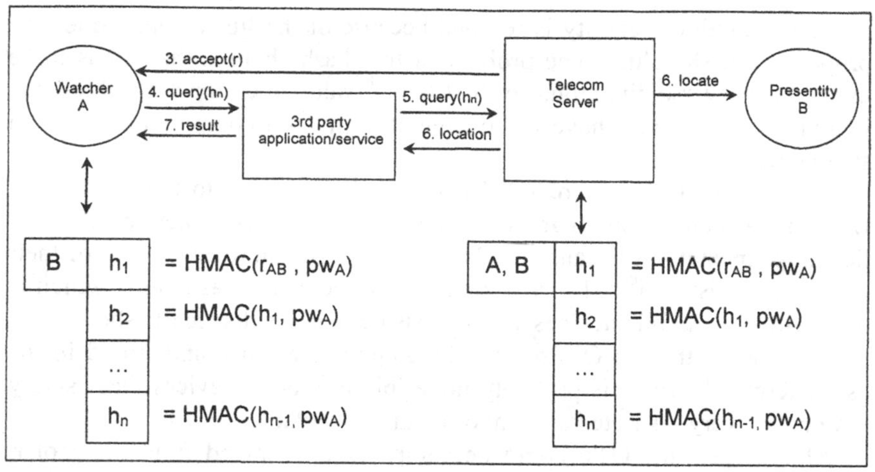

Figure 3. Creation of chained hash values based on HMAC

After subscription, the watcher sends the first hash value as argument of a location request to the 3rd party application. The request is forwarded via the standardized interface to the Telecom Server where the real identity of the presentity and its location is determined.

The Telecom Server maintains a hashed access control list, such that exactly one access to a database table is required to resolve authorization rights, whereas an implementation based on standard procedures would have to first fetch the presentity data from a user table and then to look up the watcher (and/or application) in an 'authorized watcher' table and to check the allowed operations. Thus, the performance gain for the access control look-up is at least a factor of 2.

Finally, if authorizations have to be revoked, the Telecom Server has just to delete the stored hash value from the access control list, leading to a new subscription at the watcher by the time the next localization requests arrives.

3.2 Protocol error handling

The correct operation of PRIVES relies on the synchronization of hash value creation at watcher and server side. However, in case of erroneous transmission channel, unexpected crashes of client applications due to loss of battery power, synchronization is lost.

Therefore, each hash value has to be stored persistently on the user's device after a location request. Each time the application starts, the stored hash value and the user's password are used for calculation of subsequent hash values.

If for any reason the Telecom Server cannot process a received hash value, it returns coordinates (0, 0) as result, which indicates an error. To recover from the error, the client sends a new `subscribe` message for the respective presentity. Since the watcher subscribed this presentity already, the server computes and issues a new root which reinitializes the hash values and allows the watcher to request the position again.

The use of hash values in PRIVES increases also the efficiency of the Telecom Server that in normal case has to check if the watcher, and the 3rd party application are authorized to locate the presentity .

4. PROTOTYPE REALIZATION

In order to validate the architecture and measure the performance of the system, we implemented in the lab a "Telecom Server" that obtains real location information from a Parlay-X location service. Thus, several mobile phones (the presentities) can be localized. The watchers are currently implemented on a J2ME Personal Profile 1.0 [12] platform. A simple application is being currently implemented for demonstration purposes: it tracks the user and calculates periodically the distance between successive retrieved locations of that user. The scenario is a bit more complex than that described in Fig. 3: it starts with the watcher (which is identical with the presentity in this case) sending the application one single startRoute() request with the pseudonym as parameter. The application requests from the Telecom Server to be notified with the new position of the user every T minutes (using the pseudonym as parameter). This architecture has the advantage that the application can communicate much more efficiently with the Telecom Server than the wireless client and it relieves the client from transmitting a high amount of data. When the terminal leaves a certain geographical area, the total (direct) distance is summed up, the user is notified and the service session terminates.

4.1 Performance and security considerations

The performance gain is determined by the following operations:

- Hash value calculation at watcher and server
- Checking a certain hash-value at the server

One reason for using hash techniques is their computational efficiency that makes them suitable for today's wide spread mobile devices. Our performance measurements were undertaken for MD5, SHA-1, MD5/HMAC and SHA-1/HMAC, carried out on a mobile emulator of the J2ME Wireless Toolkit 2.0 with preset VM speed emulation of 100 byte codes/millisecond.

All results in Table 1 and Table 2 are mean values based upon calculation of 100 hash values. Our performance analysis first concentrated on an implementation of the OTP (one time password) System designed to allow only up to n=10 authentications before the system needs to be reinitialized. This would require n*(n+1)/2=55 hash value calculations (see Table 1).

Table 1. Mean time in ms for single hash value calculations

function	mean time
MD5	51ms
SHA-1	139ms
MD5/HMAC	164.1ms (0.164sec)
SHA-1/HMAC	448ms (0.448sec)

To keep computing effort low, n has to be small which however results in frequent re-initializations. From Table 2 we see that an implementation based on chained hash calculations as it is done in the OTP System is not feasible on mobile devices with low processing power.

The calculations based on MD5/HMAC and SHA-1/HMAC take longer than those based on corresponding MD5 and SHA-1, but since the hash value can be safely calculated from a previous one (n=1), the HMAC procedure is fast enough for our purposes.

Table 2. One-Time Password scheme and expected mean time needed for calculation

function	Total computation time for 10 authentications	Total computation time for 100 authentications
MD5	2190ms (2.19sec)	205003ms (~3.4min)
SHA-1	6677ms (6.677sec)	701950ms (~11.7min)

Our performance analysis first concentrated on an implementation of the hash chain scheme that calculates OTPs in reverse order. The implementation does not store the calculated hash values, but has to partially rebuild the hash chain for each authentication. Storing n 128 bit (MD5) or 160 bit (SHA-1) values for each buddy requires a memory amount that may not be available on every today's mobile devices. A preset hash chain length of n=10 allows 10 authentications before re-initialization. The first authentication requires the computation of 10 hash values, the next 9 and so on. This means that the largest delay for getting the hash chain is determined by the first authentication and grows linearly with n.

To keep computing effort low and authentication delays acceptable, n has to be small which however results in frequent re-initializations. Our calculations show that an implementation based on reverse chained hash calculations (without storing the hash chain) is not feasible on mobile devices with low processing power.

The time needed to calculate a hash value based on MD5/HMAC and SHA-1/HMAC takes longer than with MD5 and SHA-1 (see Table 1). But since only one HMAC value has to be calculated per authentication and each hash value can be safely calculated from a previous one (n=1), HMAC is better suited. We also see from Table 1 that the HMAC calculation takes about three times longer than the underlying hash function. This means that for n=4 the HMAC scheme is already faster than the reverse order scheme.

From a security point of view HMAC is secure enough given the (not so high) sensitivity of the data. If higher secrecy of data is required, SHA-1/HMAC should be preferred over MD5/HMAC, since collisions in compressing functions of MD5 have already been found [12]. In case processing power is critical, MD5/HMAC will be the better choice because it is computed approximately three times as fast as SHA-1/HMAC.

In general, it is not possible for an intruder to calculate easily hash values by eavesdropping the previous HMAC hash value. Examinations on HMAC in [11] show that finding a collision (guessing the input values that result in the same hash value) for hash values of l =128 bit length would require $2^{l/2}$ messages for each given key. It can be assumed that this is improbable to occur. As [11] further states, using the same password an attack would require approximately 250.000 years.

5. CONCLUSIONS AND FURTHER RESEARCH

In this work we propose PRIVES, a scheme that allows a third party application to receive and process user location information from a network operator without being able to identify the localized user. Monitoring users, building of location profiles or aggregation across different applications becomes impossible, a fact that would increase the user acceptance for location based services. Since the presentity grants the watcher the access to location information no additional security responsibility has to be taken by the LBS to check the authorization of the watcher. Furthermore, PRIVES retrieves data tuples from access control lists in an efficient way.

As further research directions, we will investigate, whether PRIVES can be extended to other location operations, such as triggered location or periodical notifications, or to other privacy relevant services, like presence.

6. ACKNOWLEDGEMENT

This work has been done at the Telecommunications Research Center Vienna (http://www.ftw.at) within the project "Service Platforms beyond OSA", and partially funded by the Austrian Kplus Program.

REFERENCES

1 Directive 2002/58/EC of the European Parliament and of the Council concerning the processing of personal data and the protection of privacy in the electronic communications sector (Directive on privacy and electronic communications), 12 July 2002, Official Journal of the European Communities L 201/37

2 "Common Criteria for Information Technology Security Evaluation , Part 2: *Security functional requirements*, January 2004, Version 2.2, CCIMB-2004-01-002, aligned with ISO 15408, http://www.commoncriteria.de/it-security_english/ccinfo.htm

3 RAPID - Work Package 2 - Stream 6: *PETs in Infrastructure*, FP5 IST Roadmap project, RAPID – IST-2001-38310: Roadmap for Advanced Research in Privacy and Identity Management, http://www.ra-pid.org/

4 J. Cuellar, J. Morris, D. Mullignn, J. Peterson and J. Polk, Geopriv Requirements, IETF RFC 3693, Feb. 2004

5 3GPP, *Enhanced User Support for Privacy in Location Services*, 3GPP TR 23871 V 5.0.0

6 C. Hauser, M. Kabatnik, *Towards Privacy Support in a Global Location Service*, Proceedings of the IFIP Workshop on IP and ATM Traffic Management (WATM/EUNICE 2001), Paris, 2001

7 Ronald L. Rivest: *The MD5 Message-Digest Algorithm. RFC 1321*, April, 1992.

8 National Institute of Standards and Technology: *Secure Hash Standard*, Federal Information Processing Standards (FIPS) Publication 180-2, 2002

9 Leslie Lamport, *Password Authentication with Insecure Communication*, Communications of the ACM, vol. 24(11), 1981, pp. 770-772.

10 N. Haller, C. Metz, P. Nesser, M. Straw, *A One-Time Password System*, RFC 2289, 1998

11 Mihir Bellare, Ran Canetti, Hugo Krawczyk, Message Authentication using Hash Functions – The HMAC Construction, CryptoBytes, Vol. 2, No. 1, 1996, http://www.cs.ucsd.edu/users/mihir/papers/hmac-cb.pdf

12 Hans Dobbertin: *Cryptoanalysis of MD5 Compress*, Announcement on Internet, May, 1996, http://citeseer.ist.psu.edu/dobbertin96cryptanalysis.html

13 J2ME Personal Profile, Version 1.0, http://java.sun.com/products/personalprofile/index.jsp

NETWORK SECURITY MANAGEMENT: A FORMAL EVALUATION TOOL BASED ON RBAC POLICIES

Romain Laborde, Bassem Nasser, Frédéric Grasset, François Barrère, Abdelmalek Benzekri

IRIT/SIERA Université Paul Sabatier, 118 Rte de Narbonne, F31062 Toulouse Cedex04 France

Abstract: The complexity of factors to consider makes increasingly difficult the design of network security policies. Network security management is by nature a distributed function supplied by the coordination of a variety of devices with different capabilities. Formal evaluation techniques should be used to ensure that correct security network strategy are enforced. In this paper, we propose a new formal tool which allows to describe a given network security strategy, a network topology and the security goals required. The tool includes an evaluation method that checks some security properties and provides information to refine the strategy used. We introduce an example of VPN architecture which validates our approach.

Key words: Policy, Network Security, Security Management, Security Evaluation

1. INTRODUCTION

Basically, the security of distributed applications is supported by a set of network security services which are implemented by means of security mechanisms. The security administrator should determine the *security services to use* and the *security mechanisms configurations to apply.* End to end security (e.g. SSL based solution) is often used, but it leads to conceal the underlying network. If such solutions can provide confidentiality, integrity, non repudation and authenticity properties, it is not suitable regarding the availability, anonymity property (e.g. deny of service, non accessibility) or regarding networks characteristics (throughput, flow control...). Then the design, the operation, and the maintenance of these *network configurations* constitute an important part of the security management task.

Network management is by nature a distributed function supplying the coordination of a variety of devices with different capabilities (PC, firewall, secure gateways, routers, etc.). Once deployed, network security often become unmanageable over time since more rules are added and there is a real difficulty in retrieving, managing and getting rid of old unnecessary rules.

The first category of security management problems is the *security mechanisms inconsistency*. It can be divided into two sub-groups: the *atomic* security mechanisms inconsistency and the *distributed* security mechanisms inconsistency. The atomic inconsistency problem considers that two or more configuration rules on the same device can be incompatible. For example, one rule states that data flows with the source IP addresses in the range 10.0.0.0 can pass through the firewall and another rule on the same firewall states that the data flow with the source IP address 10.20.30.4 is denied. Several techniques[1] can be used to solve it, for example:

- *Denials take precedence* : negative authorizations take precedence,
- *Most specific take precedence* : the authorization that is most specific w.r.t. a partial order wins,
- *Positional*: the priority of the authorization depends on which they appear in the authorization list,
- *Priority level* : each authorization is assigned with a priority level, the authorization with the highest priority wins,
- Etc.

The distributed inconsistency concerns incompatible rules mapped on different devices. Thus, the administrator should pay a special attention to all dependency relations between rules present on different devices. For instance, an IPsec tunnel is correctly configured between two VPN gateways and a firewall between them blocks their IPsec data flows. Some works provide a partial solution considering only one kind of device, for example firewalls[2-4], IPsec gateways[5] or filtering IPsec gateways[6].

Nevertheless, security mechanisms consistency does not imply that the security objectives are achieved (i.e. the administrator has chosen the good security services). The latest management paradigms[7-9] aim to automate the management tasks. In this context, policy based management approach[7,10,11] considers abstract security policies that can represented at different levels[12,13] ranging from the business goals to the devices-specific configurations. The process that transforms a definite goal into the corresponding configurations is called derivation process[14]. Thus, it tries to define the relation between the objectives and mechanisms configuration.

As there is no formal and automatic evaluation method of the couple services/mechanisms against the objectives yet, we are working on the definition of a general framework for the specification and the evaluation of network security mechanisms/services against network security goals. In our approach we have defined a language that allows the expression of the network security objectives, the network security services and the network security mechanisms with their configuration. Moreover, it brings the ability to specify the network topology because the efficiency of the security mechanisms depends on. It also includes a formal evaluation process.

This article only presents our specification language, its expressiveness and our evaluation tool. The formal definition of the language and the formal evaluation process is presented by Laborde et al[15]. Section 2 exposes our way of defining the network security objectives. In section 3, we define our network model and our specification language and briefly comment the evaluation process. In section 4, we present our tool that automates the evaluation task by a simplistic example. Finally, in section 5, we conclude and introduce our plans for future works.

2. DEFINITION OF NETWORK SECURITY OBJECTIVES

Traditionally, the network security officer addresses security problems using an emperical approach where each problem is considered one after the other. Such a point of view does not permit to determine correct security objectives because they are not integrated in a global security management process. Management models, like the TMN[16] one, show clearly that networks provide services to applications. So, the requirements of the applications constitute the network objectives. Thus, in this section, we formalize this dependency in the security context.

2.1 The relationship between an application security policy and a network security policy

When a user accesses a service, a set of data flow is exchanged between the device from which the user launches the service and the devices supporting the service execution (fig. 1). So, a relation between a network security policy and an application security policy can be distinguished. For example, if the application security policy states that user "u_1" can read object "o_1"- noted (u_1, o_1, +read), then it implies that a corresponding data flow *flow(o_1,+read)* between the device of user "u_1" and the device of "o_1" can exists on the network. Consequently, the associated network security policy must allows the data flows *flow(*o_1*,+read)* between these two devices – noted (device(u_1) ↔ device(o_1), +flow(o_1, read)). Conversely, if the application security policy states that user u_2 cannot read object o_2 noted (u_2, o_2, - read), there no flow *flow(*o_2*,read)* between the devices of u_2 and o_2. Therefore, the network security policy must forbid *flow(*o_2*, read)* between the devices of u_2 and o_2, i.e., (device(u_2) ↔ device(o_2), - flow(o_2, read)). We thus obtain the derivation relation noted "$\Rightarrow^d$" as $\forall u \in$ USERS, $\forall$ o $\in$ OBJECTS, $\forall$ a $\in$ ACTIONS, (u, o, ±a) $\Rightarrow^d$ (device(u) ↔ device(o), ± flow(o, a)).

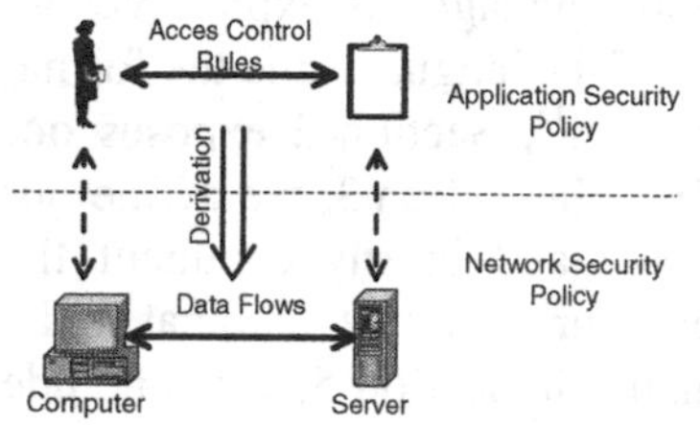

Figure 1. Security policy derivation

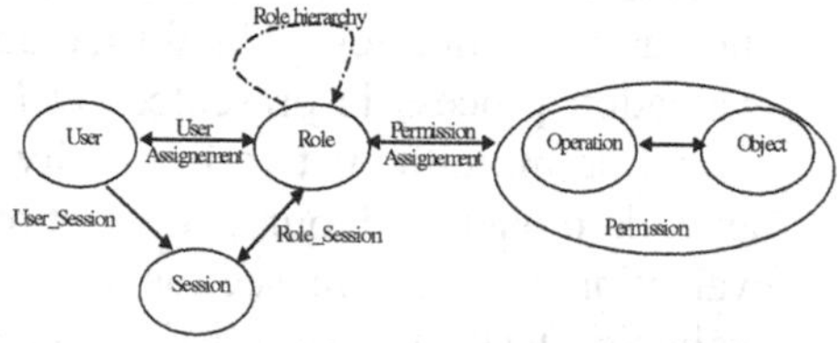

Figure 2. The NIST RBAC Model

2.2 The NIST RBAC Model

Access control is the process of mediating every request to resources and data maintained by a system and determining whether the request should be granted or denied. Access control models[1] provide a formal representation of the access control security policy and its working. The formalization allows the proof of properties[1,17-20] on the security provided by the access control system being designed.

Among the access control models, we have chosen the NIST RBAC model[17] because it simplifies the management tasks. Actually, the role concept allows aggregating the users' permissions and then it facilitates the users' rights modifications made by an administrator. Moreover, the hierarchies between roles represent a good tool for modeling an organization according to different points of view.

The NIST group proposes the standardization of the RBAC model. It is made up of two sub-models: the core model and the hierarchical model (fig.2).

The core model includes five sets of basic data elements:

- A *user* is an active entity, i.e., human or intelligent agent.
- A *role* is a job function within the context of an organization with some associated semantic regarding the authority and responsibility on the user assigned to the role. We can notice that the definition is very vague.
- A *permission* is an approval to perform an operation on one or more protected objects.
- An *operation* is an executable image of a program, which upon invocation executes some function on behalf of the user.
- An *object* is an entity that contains or receives information.

Finally, a set of roles is assigned to a user, and a set of permissions is assigned to a role. A session is a mapping of one user to a set of authorized roles.

The hierarchical model adds relations for supporting role hierarchies. There exist different approaches for constructing a role hierarchy: based on privileges[20] or based on users' job functions[21,22].

2.3 Towards an "RBAC network security policy"

Users are considered in an RBAC system by their assigned role. Consequently, the derivation relation becomes: $\forall$ r $\in$ ROLES, $\forall o_i \in$ OBJECTS, $\forall$ op_j $\in$ OPERATIONS, $\forall u \in$USERS, $\forall u' \in$USERS • (r, {(op_j, o_i)}) $\wedge$ Assigned(u,r) $\wedge$ $\neg$Assigned(u',r) $\Rightarrow^d$ (device(u) $\leftrightarrow$ device(o_i), +flow(o_i, op_j)) $\wedge$ (device(u') $\leftrightarrow$ device(o_i), - flow(o_i, op_j)).

Thereafter, we consider that there is *no hierarchy* and that *roles have disjoint privileges* (if this is not the case then we may create a partition of this set): such a constraint will help us to group data flows based on the permissions assigned to one role and then identifying them by the role. Afterward, we note by the name of the role the set of flows corresponding to the permissions assigned to the role.

According to these definitions, we present our language which is able to express the network security objectives, i.e. the RBAC information, the network security mechanisms and the network topology.

3. NETWORK ARCHITECTURE MODEL AND NETWORK SECURITY SPECIFICATION

Each communication generates data flows between a source and a destination system. Our approach consider that the applicable treatments on data flow can be brought together into four basic functionalities. Devices are modeled while interconnecting these basic functionalities:

- Mechanisms that *consume/produce* data flows such as the end-systems,
- Mechanisms that *propagate* data flows such as physical supports and associated devices,
- Mechanisms that *transform* data flows into another one such as the security protocols,
- Mechanisms that *filter* data flows such as the firewall ones.

In our model (fig. 3), there are a set of active entities and a set of passive entities, and a set of functionalities (end-flow, channel, transform and filter) which act on information flows. An active entity corresponds to a user in the RBAC model, and a passive entity is a set of objects in the RBAC model.

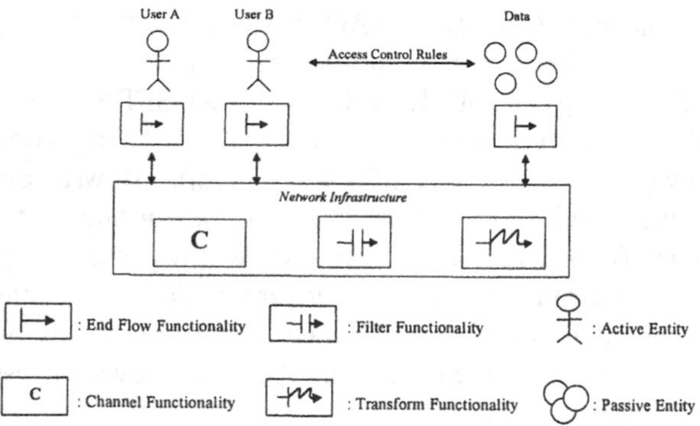

Figure 3. The network security and topology model

3.1 Definition of the functionalities

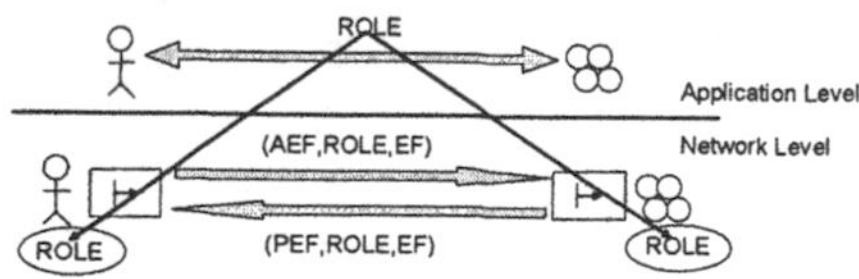

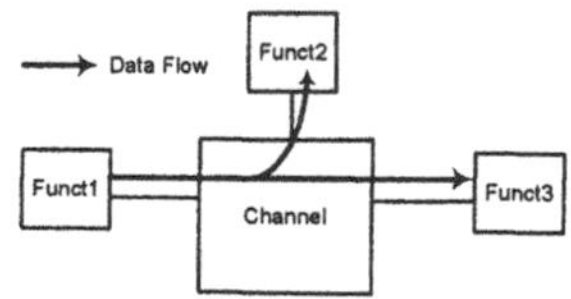

Figure 4. End-flow functionality

Figure 5. Channel functionality

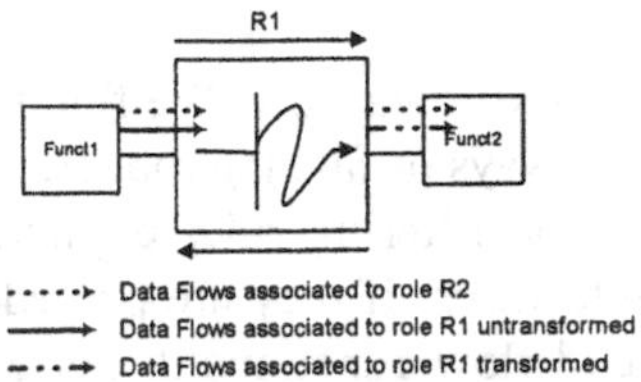

Figure 6. Transform functionality

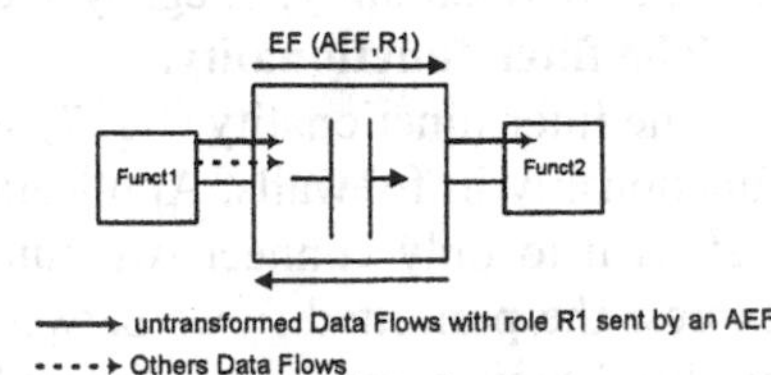

Figure 7. Filter functionality

We have modeled data flows and all basic functionalities using Colored Petri Nets. The formal definition is given in Laborde et al[15]. We just present here a non formal definition of each functionality:

The end-flow functionality.

An end-flow (EF) is a functionality that is specific to end-systems, i.e., data and application servers as well as the workstations. It constitutes the link between the application level security model, i.e. RBAC, and our network model. Hence, we consider two types of end-flow functionalities:

- *Active End Flow functionality* (AEF): An EF is said active if any active entity is connected to this EF.
- *Passive End Flow functionality* (PEF): An EF is said passive if any passive entity is connected to this EF.

We append a list of roles to each EF for indicating the flows that the EF can produce. The list corresponds to the set of roles assigned to the user representing the connected active entity for an AEF. In the case of a PEF, it is the set of roles assigned to the permissions that concern an object of the connected passive entity. When the users launch their authorized services, it implies a communication between all the AEF and the PEF with the same role (fig. 4). The flow produced by an AEF (resp. PEF) with role R is noted (AEF,R,EF) (resp. (PEF,R,EF)). It allows the expression of the network security objectives.

The channel functionality.

The channel functionality models the physical network. It receives the flow from one of the connected functionalities and retransmits it to all the others connected functionalities (fig. 5).

The transform functionality.

The transform functionality receives a data flow (ex: (AEF,R,EF)) from one of its two interfaces, according to transformation rules represented by a list of roles which identifies the data flows that must be transformed, and sends to the other interface the same data flow or the data flow transformed represented by the parameter TR (ex: (AEF,R,TR)) (fig. 6). This new flow has the confidentiality, integrity and authenticity properties.

The filter functionality.

The filter functionality (fig. 7) stops or forwards a data flow. We find this functionality in firewalls, Application Level Gateways or filtering routers. We restrict it to only connect two functionalities. The filtering rules explicitly express the permitted flows between its two interfaces; if they are preceded by "EF" they come untransformed from an end-flow functionality, else if they are preceded by "TR" then they have been modified by a transform functionality.

3.2 Security analysis

The CPN model associated to each specification produces a reachability graph. It is analyzed with the set of security properties described here after. The formal properties definitions, the analysis process, its applicability in complex studies are given in Laborde et al[15].

Property of confidentiality.

Basically, the property of confidentiality limits protects the data from unauthorized disclosure. Thus, in our model, it prohibits an end-flow functionality from receiving at any time a untransformed data flow with any unassigned role.

Property of integrity.

Classically, the property of integrity prohibits non granted entities from any creation, modification or destruction of objects. Then, in our model, this property lay down that an end-flow functionality can only generates data flows with its assigned roles.

Property of availability.

This property stipulates that all the granted services must be available to all the authorized entities. In the network environment, the data flows corresponding to this must be able to travel between both devices. Consequently, its translation in our model is all active (resp. passive) end-flow functionalities must be able to consume all the data flows with an assigned role sent by every passive (resp. active) end-flow functionalities.

As we intend to address devices configurations, we complete these classical security properties with new ones:

Property of partitioning.

It is used to limit to the propagation of data flows. It declares that a data flow can only pass a filter functionality if it is situated between the data flow source and a possible correct destination.

Non productive filtering rule.

It is used to eliminate unnecessary filtering rules. Let f, a filter functionality connected to the functionalities fct_1 and fct_2. We say that the filtering rule which let pass a data flow from fct_1 to fct_2 is non productive if this flow never try to pass through the filter functionality.

Non productive transform rule.

This one is used to eliminate unnecessary transform rules. A transform rule tf that transforms the data flows with the role "r" from fct_1 to fct_2 is non productive, if any flow with the role "r" pass through the transform functionality in the direction fct_1 to fct_2 at any time.

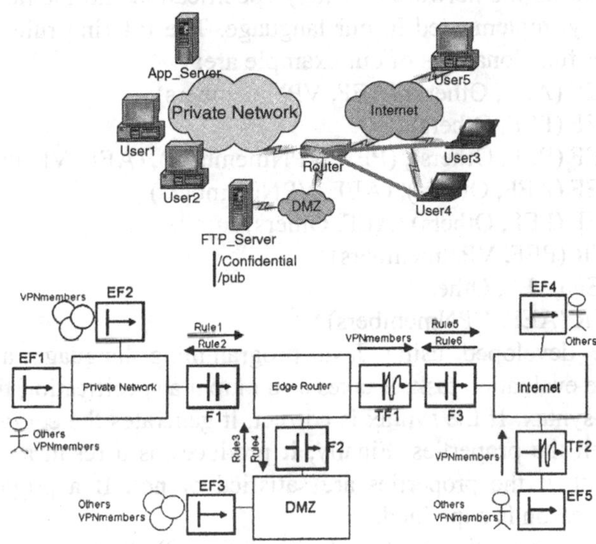

Figure 8. Architecture and graphical specification of our VPN example

4. A NETWORK SECURITY POLICY EVALUATION EXAMPLE

Like in traditional enterprise network, this example considers an edge router interconnecting a private network and a DMZ. An "App_Server" server and an FTP server are respectively installed in the private network and in the DMZ (fig. 8). The application level security policy is a RBAC one, without hierarchy, where two user groups "*VPNmembers*" and "*Others*"

are defined. This organization is only based on the granted privileges. The "App_Server" server is dedicated only to the services usable by the *VPNmembers* group. The FTP_Server has two directories: /confidential and /pub. The directory "confidential" contains data only accessible to the *VPNmembers* users group. Data of the "pub" directory is accessible to everyone. $User_1$, $User_2$, $User_3$ and $User_4$ belong to *VPNmembers* and *Others* groups. $User_5$ is only member of the *Others* group.
The application level security policy can be expressed as:

Permissions(VPNmembers) = {(+all_access, FTP_Server/confidential),
(+all_access, App_Server)}

Permissions(Others) = {(+all_access, FTP_Server/pub)}

Figure 8 shows the network topology specification and the network level security policy implemented in our language. The filtering rules associated with the filter functionalities of our example are:

- Rule1 = EF (AEF, Others) (AEF, VPNmembers)
- Rule2 = EF (PEF, Others)
- Rule3 = EF (PEF, Others), (PEF, VPNmembers), (AEF, VPNmembers)
- Rule4 = EF (AEF, Others), (AEF, VPNmembers)
- Rule5 = EF (PEF, Others) (AEF, Others)
  TR (PEF, VPNmembers)
- Rule6 = EF (AEF, Others)
  TR (AEF, VPNmembers)

We have developed using Java programming language a tool that automates the evaluation task. It takes as an input a specification file. First, it analyzes the syntax. If the syntax is correct, it generates the equivalent CPN and checks all the properties. Finally, it produces as a result a file (fig. 9) that deals with if the properties are satisfied or not. If a property is not satisfied, the reason is explained.

In our example, the tool indicates (fig. 9) that the property of confidentiality is satisfied and there is no non productive transform rule. Nevertheless, the availability is not satisfied because ef_2 cannot receive any flow with the role VPNmembers from ef_5, ef_1 cannot receive any flow with the role VPNmembers from ef_3 and ef_5 cannot receive any flow with the role VPNmembers from ef_2. The partitioning properties is not satisfied on account of the rule EF (AEF,Others) from tf_1 to Internet in the filter functionality f_3. And finally, the filtering rule EF (AEF, VPNmembers) from dmz to edge_router in the filter functionality f_2 is non productive. To resume, this specification is not secure.

```
Property of Confidentiality :
---------------------------------
ef5 : OK
ef4 : OK
ef1 : OK
=> The property of confidentiality is satisfied

Property of Availability :
------------------------------
ef5 :
   no flow with the role vpn-members from ef2
ef4 : OK
ef1 :
   no flow with the role vpn-members from ef3
ef3 : OK
ef2 :
   no flow with the role vpn-members from ef5
=> The property of availability is not satisfied

Partitioning Property :
---------------------------
f3 :
   Rule 1 -> 2 :
     [ EF (AEF ,others) ]
   Rule 2 -> 1 : OK
f2 :
   Rule 1 -> 2 : OK
   Rule 2 -> 1 : OK
f1 :
   Rule 1 -> 2 : OK
   Rule 2 -> 1 : OK
=> There is one or more partitioning problem

Non Productive Transform Rules :
------------------------------------
tf2 :
   rules 1 -> 2 : OK
   rules 2 -> 1 : OK
tf1 :
   rules 1 -> 2 : OK
   rules 2 -> 1 : OK
=> There is no non productive rule

Non Productive Filtering Rules :
------------------------------------
f3
   rules 1 -> 2 :  OK
   rules 2 -> 1 :  OK
f2
   rules 1 -> 2 :
       [ EF (AEF, vpn-members) ],
   rules 2 -> 1 :  OK
f1
   rules 1 -> 2 :  OK
   rules 2 -> 1 :  OK
=> There is one or more non productive rule
```

Figure 9. Evaluation result file

5. CONCLUSION

The work presented here combine different levels of policy abstraction and security analysis coming from new management approaches and the formal modeling and evaluation techniques. The quiet simple language that we have proposed allows to formally evaluate the network security policy, while using the underlying CPN powerful.

At present, we are testing our approach through different case studies to enhance our method. We are focussing on validating the real configurations on devices. As our tool is independent from the security technologies implemented on the devices, it confines itself to only validate security mechanisms constraints. The next usefull step is to bridge this gap thanks to the Common Information Model[23] defined by the DMTF task force to harmonize the management systems. Hence, we could interconnect our work with management platforms.

REFERENCES

1. Samarati P., De Capitani di Vimercati S., "Access Control: Policies, Models and Mechanisms", Foundations of Security Analysis and Design, R. Focardi and R. Gorrieri (eds), LNCS 2171, Springer-Verlag. 2001.
2. Guttman J., "Filtering postures : Local enforcement for global policies", IEEE Symposium on Security and Privacy, Oakland CA, USA, 1997.
3. Ehab Al-Shaer and Hazem Hamed, "Discovery of Policy Anomalies in Distributed Firewalls", in IEEE INFOCOMM'04, March 2004.
4. Y. Bartal., A. Mayer, K. Nissim and A. Wool. "Firmato: A Novel Firewall Management Toolkit." proceedings of 1999 IEEE Symposiumon Security and Privacy, May 1999.
5. Guttman J., Herzog A., Thayer F., "Authentication and confidentiality via IPsec", 6th European Symposium in Computer Security ESORICS, Toulouse, France, 2000.
6. Z. Fu, F. Wu, H. Huang, K. Loh, F. Gong, I. Baldine and C. Xu."IPSec/VPN Security Policy: Correctness, Conflict Detection and Res-olution." Proceedings of Policy'2001 Workshop, January 2001.
7. Yavatkar R., Pendarakis D., Guerin R., "A Framework for Policy-based Admission Control", RFC 2753, January 2000.
8. Jennings N.R., Bussmann S., "Agent based Control Systems, why are they suited to engineering complex systems? IEEE Control Systems Magazine, vol 23, No 3, June 2003.
9. IBM Corporation, An Architectural Blueprint for Autonomic Computing, IBM white papers, April 2003.
10. http://www.solsoft.com
11. Hinrichs S., "Policy Based Management : bridging the gap", in 15th Annual Computer Security Applications Conference (ACSAC 99), December 1999.
12. Westerinen A., Schnizlein J., Strassner J., Scherling M., Quinn B., Herzog S., Huynh A., Carlson M., Perry J., Waldbusser S., "Terminology for Policy-Based Management", RFC 3198, November 2001.
13. Moore B., Ellesson E., Strassner J., Westerinen A., "Policy Core Information Model -- Version 1 Specification", RFC 3060, February 2001.
14. Arosha K Bandara, Emil C Lupu, Jonathan Moffet, Alessandra Russo,. "A Goal-based Approach to Policy Refinement", in: Policy 2004, June 2004.
15. Laborde R., Nasser B., Grasset F., Barrère F., Benzékri A., "A formal approach for the evaluation of network security mechanisms based on RBAC policies", In WISP'04, Electronic Notes in Theoretical Computer Science, Elsevier, to appear.
16. "Principles for a Telecommunications Management Network", ITU-T, M3010, May 1996.
17. " Role-Based Access Control", ANSI/INCITS 359-2004, February 2004.
18. R. Peri, "Specification and verification of security policies", PhD Dissertation, University of Virginia, January 1996.
19. Wijesekera D., Jajodia S., "A propositional policy algebra for access control", ACM Transactions on Information and System Security (TISSEC), vol 6,2003.
20. Nyanchama M., Osborn S., "The role graph model and conflict of interest", ACM Transactions on Information and System Security (TISSEC), vol. 2, 1999.
21. Moffett J. D., "Control Principle and Role Hierarchies", 3rd ACM Workshop on Role Based Access Control, Fairfax, VA, 1998.
22. Crook R., Ince D., Nuseibeh B, "Modeling Access Policies using Roles in Requirements Engineering", Information and Software Technology, 2003, Elsevier
23. http://www.dmtf.org/standards/cim

Part Three: Quality of Service

A DYNAMIC CROSS LAYER CONTROL STRATEGY FOR RESOURCE PARTITIONING IN A RAIN FADED SATELLITE CHANNEL WITH LONG-LIVED TCP CONNECTIONS(*)

Nedo Celandroni[1] Franco Davoli[2] Erina Ferro[1], Alberto Gotta[1]

[1]ISTI-CNR, Area della Ricerca del C.N.R., Via Moruzzi 1, I-56124 Pisa, Italy, {nedo.celandroni,erina.ferro,alberto.gotta}@isti.cnr.it; [2]Italian National Consortium for Telecommunications (CNIT), University of Genoa Research Unit, Via Opera Pia 13, 16145 Genova, Italy; National Laboratory for Multimedia Communications, Via Diocleziano 328, Napoli, Italy, franco.davoli@cnit.it

Abstract: The paper aims at devising a control system for dynamic resource allocation in a packet-oriented satellite network. The traffic to be served is represented by TCP long-lived connections (elephants). A Master Station adaptively assigns bandwidth and transmission parameters (bit and coding rate) to TCP buffers at the earth stations, grouping connections characterized by the same source-destination pair. The assignment is effected according to each pair's traffic load and fading conditions, in order to reach a common goal. The latter may consist of maximizing the overall TCP goodput, of equalizing the connections' goodput for global fairness, or a combination thereof. Three different allocation strategies are devised, and their respective performance is compared, under a realistic link budget.

Key words: Satellite Communications, Fade Countermeasures, Cross-Layer Optimization, TCP, Dynamic Bandwidth Allocation.

(*) Work supported by MIUR (Ministero dell'Istruzione, Università e Ricerca) in the framework of the "*TANGO*" and "*DIDANET*" projects, by the Italian National Research Council (CNR), under the *IS-MANET* project, and by the European Commission in the framework of the *SatNEx* NoE project (contract N. 507052).

1. INTRODUCTION

Satellite systems not only have to face variable load multimedia traffic, but also variable channel conditions with large propagation delay. The variability in operating conditions is due both to changes in the traffic loads and to the signal attenuation on the satellite links, because of bad atmospheric events, which particularly affect transmissions in the Ka band (20-30 GHz). It is therefore stringent to make use of adaptive network management and control algorithms to maintain the Quality of Service (QoS) of the transmitted data. Our study considers the following scenario. A geo-stationary satellite network consists of *N* stations, among which a master station exerts, in addition, the control on the access to the common resource, i.e., the satellite bandwidth, and *N*-1 traffic stations exchange non-real-time (bulk) traffic. The traffic consists of a number of long-lived TCP connections (also called *elephants*[1]) any station may have with any other station in the network. In Celandroni[2,3] the application of adaptive FEC (Forward Error Correction) was investigated, to optimize the efficiency of TCP connections over AWGN (additive white Gaussian noise) links with high delay-bandwidth product. This case well matches transmissions over rain faded geo-stationary satellite channels, with fixed user antennas. In our study, we thus adopt the same philosophy and operate at the physical level, by trading the bandwidth of the satellite link for the packet loss rate due to data corruption. In fact, over wireless links, any gain in the *bit error rate* (BER) (i.e., in the packet loss) is generally obtained at the expenses of the *information bit rate* (IBR), and the end-to-end transfer rate of a TCP connection (also called *goodput*) increases with the IBR and decreases with the BER. The adaptation techniques adopted do not interfere in any way with the normal behavior of the TCP stack; the end-to-end protocols are thus left unaltered, while the transmission parameters of the satellite link are appropriately tuned-up.

In this paper, we study bandwidth, bit and coding rate allocation methods for TCP long-lived connections that are in effect on a number of source-destination (SD) pairs over satellite links, which may be generally subject to different fading conditions. In this situation, connections belonging to the same SD pair feed a common buffer at the IP packet level in the earth station, which "sees" a transmission channel with specific characteristics; the latter may generally be different from those of other SD pairs originating from the same station or from other stations. The bandwidth allocated to serve such buffers is shared by all TCP connections in that group, and, once fixed, it determines the "best" combination of bit and coding rates for the given channel conditions. The goal of the allocation is to satisfy some global optimality criterion, which may involve goodput, fairness among the connections, or a combination thereof. Therefore, in correspondence of a specific situation of channel conditions, determined by the various up- and down-link fading patterns, and a given traffic load, we face a possible two-criteria optimization problem, whose decision variables are the service rates of the above mentioned IP buffers for each SD pair, and the corresponding transmission parameters. We will refer to these allocation strategies as TCP-CLARA(*C*ross *L*ayer *A*pproach for *R*esource *A*llocation).

Though there is a vast literature on performance aspects related to the adaptation of TCP congestion control mechanisms over large bandwidth-delay product and error-prone satellite channels (see, Kota[4] and references therein and Jamalipour[5]), as well as on resource allocation and Quality of Service (QoS) control in broadband packet networks[6], even in the satellite environment[4, 5, 7], to the best of the authors' knowledge, this is the first attempt to conjugate TCP over satellite performance and resource allocation, in the framework of a cross-layer approach.

2. GOODPUT ESTIMATION OF LONG-LIVED TCP CONNECTIONS

When a number of long-lived TCP sources share the same bottleneck-rate link, it was empirically observed in Lakshman[8] (by making use of simulation) that, if all connections have the same latency, they obtain an equal share of the link bandwidth. This is strongly supported by our simulations as well (obtained by using Network Simulator[9] – ns2), where we suppose the bottleneck is the satellite link. As the latency introduced by a geo-stationary satellite is quite high (more than half a second) it is reasonable to assume that the additional latency introduced by the satellite access network in the entire link path is negligible with respect to the satellite one, and that all connections have the same latency.

In order to avoid time consuming simulations, reasonable estimations can be constructed for the goodput of a TCP Reno agent. A first relation that can be used is the one taken from Padhye[10], which is estimated for infinite bottleneck rate, and thus it is valid far apart the approaching of the bottleneck rate itself. Let μ be the bottleneck (the satellite link) rate expressed in segments/s, n the number of TCP sources, and τ the delay between the beginning of the transmission of a segment and the reception of the relative acknowledgement, when the satellite link queue is empty. Moreover, assume the segment losses to be independent with rate q. We have $\tau = c_l + 1/\mu$, where c_l is the channel latency. The TCP connections that share the same link also share an IP buffer, inserted ahead the satellite link, whose capacity is at least equal to the product $\mu\tau$. Let also b be the number of segments acknowledged by each ACK segment received by the sender TCP, and T_0

the timeout estimated by the sender TCP. Then, by exploiting the expression of the send rate derived in Padhye[10], dividing it by $\frac{\mu}{n}$ for normalization and multiplying it by $1-q$ for a better approximation, the relative (normalized to the bottleneck rate) goodput can be expressed as

$$T_g = \frac{1-q}{\frac{\mu}{n}\left[\tau\sqrt{\frac{2bq}{3}} + T_o \min\left(1,\ 3\sqrt{\frac{3bq}{8}}\right) q(1+32q^2)\right]} \tag{1}$$

Relation (1) is rather accurate for high values of q, i.e., far apart the saturation of the bottleneck link. For low values of q, it is found by simulation that, given a fixed value of c_l, for fixed values of the parameter $y = q\left(\frac{\mu}{n}\right)^2 \tau^5$, the goodput has a limited variation with respect to individual variations of the parameters q, μ and n. Owing to the high number of simulations needed to verify this observation, a fluid simulator[11] has been employed, which was validated by means of ns2, for values of $y \le 1$. Simulation results have been obtained for the goodput estimation, with a 1% confidence interval at 99% level, over a range of values of $\frac{\mu}{n}$ between 20 and 300, and n between 1 and 10. For $0 \le y \le 1$, goodput values corresponding to the same y never deviate for more than 8% from their mean. We then interpolated such mean values with a 4-th order polynomial approximating function, whose coefficients have been estimated with the least squared errors technique. Assuming a constant c_l, equal to 0.6 *s* (a value that takes into account half a second of a geostationary satellite double hop, plus some delay for terrestrial links and some processing time), in the absence of the so-called *Delayed ACKs* option (b=1), the polynomial interpolating function results to be

$$T_g = a_o + a_1 y + a_2 y^2 + a_3 y^3 + a_4 y^4 \ ; \qquad y \le 1, \tag{2}$$

where

$a_o = 0.995$; $a_1 = 0.11\ [s^{-3}]$; $a_2 = -1.88\ [s^{-6}]$; $a_3 = 1.98\ [s^{-9}]$; $a_4 = -0.63\ [s^{-12}]$. For $y = 1$, $T_g = 0.575$. For $y > 1$, we adopt instead relation (1), with $b = 1$.

We assume to operate on an AWGN channel, a reasonable approximation for geo-stationary satellites and fixed earth stations. The segment loss rate q can be computed as in Celandroni[2]:

$$q = 1 - (1 - p_e / l_e)^{l_s}, \tag{3}$$

where p_e is the bit error probability (BER), l_s is the segment bit length and l_e is the average error burst length (*ebl*). We took p_e data from the Qualcomm Viterbi decoder data sheet[12] (standard NASA 1/2 rate with constraint length 7 and derived punctured codes), while l_e was obtained through numerical simulation in Celandroni[2]. (channel bit energy to one-sided noise spectral density ratio).

In order to make q computations easier, we interpolated data taken from Celandroni[2] and expressed p_e and l_e analytically as functions of the coding rate and the E_c / N_0 (channel bit energy to one-sided noise spectral density ratio) ratio. We have

$$\begin{aligned}
&p_e(1/2) = 10^{-(1.6E_c/N_0+3)} \;;\quad 0 \le E_c/N_0 \le 5\ dB \\
&p_e(3/4) = 10^{-(1.6E_c/N_0-2.04)} \;;\quad 4 \le E_c/N_0 \le 8\ dB \\
&p_e(7/8) = 10^{-(1.6E_c/N_0-5)} \;;\quad 6 \le E_c/N_0 \le 10\ dB \\
&l_e(1/2) = e^{-0.32E_c/N_0+1.87} \;;\quad 0 \le E_c/N_0 \le 5\ dB \\
&l_e(3/4) = e^{-0.4E_c/N_0+3.45} \;;\quad 4 \le E_c/N_0 \le 8\ dB \\
&l_e(7/8) = 63.6 - 19(E_c/N_0) + 1.94(E_c/N_0)^2 - 0.067(E_c/N_0)^3 \;;\quad 6 \le E_c/N_0 \le 10\ dB
\end{aligned} \tag{4}$$

For the uncoded case[13] we have

$$p_e(1/1) = \frac{1}{2}\, erfc\left(10^{(E_c/N_0)/20}\right); \qquad l_e(1/1) = 1; \tag{5}$$

The TCP goodput relative to the bottleneck rate is a decreasing function of the segment loss rate q, which, in its turn, is a decreasing function of the coding redundancy applied in a given channel condition C / N_0 (carrier power to one-sided noise spectral density ratio; see Section 4 for the relation between C / N_0 and E_c / N_0) and for a given bit rate b_r. The combination of channel bit rate and coding rate gives rise to a "redundancy factor" $r_{cs} \ge 1$, which represents the ratio between the IBR in clear sky and the IBR in the specific working condition.

The absolute goodput of each TCP connection $\hat{T}_g$ is obtained by multiplying the relative value by the bottleneck rate, i.e.,

$$\hat{T}_g = T_g \frac{\mu}{n} = T_g \frac{1}{n} \cdot \frac{B}{r_{cs}} \tag{6}$$

where B is the link rate in segments/s in clear sky conditions.

In Celandroni[2] it is shown that, for a given hardware being employed (modulation scheme/rate, FEC type/rate), a set of transmission parameters maximize the absolute goodput for each channel condition. Given B and C/N_0 (which results from a given link budget calculation), and for all possible bit rates, we must compute T_g for all allowable coding rates. The actual values of the goodput are obtained as mentioned above: then, the maximum value is selected. The value of T_g is taken from (1) or (2), and q is computed with (3) and (4) or (5). Numerical examples for the link budget corresponding to the Eutelsat satellite Hot Bird 6 are given in Section 4.

3. THE BANDWIDTH ALLOCATION PROBLEM

Consider now a satellite network, consisting of a bent-pipe (transparent) geo-stationary payload and N-1 traffic stations. We assume to operate in single-hop, so that the tasks of the master station are limited to resource assignment and synchronization. Note that, in this respect, we may consider a private network, operating on a portion of the total available satellite capacity, which has been assigned to a specific organization and can be managed by it (as a special case, this situation might also represent a service provider, managing the whole satellite capacity).

The problem we address in the following is the assignment of bandwidth, bit and coding rates to the IP buffers serving each specific link, given the fading conditions and the load of the network.

We make the following assumptions.

1. The end-to-end delay of the TCP connections is the same for each station.
2. In each station, there is an IP buffer for each SD pair, and we say that the TCP connections sharing it belong to the same class; obviously, they experience the same up-link and destinations' downlink conditions.
3. In the present paper, we consider the bandwidth assignment in static conditions. In other words, given a certain number of ongoing connections, distributed among a subset F of SD pairs, characterized by a certain fading attenuation, we find the optimal assignment as if the situation would last forever. Possible ways of adaptive allocation in a dynamic environment are the matter of ongoing investigation.

We assume that, if the fading conditions of an active class i ($i = 1,2,..,F$) are such that a minimum goodput $T_{g,thr}^{(i)}$ cannot be reached by its connections, the specific SD pair would be considered in outage, and no bandwidth would be assigned to it.

Let $B_i \in [0,W]$ (where W is the total bandwidth to be allocated, expressed in segments/s), $r_{cs}^{(i)} \in \{R_1,...,R_P\}$, and $n_c^{(i)}$, $i=1,...,F$, be the bandwidth, the redundancy factor, chosen in the set of available ones (each value R_i corresponds to a pair of bit and coding rates), and the number of connections, respectively, of the i-th SD pair. Note that, in the cases where different bit and coding rates yield the same redundancy factor, the pair will be selected that gives rise to the minimum BER.

We consider two basically opposite ways of assigning the bandwidth (which corresponds to setting the parameters of the scheduler serving the buffers that use a given station's up-link), together with the transmission parameters:

G1) To maximize the global goodput, i.e.,

$$\max_{B_i \in [0,W],\ r_{cs}^{(i)} \in \{R_1,...,R_P\}\ i=1,...,F} \sum_{j=1}^{F} n_c^{(j)} \ddot{T}_g^{(j)}, \tag{7}$$

$$\text{subject to } \sum_{i=1}^{F} B_i = W \tag{8}$$

G2) To reach global fairness, i.e., to divide the bandwidth (and assign the corresponding transmission parameters, namely, channel bit and coding rate) in such a way that all SD pairs achieve the same goodput.

Note that, even though the goodput optimization formulas of Section 2 are applied in both cases, the two goals are different and will generally yield different results in the respective parameters: maximizing the global goodput may result in an unfair allocation (in the sense that some SD pairs may receive a relatively poor service), whereas a fair allocation generally does not achieve globally optimal goodput.

As far as the single goals are concerned, the relative calculations may be effected as follows.

The maximization in (7) is over a sum of separable nonlinear functions (each term in the sum depending only on its specific decision variables, coupled only by the linear constraint (8)). As such, it can be computed efficiently by means of Dynamic Programming[14, 15], if the bandwidth allocations are expressed in discrete steps of a *minimum bandwidth unit* (*mbu*), which is the minimum granularity that can be achieved.

As regards the goodput equalizing fair allocation, it can be reached by starting from an allocation proportional to the number of TCP connections per SD pair, by computing the average of the corresponding optimal (in the choice of transmission parameters) goodputs, then changing the *mbu* allocations (under constraint (8)) by discrete steps, in the direction that tends to decrease the absolute deviation of each SD pair's goodput from the average, and repeating the operation with the new allocations. A reasonable convergence, within a given tolerance interval, can be obtained in few steps.

As usually one may want to achieve what one believes to be a reasonable combination of goodput and fairness, we propose the following two strategies, which will then be evaluated numerically in the next Section.

Tradeoff Strategy. The following steps are performed:

Compute the pairs $\left(B_i^*, r_i^*\right)$ $i = 1,...,F$, maximizing the global goodput (7), under constraint (8);

Compute the pairs $\left(\bar{B}_i , \bar{r}_i\right)$ $i = 1,...,F$, corresponding to the goodput equalizing fair choice;

Calculate the final allocation as $\tilde{B}_i = \bar{B}_i\, \rho + B_i^*(1-\rho)$, $i = 1,...,F$, where $0 \le \rho \le 1$ is a tradeoff parameter, along with the corresponding bit and coding rates.

Range Strategy. The following steps are performed:

Compute the pairs $\left(\bar{B}_i , \bar{r}_i\right)$ $i = 1,...,F$, corresponding to the goodput equalizing fair choice;

Choose a "range coefficient" $\beta \ge 0$;

Compute the global goodput maximizing allocation, by effecting the constrained maximization in (7), with $\bar{B}_i$ varying in the range $\left[\max\left(\bar{B}_i(1-\beta), 0\right), \min\left(\bar{B}_i(1+\beta), W\right)\right]$, instead of $\left[0, W\right]$, $i = 1,...,F$.

As a term of comparison, we will also consider another possible strategy (termed *BER Threshold* in the following), which only assigns the transmission parameters (bit and coding rate) to each SD pair, in order to keep the BER on the corresponding channel below a given threshold. The bandwidth assignment is done proportionally to the number of connections per link, multiplied by the corresponding redundancy.

In order to comparatively evaluate the different options, we define the following terms of comparison:

Goodput Factor:
$$f_g = \frac{\sum_{i=1}^{F} n_c^{(i)} \hat{T}_g^{(i)}\left(B_i, r_i\right)}{\sum_{i=1}^{F} n_c^{(i)} \hat{T}_g^{(i)}\left(B_i^*, r_i^*\right)} \tag{9}$$

where (B_i, r_i) is a generic choice and (B_i^*, r_i^*) is the goodput-maximizing one.

$$\text{Fairness Factor:} \qquad f_f = 1 - \frac{\sum_{j=1}^{L} \left| \hat{T}_g^{(j)} - \overline{T}_g \right|}{2\overline{T}_g (L-1)} \tag{10}$$

where $L = \sum_{i=1}^{F} n^{(i)}$ is the total number of ongoing TCP connections, and $\overline{T}_g = \frac{1}{L} \sum_{k=1}^{L} \hat{T}_g^{(k)}$ is the average goodput. Note that $f_f = 1$ when all goodputs are equal, and $f_f = 0$ when the unbalance among the connections' goodputs is maximized, i.e., the goodput is $\overline{T}_g \cdot L$ for one connection and 0 for the others (yielding a deviation from the average $(\overline{T}_g \cdot L - \overline{T}_g) + |0 - \overline{T}_g| \cdot (L-1) = 2\overline{T}_g \cdot L - 2\overline{T}_g$, which is the denominator of (10)).

4. NUMERICAL RESULTS

We considered a fully meshed satellite network that uses bent-pipe geostationary satellite channels. This means that the satellite only performs the function of a repeater and it does not make any demodulation of data. The system operates in TDMA mode. The master station maintains the system synchronization, other than performing capacity allocation to the traffic stations. The master station performance is the same as the others; thus, the role of master can be assumed by any station in the system. This assures that the master normally operates in pretty good conditions, because when the current master's attenuation exceeds a given threshold, its role is assumed by another station that is in better conditions. To counteract the signal attenuation the system operates bit and coding rates changing. Traffic stations transmit in temporal slots assigned by the master.

In order to compute the link budget, we considered a portion of the Ka band (20/30 GHz) transponder of the Eutelsat satellite Hot Bird 6, and took data from the file "*Hot Bird 6 data sheet.fm*", which is downloadable from the web[16].

We consider exploiting 1/4 of the transponder power. Our carrier is modulated in QPSK (quadrature phase shift keying) at 5, 2.5 or 1.25 Msymbols/s; thus, the resulting uncoded bit rates range from 10 to 2.5 Mbit/s. A

1/2 convolutional encoder/Viterbi decoder is employed, together with the punctured 3/4 and 7/8 codes for a possible total 12 combinations of bit/coding rates. The net value of about 7.5 dB of E_c/N_0 (C/N_0=77.5 dB), with the maximum modulation rate and the 7/8 coding rate, is assumed as the clear sky condition. In clear sky, after the Viterbi decoder, the bit error rate is about 10^{-7}. The *mbu* size, i.e., the minimum bandwidth unit that can be allocated, has been taken equal to 5 kbit/s; this value is referred to clear sky conditions.

In order to compute the resulting net values of E_c/N_0 at the earth station's receiver input we used relation (11) below.

$$E_c/N_0 = C/N_0 - 10Log_{10}b_r - m_i, \tag{11}$$

where b_r is the uncoded data bit rate in bit/s and m_i is the modem implementation margin (taken equal to 1 dB). We have assumed b=1 (no *Delayed ACKs* option) and T_o =1.5 s when using relation (1). We also considered l_s = 4608 (576 bytes), which is the default segment length assumed by sender and receiver TCPs, when no other agreement has been possible.

Actually, not all combinations of bit and coding rates must be probed to find the maximum goodput, because some of them result inefficient (i.e., they yield higher BER with the same redundancy). The possible cases are then limited to the following 7 ones: 10 Mbits/s, with code rates 7/8, 3/4, and 1/2; 5 Mbits/s, with code rates 3/4, and 1/2; 2.5 Mbits/s, with code rates 3/4 and 1/2. The uncoded case results inapplicable with the values of C/N_0 available in our situation, even in clear sky conditions.

Table 1. Configuration of the 3 tests (n_c is the number of TCP connections in the relevant class).

Record #	Class 1	Class 2	Class 3	Class 4	Class 5
	$C/N_0;n_c$	$C/N_0;n_c$	$C/N_0;n_c$	$C/N_0;n_c$	$C/N_0;n_c$
1	78.0;2	77.0;3	76.1;2	68.9;4	74.0;2
2	78.0;2	68.0;3	76.1;2	68.9;4	73.0;2
3	78.0;2	72.0;3	76.1;2	69.9;4	73.0;2
Record #	Class 6	Class 7	Class 8	Class 9	Class 10
1	75.0;3	76.5;1	71.8;4	76.3;3	76.6;6
2	73.0;3	75.0;1	72.0;4	76.3;3	76.6;6
3	75.0;3	76.5;1	71.8;4	76.3;3	63.0;6

Table I shows the configurations of the three tests carried on, denoting the link status (C/N_0 [dB]) and the number of TCP connections in each class.

The following Figs 1 and 2 have been obtained, by using the above data, by means of the TEAM (*T*CP *E*lephant bandwidth *A*llocation *M*ethod) modeling and optimization software, which was specifically developed to implement the mechanisms proposed in this paper and other, more general ones, which will be the subject of forthcoming research.

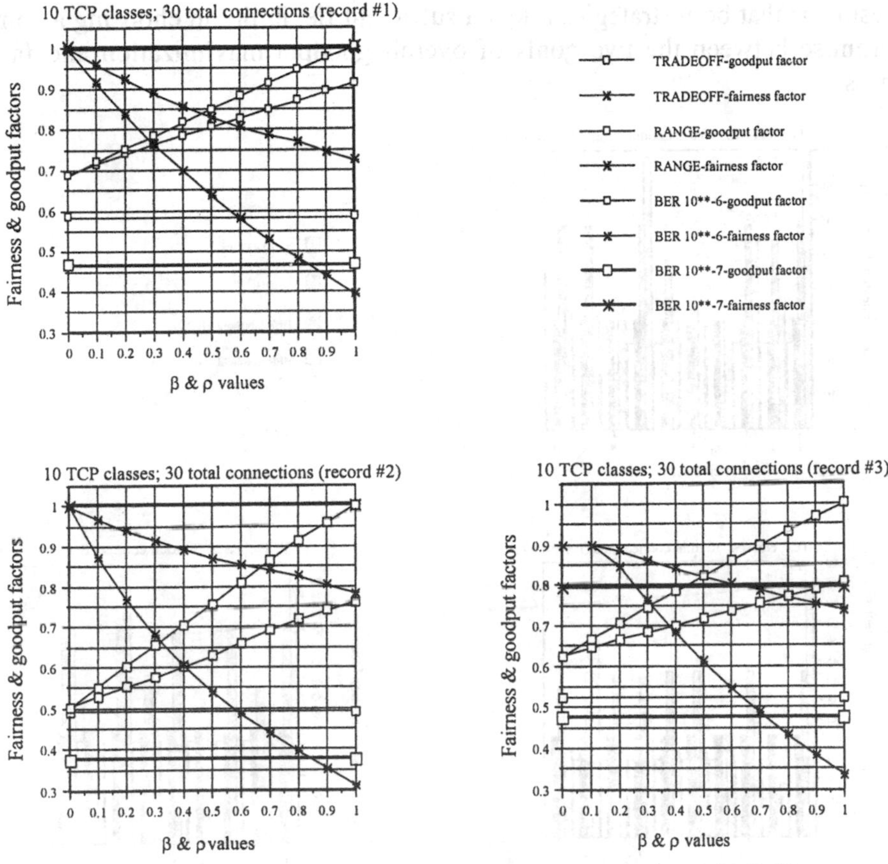

Figure 1 Goodput and fairness indexes for the data in record #1 #2 #3.

For each record of values in Table I, the figures depict the behavior of the goodput and fairness factors, respectively, for the *Tradeoff*, *Range*, and *BER Threshold* strategies that have been defined in the previous Section, for values of the parameters ρ and β between 0 and 1. It can be noted that constantly keeping the BER below a given threshold lowers the goodput and not always maximizes the fairness (as can be seen in Fig. 1 record #3); moreover, enforcing a constant lower BER (10^{-7}) lowers the goodput, without any appreciable gain in fairness.

The *Tradeoff* and *Range* strategies have a similar behavior, though they span different values of goodput and fairness factors, depending on the system parameters. In all cases, as expected, the goodput factor increases and the fairness factor decreases with increasing ρ and β. In general, the span of the *Tradeoff* strategy's goodput and fairness index values is wider in the interval [0, 1] than that of the *Range* strategy, but it must be noted that the parameter β could be increased beyond 1, within the limits imposed by the total bandwidth available. A conclusion that can be drawn from the obtained results is that both strategies allow a sufficient flexibility in choosing a compromise between the two goals of overall goodput maximization and fairness.

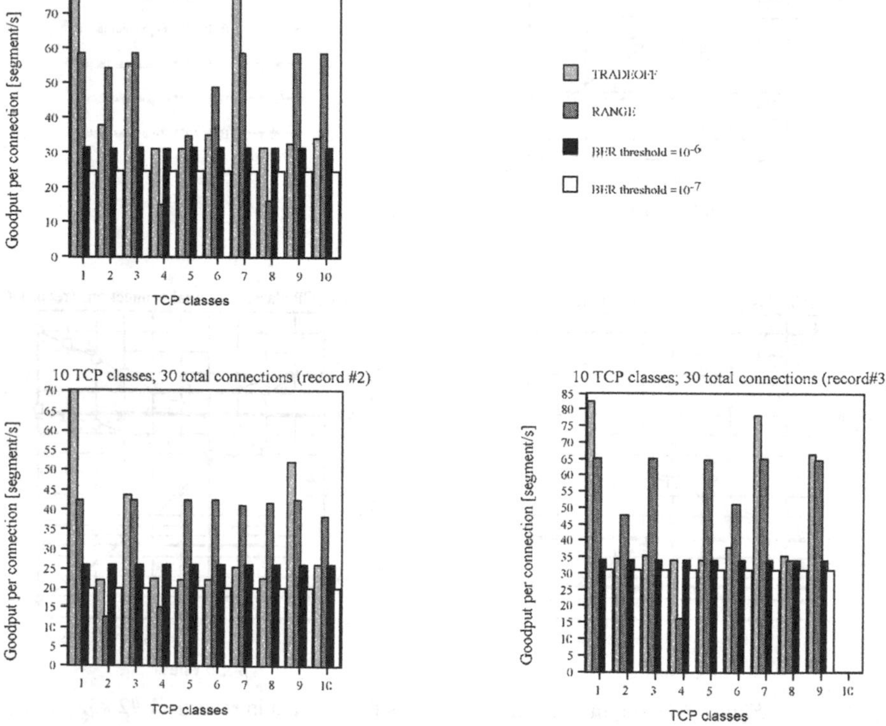

Figure 2. Goodput per TCP class for the data in record #1 #2 #3. (class 10 of record #3 is in outage for all strategies).

The total goodput of the system [segments/s] calculated by the TEAM modeling and optimization software, is in total agreement with simulation results of ns2. The latter have been obtained by running the TCP connections under the bandwidth partitions provided by the TEAM software, and confirm the good prediction properties of the TCP model used in the calculations. The values of the coefficients ρ and β (0.16 and 0.6, respectively) have been chosen with the criterion of maintaining a fairness factor always higher than 0.8 in the three cases. Under these conditions, it can be noted that the *Range* strategy always exhibits a higher total goodput (anyway, this is not true in general).

As a final comparison, Fig. 2 show the goodput per connection in the three cases considered, for all classes, with the two selected values ρ=0.16 and β=0.6, and for all strategies. In Fig. 2 record #1 and #3, the goodputs corresponding to the *Tradeoff* strategy are always higher than those of the *BER Threshold* strategies. However, it is not so in Fig. 2 record #2, which makes it difficult to draw a general conclusion in this sense. In perspective, an adaptive choice of the coefficients, even within possible predefined limits (such as imposing a minimum threshold on the fairness, like we have done), may turn out to be advisable. This is the matter of current investigation. What can be further observed is that, in all cases considered, the range strategy tends to put a stronger penalization on stations suffering from heavy fading (as is the case with station # 4), and to more evenly favor the other stations.

5. CONCLUSIONS

We have considered a problem of cross-layer optimization in bandwidth assignment to TCP connections, traversing different links in a geostationary satellite network, characterized by differentiated levels of fading attenuation. On the basis of the observation that there exists a tradeoff between bandwidth and data redundancy that influences TCP goodput, we have proposed optimization mechanisms that can be used to control the Quality of Service, in terms of goodput and fairness, of the TCP connections sharing the satellite bandwidth. The performance analysis of the methods proposed, conducted on a few specific cases with real data, by means of the modeling and optimization software developed for this purpose, has shown that relevant gains can be obtained with respect to fade countermeasures that only attempt to constrain the BER below a given threshold, and that a good range of flexibility can be attained in privileging the goals of goodput or fairness. Further research is currently ongoing, to apply the proposed strategies in a dynamic environment, where fading levels and number of TCP connections in the system change over time in unpredictable fashion, as well as in traffic engineering for multiservice satellite networks.

REFERENCES

1. M. Ajmone Marsan, M. Garetto, P. Giaccone, E. Leonardi, E. Schiattarella, A. Tarello, "Using partial differential equations to model TCP mice and elephants in large IP networks", *Proc. IEEE Infocom 04*, Hong Kong, March 2004.
2. N. Celandroni, F. Potortì, "Maximising single connection TCP goodput by trading bandwidth for BER", *Internat. J. Commun. Syst.*, vol. 16, no. 1, pp. 63-79, Feb. 2003.
3. N. Celandroni, E. Ferro, F. Potortì, "Goodput optimisation of long-lived TCP connections in a faded satellite channel", *Proc. IEEE Vehic. Technol. Conf.*, Milan, Italy, May 2004.
4. S. Kota, M. Marchese, "Quality of Service for satellite IP networks: a survey", *Internat. J. Satell. Commun. Network.*, vol. 21, no.4-5, pp. 303-348, July-Oct. 2003.
5. A. Jamalipour, M. Marchese, H.S. Cruickshank, J. Neale, S.N. Verma (Eds.), Special Issue on "Broadband IP Networks via Satellites – Part I", *IEEE J. Select. Areas Commun.*, vol. 22, no. 2, Feb. 2004.
6. H. J. Chao, X. Guo, *Quality of Service Control in High-Speed Networks*, John Wiley & Sons, New York, NY, 2002.
7. F. Alagoz, D. Walters, A. Alrustamani, B. Vojcic, R. Pickholtz, "Adaptive rate control and QoS provisioning in direct broadcast satellite Networks," *Wireless Networks*, vol. 7, no. 3, pp. 269-281, 2001.
8. T. V. Lakshman, U. Madhow, "The performance of TCP/IP for networks with high bandwidth-delay products and random loss", *IEEE/ACM Trans. Networking*, vol. 5, no. 3, pp. 336-350, June 1997.
9. The Network Simulator – ns2. Documentation and source code from the home page: http://www.isi.edu/nsnam/ns.
10. J. Padhye, V. Firoiu, D. F. Towsley, J. F. Kurose, "Modeling TCP Reno performance: a simple model and its empirical validation", *IEEE/ACM Trans. Networking*, vol. 8, pp. 133-145, 2000.
11. http://wnet.isti.cnr.it/software/tgep.html.
12. "Q1401: K=7 rate 1/2 single-chip Viterbi decoder technical data sheet", Qualcomm, Inc., Sept. 1987.
13. J. G. Proakis, "Digital communications", McGraw Hill, New York, NY,1995.
14. K. W. Ross, *Multiservice Loss Models for Broadband Telecommunication Networks*, Springer, London, UK, 1995.
15. N. Celandroni, F. Davoli, E. Ferro, "Static and dynamic resource allocation in a multiservice satellite network with fading", *Internat. J. Satell. Commun. Network.*, vol. 21, no. 4-5, pp. 469-487, July-Oct. 2003.
16. http://www.eutelsat.com/satellites/13ehb6.html

CONTENT LOCATION AND DISTRIBUTION IN CONVERGED OVERLAY NETWORKS

Oren Unger[1,2] and Israel Cidon[1]
[1] *Department of Electrical Engineering, Technion - Israel Institute of Technology;* [2] *Zoran Microelectronics*

Abstract: A major challenge for organizations and application service providers (ASP) is to provide high quality network services to geographically dispersed consumers at a reasonable cost. Such providers employ content delivery networks (CDNs) and overlay networks to bring content and applications closer to their service consumers with better quality.

Overlay networks architecture should support high-performance and high-scalability at a low cost. For that end, in addition to the traditional unicast communication, multicast methodologies can be used to deliver content from regional servers to end users. Another important architectural problem is the efficient allocation of objects to servers to minimize storage and distribution costs.

In this work, we suggest a novel hybrid multicast/unicast based architecture and address the optimal allocation and replication of objects. Our model network includes application servers which are potential storage points connected in the overlay network and consumers which are served using multicast and/or unicast traffic. General costs are associated with distribution (download) traffic as well as the storage of objects in the servers.

An optimal object allocation algorithm for tree networks is presented with computational complexity of $O(N^2)$. The algorithm automatically selects, for each user, between multicast and unicast distribution. An approximation algorithm for general networks is also suggested. The model and algorithms can be easily extended to the cases where content is updated from multiple locations.

Keywords: Content Distribution; Location Problems; Hybrid networks; Overlay Networks; Tree Networks; Quality of Service

1. INTRODUCTION

Recent years have witnessed tremendous activity and development in the area of content and services distribution. Geographically dispersed consumers and organizations demand higher throughput and lower response time for accessing distributed content, outsourced applications and managed services. In order to enable high quality and reliable end-user services, organizations and applications service providers (ASPs) employ content distribution networks (CDN) and overlay networks. These networks bring content and applications closer to their consumers, overcoming slow backbone paths, network congestions and physical latencies. Multiple vendors such as Cisco[1], Akamai[2] and Digital Fountain[3] offer CDN services and overlay technologies.

An overlay network is a collection of application servers that are interconnected through the general Internet Infrastructure. Efficient allocation of information objects to the overlay network servers reduces the operational cost and improves the overall performance. This becomes more crucial as the scale of services extend to a large number of users over international operation where communication and storage costs as well as network latencies are high.

The popularity of multicast for distribution of such content is increasing with the introduction of real-time and multimedia applications that require high QoS (high bandwidth, low delay loss and jitter) and are delivered to large groups of consumers. Although multicast is efficient for large groups, its high deployment and management cost makes unicast a better solution for small groups, especially for a sparse spread or when data requirements are diverse.

Hybrid overlay networks are overlay networks that use both multicast and unicast as the transport protocol. The new approach suggested in this paper is to combine the replication used in CDNs with multicast/unicast based distribution and achieve better scalability of the service while maintaining a low cost of storage and communication. The novel hybrid approach for data distribution is based on the understanding that in some cases, it is more efficient to use unicast since it saves bandwidth or computational resources.

Our initial model is a tree graph that has a potential server located at each of its vertices. The vertices may also include local consumers. Each server is assigned with a storage cost and each edge is assigned with distribution communication costs. The distribution demands of the consumers are given. The consumers are served from servers using multicast and/or unicast communication. The costs can also be interpreted as QoS related costs or as loss of revenue resulted from reduced performance.

Our goal is to find an optimal allocation, e.g., the set of servers which store an object, with the minimum overall (communication and storage) cost. Each consumer is served by exactly one server for an object. There is an obvious tradeoff between the storage cost that increases with the number of copies and the distribution cost that decrease with that number.

In this work we present an optimal allocation algorithm for tree networks with computational complexity of $O(N^2)$. We solve the case where the mode of operation per consumer (multicast or unicast) is automatically optimized by the algorithm itself. We also suggest an approximation algorithm for general networks. The model and algorithms can be easily extended to the case where server content is dynamic and needs to updated from media sources via multicast or unicast means[4].

1.1 Related work

Application level multicast and overlay multicast protocols have been studied in recent years. Most of the works are focused on the structure of the overlay topology for a single tree[5–8]. We focus on the way an existing overlay network should be partitioned to multiple regional multicast/unicast trees while optimizing the communication and storage cost. Our earlier work[9] presents an optimal allocation algorithm for the multicast only distribution on trees.

The object allocation problem, also referred as the file allocation problem in storage systems[10] or data management in distributed databases has been studied extensively. When looking only at the unicast distribution model, we end up with the classical "uncapacited plant location problem" (UPLP)[11] with facilities replacing servers and roads replacing communication lines. The problem has been proved to be NP-complete for general graphs[11]. It was solved for trees in polynomial time[12, 13]. The UPLP model was mapped to content delivery networks[14]. There are additional works that address the severs/replicas placement problem for the unicast only distribution model[15–17].

2. THE MODEL

2.1 Objects

For each object o of the objects set O, we determine the set of vertices which store a copy of o. The algorithm handles each object separately, so the costs described below are defined (and can be different) for each object o.

2.2 The tree network

Let $T = (V, E)$ be a tree graph that represents a communication network, where $V = \{1, \ldots, N\}$ is the set of vertices and E is the set of edges. The tree is rooted at any arbitrary vertex r (r=1). Each vertex represents a network switch and a potential storage place for copies of o. Each vertex is also an entry point of content consumers to the network. Distribution demands of consumers connected to vertex i are satisfied by the network from the closest vertex (or the closest multicast tree rooted at) j which stores a copy of o. Consumers

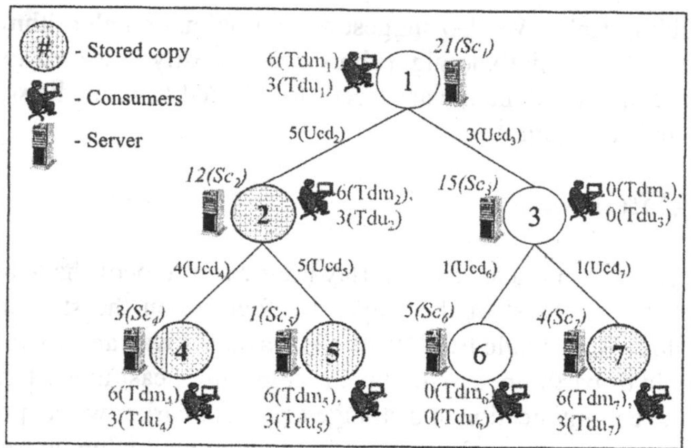

Figure 1. An example of a tree network and various costs

connected to a vertex are served by one of unicast or multicast. The selection is done automatically by the system in order to optimize the overall cost.

Denote the subtree of T rooted at vertex i as T_i; the parent of vertex i in T ($i \neq r$) as P_i; the edge that connects vertex i to its parent in T, (i, P_i) as e_i ($e_r = \emptyset$); the set of edges in $T_i \cup e_i$ as E_i ($E_r \equiv E$); the set of vertices in T_i as V_i ($V_r \equiv V$); the set of children of vertex i in T as Ch_i (For a leaf i, $Ch_i = \emptyset$).

Figure 1 displays a tree network and costs related to its vertices and edges.

2.3 Storage cost

Let the storage cost of object o at vertex i be Sc_i. Sc_i represents resources, like disk space, computational power and relative maintenance cost. Denote Φ is the set of vertices that store o. The total storage cost of o is $\sum_{i \in \Phi} Sc_i$.

2.4 Distribution traffic cost

Denote the cost per traffic unit at edge e_i as Ucd_i ($Ucd_i > 0$). Since $e_r = \emptyset$, $Ucd_r \equiv 0$. Ucd_i represents the residual cost of traffic in a physical line or the relative cost of the connection to a public network. The cost per traffic unit along a path between vertices i and j is $Dd_{i,j} = \sum_{e \in P_{i,j}} Ucd_e$, where $P_{i,j}$ is the set of edges that connect vertex i to vertex j. We define $P_{i,i} \equiv \emptyset$ and $Dd_{i,i} \equiv 0$. Since the graph is undirected, $P_{i,j} = P_{j,i}$.

The mutual exclusive hybrid model automatically selects between the unicast/muticast traffic provided to each vertex. The advantage of multicast over unicast is the aggregation of multiple streams into a single stream. On the other hand, unicast is much easier to control (in terms of flow control). The effective bandwidth required by a unicast stream is smaller than a multicast stream.

The multicast traffic provided to vertex i, Tdm_i, is either Td or 0. Td is used when at least one consumer connected to i requires the object and 0 is used when no consumers connected to i require the object. Td may be the bandwidth requirement, or other QoS related parameters†. Since unicast traffic require less bandwidth, the unicast traffic demand in vertex i, Tdu_i is $Tdu_i = q \cdot Tdm_i, 0 < q < 1$.

Denote the set of vertices which are served using unicast (and are not served by unicast) as V_{uc}; the set of edges in the multicast tree rooted at vertex i as Dmt_i. If $i \notin \Phi$, or i does not serve by multicast, $Dmt_i = \emptyset$.

The total multicast traffic cost is $\sum_{i\in\Phi} Td \cdot (\sum_{e\in Dmt_i} Ucd_e)$. The total unicast traffic cost is $\sum_{i\in V_{uc}} Tdu_i \cdot \min_{j\in\Phi} Dd_{i,j}$. (If $\exists j, k \in \Phi$ s.t. $Dd_{i,j} = Dd_{i,k}$ and $j < k$ then j, the smallest, is taken).

3. THE PROBLEM

The optimization problem is to find an object allocation that minimizes the total cost (storage and traffic):

$$\sum_{i\in\Phi} Sc_i + \sum_{i\in\Phi} Td \cdot \left(\sum_{e\in Dmt_i} Ucd_e \right) + \sum_{i\in V_{uc}} Tdu_i \cdot \min_{j\in\Phi} Dd_{i,j}$$

We developed an optimal algorithm called MX-HDT - Mutual eXclusive Hybrid Distribution for Trees, with computational complexity of $O(N^2)$.

4. MX-HDT VS. MDT/UDT RESULTS

As stated in the sub-section 2.4, the MX-HDT algorithm attempts to leverage the advantages of both MDT (multicast only distribution in trees graphs[9]) and UDT (unicast only distribution, also termed UPLP, on trees[13]) by switching between multicast and unicast in order to minimize the overall communication and storage costs. The MX-HDT algorithm always performs better than UDT/MDT and the overall costs will be equal or smaller than the minimum between UDT/MDT.

Figure 2 presents a comparison between the average results of running MX-HDT, MDT, UDT on various trees in various shapes and sizes, while the ratio between the multicast bandwidth demands and the unicast bandwidth demand is 2 : 1 (i.e. $q = 0.5$). It can be seen that MX-HDT performs better than both algorithms in all cases.

†The reason for using the same traffic rate for all vertices in multicast is the fact that the server determines the transmission rate, not each customer as in the unicast case.

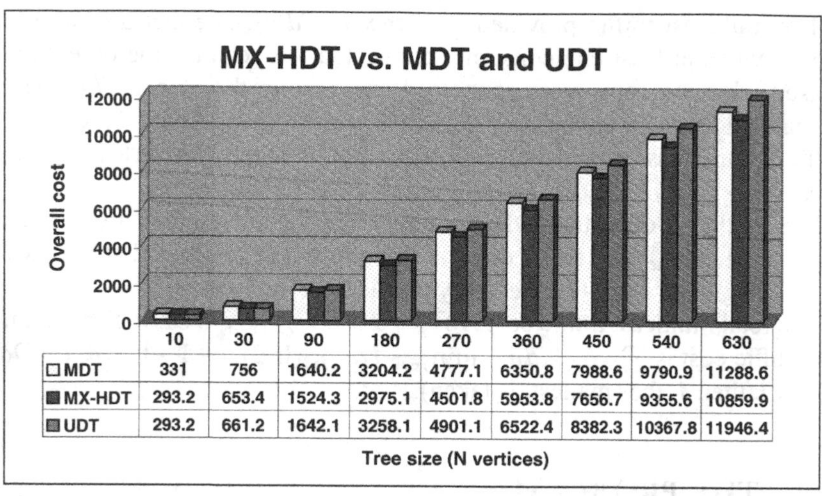

	10	30	90	180	270	360	450	540	630
MDT	331	756	1640.2	3204.2	4777.1	6350.8	7988.6	9790.9	11288.6
MX-HDT	293.2	653.4	1524.3	2975.1	4501.8	5953.8	7656.7	9355.6	10859.9
UDT	293.2	661.2	1642.1	3258.1	4901.1	6522.4	8382.3	10367.8	11946.4

Figure 2. MX-HDT vs. MDT/UDT

5. OPTIMAL ALLOCATION PROPERTIES

LEMMA 1 *In case of unicast traffic, if vertex i is served by vertex j, which satisfies $\min_{j\in\Phi} Dd_{i,j}$, and i is served through vertex k (i.e. $P_{i,j}=P_{i,k}\cup P_{k,j}$), then if vertex l is served by unicast through vertex k, l must also be served from j (k itself may not be served by unicast).*

PROOF $P_{i,j}=P_{i,k}\cup P_{k,j} \Rightarrow Dd_{i,j}=Dd_{i,k}+Dd_{k,j}$. *Suppose vertex l is not served by j, but from a different vertex m. $P_{l,j}=P_{l,k}\cup P_{k,m} \Rightarrow Dd_{l,j}=Dd_{l,k}+Dd_{k,m}$. Since the solution is optimal there must exist $Dd_{k,j}=Dd_{k,m}$. And if $\exists j,m\in\Phi$ s.t. $Dd_{k,j}=Dd_{k,m}$ and $j<m$ then j, the smallest, is taken. So j and m must be the same vertex.*

LEMMA 2 *Each vertex i can only belong to at most one multicast tree.*

PROOF *Suppose a vertex i belongs to more than one multicast tree, then by removing it from the other trees and keeping it connected to only one multicast tree we reduce the traffic in contradiction to the optimality of the cost.*

LEMMA 3 *If vertex i is served through its neighbor k in T (either parent or child), then i and k are served by the same server.*

PROOF *A direct result of lemmas 1 , 2.*

LEMMA 4 *If there is multicast traffic through vertex i, then vertex i must belong to a multicast tree (this property is not correct for unicast).*

PROOF *Suppose there is multicast traffic through vertex i, and i is served by unicast. In this case i belongs to both kinds of trees, and this is a contradiction of the mutual exclusive traffic condition.* ‡

COROLLARY 5 *The optimal allocation is composed of a subgraph of T which is a forest of unicast and multicast subtrees. Each subtree is rooted at a vertex which stores a copy of o. Each edge and vertex in T can be part of at most one unicast and at most one multicast subtree. If a vertex belongs to a multicast tree, it may still pass unicast traffic through its uplink edge.*

6. THE MX-HDT OPTIMAL ALGORITHM

The algorithm calculates the optimal object allocation cost as well as the set of servers that will store the object o.

6.1 The technique

The main idea behind the algorithm is the observation that in tree graphs, since there is only one edge from each vertex i to its parent, and due to lemma 3, if we consider the influence of the optimal allocation outside T_i on the optimal allocation within T_i, it is narrowed to few possibilities, and it is fairly easy and straight forward to calculate the optimal allocation for vertex i and T_i based on the optimal allocation calculated for each c and T_c, where $c \in Ch_i$.

As a result, MX-HDT is a recursive algorithm that finds the optimal allocation for a new problem which is a subset of the original problem, for vertex i and T_i, based on the optimal allocation computed by its children Ch_i.

The algorithm is performed in two phases. The first phase is the cost calculation phase which starts at the leaves and ends at the root, while calculating the optimal allocation and its alternate cost for each vertex pair i, j for each scenario. The second phase is a backtrack phase which starts at the root and ends at the leaves where the algorithm selects the actual scenario in the optimal allocation and allocates the copies in the relevant servers.

The new optimization problem is defined as follows: Find the optimal allocation and its alternate cost in $T_{i,j}$, for the scenarios described in subsection 6.2 that are possible for each vertex pair i, j. These scenarios cover all the possible external influences on the optimal allocation within T_i.

Figure 3. demonstrates the distribution forest (Described in corollary 5) with the different possible scenarios of the vertices and edges in T.

‡Note: the opposite is not a contradiction, i.e. even if there is unicast traffic through vertex i (i.e. - another vertex k is served by unicast through i), then vertex i may still be served by multicast.

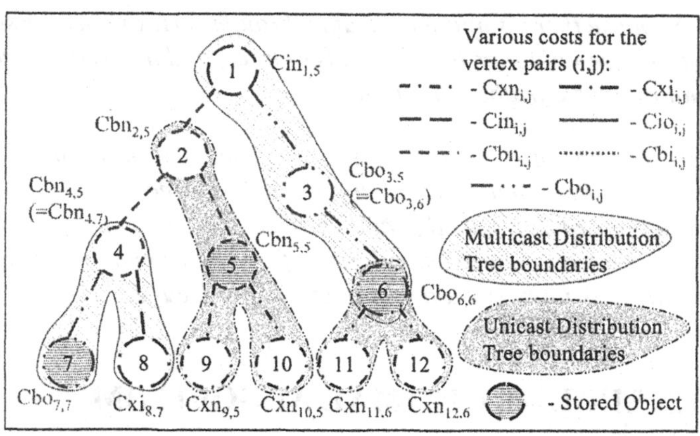

Figure 3. An allocation, scenarios and distribution forest example

6.2 The cost calculation phase

For each vertex pair i, j the algorithm calculates for $T_{i,j}$ (vertex j is assumed to allocate a copy of the object o) alternate costs for the following scenarios:

$Cxn_{i,j}$ - e**X**ternal only allocation and **N**o incoming multicast traffic. No copy of o is located inside T_i ($i \neq r$) and edge e_i may only carry unicast traffic. Legal only when $j \notin V_i$.

$Cxi_{i,j}$ - e**X**ternal only allocation and **I**ncoming multicast traffic. No copy of o is located inside T_i ($i \neq r$) and there is distribution demand in T_i that is served by multicast through edge e_i. Legal only when $j \notin V_i$.

$Cin_{i,j}$ - **I**nternal only allocation and **N**o outgoing multicast traffic. All the copies of o are located only inside T_i. Edge e_i may only carry unicast traffic. Legal only when $j \in V_i$.

$Cio_{i,j}$ - **I**nternal only allocation and **O**utgoing multicast traffic. All the copies of o are located only inside T_i and there is distribution demand outside T_i that is served by multicast through edge e_i. Legal only when $j \in V_i$.

$Cbn_{i,j}$ - **B**oth sides allocation and **N**o multicast traffic. Copies are located both inside and outside T_i. Edge e_i may only carry unicast traffic.

$Cbi_{i,j}$ - **B**oth sides allocation and **I**ncoming multicast traffic. Copies are located both inside and outside T_i and there is distribution demand inside T_i that is served by multicast through edge e_i.

$Cbo_{i,j}$ - **B**oth sides allocation and **O**utgoing multicast traffic. Copies are located both inside and outside T_i and there is distribution demand outside T_i that is served by multicast through edge e_i.

The result of the property described in lemma 4, is that for each scenario which contains multicast distribution through edge i ($xi_{i,j}$, $io_{i,j}$, $bi_{i,j}$ and $bo_{i,j}$), vertex i must be part of a multicast tree. And for each scenario which does not contain multicast distribution through edge i, vertex i may still belong to a multicast tree (as a leaf) or may be served by unicast traffic.

The algorithm calculates the costs as follows:

$$Cxn_{i,j} \leftarrow \begin{cases} \infty, & \text{if } j \in V_i \\ Tdu_i \cdot Dd_{i,j} + sum1, & \text{if } j \notin V_i \end{cases}$$

$$Cxi_{i,j} \leftarrow \begin{cases} \infty, & \text{if } j \in V_i \\ Td \cdot Ucd_i + sum2, & \text{if } j \notin V_i \end{cases}$$

$$Cin_{i,j} \leftarrow \begin{cases} \min\{sum4, sum5, sum6, sum8, min1\}, & \text{if } j \in V_k, k \in Ch_i \\ Sc_i + sum3, & \text{if } j = i \\ \infty, & \text{if } j \notin V_i \end{cases}$$

$$Cio_{i,j} \leftarrow \begin{cases} Td \cdot Ucd_i + \min\{sum5, sum8, min1\}, & \text{if } j \in V_k, k \in Ch_i \\ Td \cdot Ucd_i + Sc_i + sum3, & \text{if } j = i \\ \infty, & \text{if } j \notin V_i \end{cases}$$

$$Cbn_{i,j} \leftarrow \begin{cases} \min\{sum6, sum8, min1\}, & \text{if } j \in V_k, k \in Ch_i \\ Sc_i + sum3, & \text{if } j = i \\ \min\left\{\min_{l \in V_i} Cbn_{i,l}{}^{\star}, \min\{min2, min4\}\right\}, & \text{if } j \notin V_i \end{cases}$$

$$Cbi_{i,j} \leftarrow \begin{cases} Td \cdot Ucd_i + sum7, & \text{if } j \in V_k, k \in Ch_i \\ \infty, & \text{if } j = i \\ \min\left\{\min_{l \in V_i} Cbi_{i,l}{}^{\star}, Td \cdot Ucd_i + min3\right\}, & \text{if } j \notin V_i \end{cases}$$

$$Cbo_{i,j} \leftarrow \begin{cases} Td \cdot Ucd_i + \min\{sum8, min1\}, & \text{if } j \in V_k, k \in Ch_i \\ Td \cdot Ucd_i + Sc_i + sum3, & \text{if } j = i \\ \min\left\{\min_{l \in V_i} Cbo_{i,l}{}^{\star}, Td \cdot Ucd_i + min4\right\}, & \text{if } j \notin V_i \end{cases}$$

The cost of the optimal allocation in T is $min_{j \in V} Cin_{r,j}$.

$sum1$-$sum8$ and $min1$-$min4$ represent combinations of children scenarios ($sum1$-$sum8$ equal 0, $min1$-$min4$ equal ∞ if i is a leaf). A detailed explanation about the combinations and the proof of optimality exist in our TR[4].

*The minimum value should be calculated efficiently if $Cb?_{i,j}$, $j \in V_i$ are calculated prior to calculating any $Cb?_{i,j}$, $j \notin V_i$

6.3 Backtracking for the content allocation

While calculating the alternate costs for each vertex pair i, j, the algorithm remembers for each such cost (scenario), if a copy needs to be stored at vertex i and the relevant scenario of each child k that was used in the calculation.

The backtrack phase starts at the root and ends at the leaves of T. For each vertex i, the algorithm determines the actual scenario in the optimal allocation, if a copy should be stored at i (will happen if (i, i) pair was selected for an actual scenario) and if it is necessary to keep advancing towards the leaves of T. The algorithm uses the backtrack information that was saved earlier. The pseudo code and backtrack details of the algorithm are given in our TR[4].

6.4 Computational complexity of MX-HDT

In the cost calculation phase, each vertex in the tree $i \in V$ the algorithm calculates up to $7 \cdot N$ alternate costs. Each cost calculation requires $O(|Ch_i| + 1)$. $|V| = N$ and the total number of children in the tree is $N-1$ (only the root r is not a child). The complexity of the backtrack phase for vertex i is $O(1)$. The computational complexity of the algorithm is:

$$O_{HDT} = O(N) + \sum_{i \in V} 7N \cdot O(|Ch_i| + 1) = O\left(N + 7N \cdot \sum_{i \in V}(|Ch_i| + 1)\right)$$
$$= O(N + (7N + 1) \cdot (2N - 1)) = O(N^2)$$

The computational complexity of MX-HDT is $O(N^2)$.

7. THE MX-HDG APPROXIMATION ALGORITHM

We extended the model to general graphs. Figure 4 depicts a network with the costs related to its vertices and edges.

7.1 The optimal allocation properties

The well known Steiner tree problem[18] is defined as follows: given a (edge) weighted undirected graph and a given subset of vertices termed terminals, find a minimal weight tree spanning all terminals. Consequently, an optimal multicast tree in a general graph is a Steiner tree. The Steiner tree problem is NP-hard on general graphs[19].

The property of lemma 2 is also valid for general graphs. The optimal solution in a general graph is a forest of Steiner trees for multicast traffic and

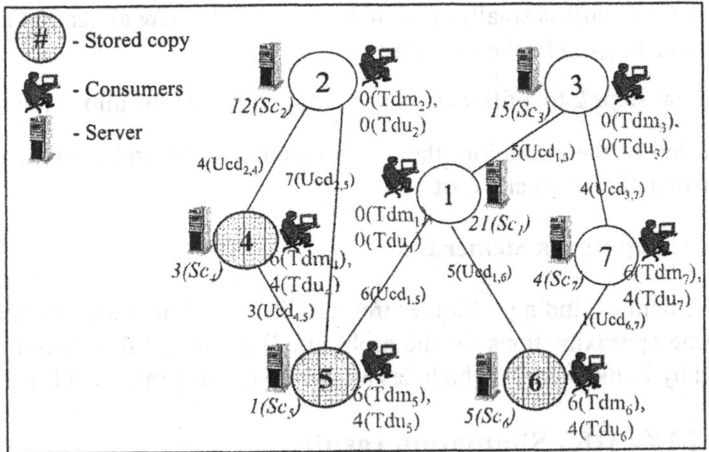

Figure 4. An example network and costs.

additional unicast paths where unicast is used. The total cost of the optimal allocation is constructed of the storage cost, the multicast Steiner trees costs and the unicast traffic cost.

Since finding a Steiner tree in a general graph is NP-hard, it is obvious that finding a forest of Steiner trees is NP-hard as well.

7.2 The approximation algorithm

As the allocation problem in general graphs is NP-hard, we use our optimal MX-HDT algorithm for trees to develop an approximation to that problem. This is an iterative algorithm that starts with an initial random or arbitrary allocation, and converges to an allocation which is optimal in an approximated Steiner tree computed for the general graph.

7.2.1 The MX-HDG algorithm steps

1 Start with a random allocation and set min_{cost} to ∞.

2 Compute a Steiner tree for the graph where the terminals are all the vertices with distribution demand and the vertices which store the object.

3 Run MX-HDT on the extracted tree. For the $Dd_{i,j}$ values use the shortest path between i and j in the graph. The result of the algorithm is a new set of vertices which store the object and a new approximated cost.

4 If the new cost is smaller than min_{cost} save the new allocation and update min_{cost} to be the new cost.

5 Repeat steps 2 to 4 till there is no improvement in the allocation cost.

At the end of the execution, the last saved allocation and min_{cost} are the approximated allocation and cost.

7.2.2 Computing a Steiner tree

The problem of finding a Steiner tree is NP-hard. There are several polynomial time approximations for the problem. We selected the approximation suggested by Zelikovsky[20], which has an approximation ratio of $11/6$.

7.3 MX-HDG Simulation results

We've generated general graphs using the Internet Model suggested by Zegura et al.[21, 22]. We've generated graphs with various node counts.

We've run our approximation MX-HDG algorithm on these graphs, and compared the results to random allocations and the multicast only (MDG) algorithm[4]. Figure 5 displays these charts.

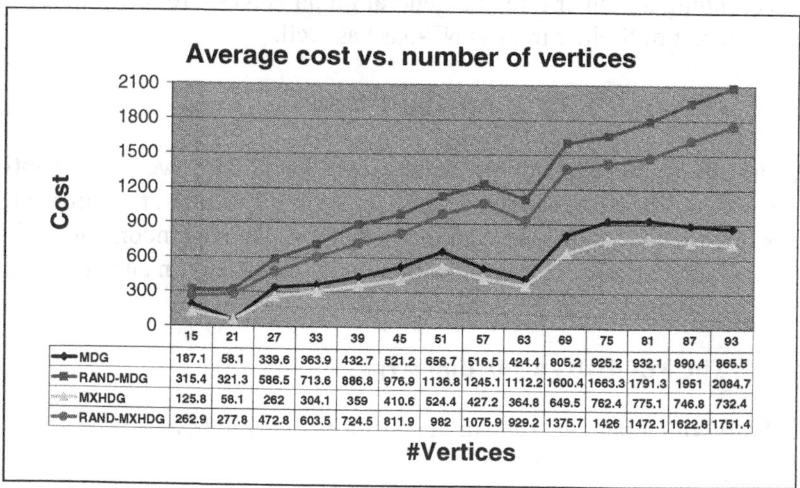

	15	21	27	33	39	45	51	57	63	69	75	81	87	93
MDG	187.1	58.1	339.6	363.9	432.7	521.2	656.7	516.5	424.4	805.2	925.2	932.1	890.4	865.5
RAND-MDG	315.4	321.3	586.5	713.6	886.8	976.9	1136.8	1245.1	1112.2	1600.4	1663.3	1791.3	1951	2084.7
MXHDG	125.8	58.1	262	304.1	359	410.6	524.4	427.2	364.8	649.5	762.4	775.1	746.8	732.4
RAND-MXHDG	262.9	277.8	472.8	603.5	724.5	811.9	982	1075.9	929.2	1375.7	1426	1472.1	1622.8	1751.4

Figure 5. The number of copies and cost of allocations in general graphs

As can be seen from the results, the average costs of MX-HDG are better than MDG (typically by 20%) and significantly better than of the results of the random allocations (the differences between RAND-MDG and RAND-MXHDG, is due the distribution model - multicast vs. hybrid).

8. CONCLUSIONS AND FUTURE WORK

In this work, we addressed the content location problem in hybrid overlay networks, while optimizing the storage and communication costs in the context of QoS provisioning.

We developed an optimal content allocation algorithm for tree networks with computational complexity of $O(N^2)$. The algorithm is recursive and is based on dynamic programming. The algorithm can easily be specified as a distributed algorithm due to the independent calculations at each vertex (only based on information from its neighbors) and due to the hierarchical data flow.

In addition to the optimal algorithm we suggested an approximation for general networks, which requires a small number of iterations, and is based on our optimal algorithm for tree networks.

In our extended work[4], that is not presented here because of space limitations, we address a more general problem where additional media sources are added and additional update communication from the media sources to the servers is considered in the optimization problem.

REFERENCES

1. Cisco, http://www.cisco.com/
2. Akamai, http://www.akamai.com/
3. Digital Fountain, http://www.digitalfountain.com/
4. O. Unger and I. Cidon, Content location in multicast based overlay networks with content updates, CCIT Report #432, Technion Press, June 2003.
5. S. Shi and J. Turner. Routing in Overlay Multicast Networks. In Proc. of IEEE INFOCOM, June 2002.
6. P. Francis. "Yoid: Extending the Internet multicast architecture", April 2000.
7. D. Helder and S. Jamin. Banana tree protocol, an end-host multicast protocol. Technical Report CSE-TR-429-00, University of Michigan, 2000.
8. J. Jannotti, D. K. Gifford, K. L. Johnson, M. F. Kaashoek and J. O'Toole. Overcast: Reliable multicasting with an overlay network. In Proceedings of the Fourth Symposium on Operating Systems Design and Implementation, pp. 197–212, October 2000.
9. I. Cidon and O. Unger, Optimal content location in IP multicast based overlay networks, In Proceedings of the 23rd ICDCS workshops, May 2003.
10. L. W. Dowdy and D. V. Foster. Comparative Models of the File Assignment Problem. ACM Computing Surveys, 14(2) pp. 287-313, 1982.
11. Pitu B. Mirchandani, Richard L. Francis. Discrete Location Theory, John Wiley & Sons, Inc. 1990.
12. Kolen A., "Solving covering problems and the uncapacited plant location problem on trees", European Journal of Operational Research, Vol. 12, pp. 266-278, 1983.

13. Alain Billionnet, Marie-Christine Costa, "Solving the Uncapacited Plant Location Problem on Trees", Discrete Applied Mathematics, Vol. 49(1-3), pp. 51-59, 1994.
14. I. Cidon, S. Kutten, and R. Sofer. Optimal allocation of electronic content. In Proceedings of IEEE Infocom, Anchorage, AK, April 22-26, 2001.
15. L. Qiu, V. N. Padmanabham, and G. M. Voelker. On the placement of web server replicas. In Proc. 20th IEEE INFOCOM, 2001.
16. Sugih Jamin, Cheng Jin, Anthony R. Kurc, Danny Raz and Yuval Shavitt. Constrained Mirror Placement on the Internet, IEEE Infocom 2001.
17. J. Kangasharju and J. Roberts and K. Ross. Object Replication Strategies in Content Distribution Networks, In Proceedings of WCW'01: Web Caching and Content Distribution Workshop, Boston, MA, June 2001.
18. S. L. Hakimi. Steiner's problem in graphs and its implications. Networks, Vol. 1 (1971), pp. 113-133.
19. M. R. Garey, R. L. Graham and D.S. Johnson. The complexity of computing Steiner minimal trees. SIAM J. Appl. Math., 32 (1977) pp. 835–859.
20. A. Z. Zelikovsky. The 11/6–approximation algorithm for the Steiner problem on networks. Algorithmica 9 (1993) 463–470.
21. Ellen W. Zegura, Ken Calvert and S. Bhattacharjee. How to Model an Internetwork. Proceedings of IEEE Infocom '96, San Francisco, CA.
22. GT-ITM: Georgia Tech Internetwork Topology Models, http://www.cc.gatech.edu/fac/Ellen.Zegura/ gt-itm/gt-itm.tar.gz

A COMMUNICATION ARCHITECTURE FOR REAL-TIME AUCTIONS

Hella Kaffel Ben Ayed, Safa Kaabi Chihi and Farouk Kamoun
CRISTAL Lab., National School of Computer Science, University of Manouba, Tunis, Tunisia

Abstract: This paper explores the possibility to use a communication protocol other than HTTP under real-time auction applications in order to provide best-suited communication services. We specify a distributed communication architecture named AHS (Auction Handling System) based on the IRC architecture to support real-time auctions. While using the suitable services provided by IRC, this architecture provides auction applications with what we define as required communication services. We also specify a communication protocol, called BSA-protocol, to support interactions between auction participants and the auctioneer in a real-time auction process. This protocol uses the services provided by the IRC-client protocol as well as the channel facilities provided by the IRC architecture for group communications. We report on the encapsulation of this protocol within the IRC-client protocol and on the implementation of a prototype. The originality of this architecture lies in the fact that it both frees auction applications from communication issues and is independent from the auction protocol.

Key words: Auctions, communication protocols, communication services, communication architecture, interactions, IRC, messages.

1. INTRODUCTION

Auctions are market mechanisms whereby one seller or one buyer – named here the initiator– is involved with many buyers or sellers called bidders. More complex forms have appeared and involve many initiators and many bidders (Beam et al., 1996; Ben Ameur,2001; Ben Youssef et al., 2001; EPRI, 2001). The way participants interact is controlled by a set of rules called the auction protocol. These interactions are managed by a mediator, called the auctioneer (e.g. the auction site), which provides the institutional setting for the auction (Beam and Segev, 1997; Klein,1997 ; Kumar and Feldman, 1998a; Strobel, 2000). They result in the exchange of various messages: bid, bid withdrawal (BW), bid admittance, bid rejection, price quote (PQ) and transaction notification (TN) (Wurman et al., 1998) .

Well known auctions (English, Dutch, Vickrey and Combinatorial Double Auction or CDA) have been variously classified (Ben Ameur, 2001;Wurman et al., 1998) (ascending/descending, sealed/outcry, single/double, etc.). Our study concerns real-time auctions. We talk about real-time when data from auction participants must be continuously monitored and processed in a timely manner to allow for a real-time decision over the Internet (Pen et al., 1998). Real-time auctions' activities are organized in rounds governed by a clock. The duration of a round depends on the auction type and the evaluation by the auctioneer of submitted bids during a round can be scheduled, or occurs, at random times or through bidders' activity/inactivity, depending on the auction protocol (Panzini and Shrivastava, 1999;Wurman et al., 1998).

Communication issues have an important impact on the process and results of real-time auctions (Pen et al., 1998). Hence, we focus on the communication protocol that supports the transfer of messages related to real-time auctions.

Current Internet sites running real-time auctions use the HTTP protocol. This is a Request/Response protocol that does not provide any additional communication service. Hence, required services for the auction process are implemented within the auction application (Wurman et al., 1998; Kumar and Feldman, 1998a; Kumar and Feldman, 1998b; Panzini and Shrivastava, 1999). This results in overloading auction applications with communication issues. This may influence the auctioneer's quality of service and cause frustrated customers (bidders or initiators) to leave the auction site (Amza et al.,2002; Cardellini et al.,2001; Pen et al., 1998).

Our objective is to explore the possibility to use another communication protocol in the application layer of the TCP/IP stack in order to provide auction applications with adequate communication services.

First, we defined in Section 2 real-time auctions' requirements in terms of communication services. We show that the IRC protocols are best suited to the defined auction requirements (Kaabi et al., 2001; Kaabi et al., 2003). As a result, we propose in Section 3 a novel communication architecture, based on IRC and named AHS, to support real-time auctions. In Section 4, we depict the specification of the BSA-protocol. Section 5 describes the encapsulation of this protocol within the IRC-client protocol and the implementation of a prototype. Finally section 6, concludes the paper and opens up prospects for future work.

2. WHICH COMMUNICATION SERVICES FOR REAL-TIME AUCTIONS?

In his section we define first the requirements of real-time auctions in terms of communication services. Then, we report on a functional comparison of: HTTP, IRC, NNTP and SMTP with regard to the provided services and underline the suitable features of IRC.

2.1 Real-time auctions' requirements

In the literature, various studies have described the typical real-time process and identified the following steps: Initiator's registration, Setting up the auction event, Bidders' registrations, Request for participation, Bidding and Settlement(Beam and Segev, 1997; Kaabi et al., 2001; Klein,1997; Kumar and Feldman, 1998b; Turban, 1997). Some of above mentioned studies have defined a number of requirements (Panzini and Shrivastava, 1999) —such as scalability, responsiveness and consistency— to be considered. We lay emphasis on the requirements that must be met by the communication protocol underlying the auction applications. We identify two sets of services: basic and optional.

Basic services:

These are always required and are considered as mandatory.

i. Synchronous and push modes: The synchronous mode enables real-time interactions between bidders/initiators and the auctioneer. The Push mode is recommended to minimize the bidders' reaction time and increase the system's reactivity (Kumar and Feldman, 1998b; Panzini and Shrivastava, 1999).

ii. Group communications: These are required to support the broadcasting of the intermediary result –PQ– and final result –TN (Maxemchuk and Shur, 2001,Panzini and Shrivastava, 1999). Group communications minimize the number of messages sent by the auctioneer and optimize the way the bandwidth is used.

Optional services:

These services are required only in some cases such as particular auction events, auction protocols or auction messages.

iii. Duration of message validity: Usually bids and price quotes are time-sensitive in that they are valid for a certain time only and then become obsolete (Wurman et al., 1998; Kumar and Feldman, 1998b). This service allows the communication layer to reject obsolete bids or other messages transparently to the auction application.

iv. Fairness: In our context, fairness means that all auction participants have the same privilege with regard to communications. Fairness may be required for both bidders and initiator or for only one side (Asokan, 1998).

v. Message-tracking: This service would allow the history of an auction event to be saved and different stages of the auction process to be kept track as proposed by Wurman (Wurman et al., 1998).

vi. Notification: A notification is an acknowledgment that confirms the reception of a message by its recipient. This allows auction participants to be informed about intermediary stages of the bidding process and contribute to provide the non repudiation of the recipient service (Asokan, 1998). We defined three notifications:
* Submission Notification (SN). It acknowledges the reception of a bid by the auctioneer.
* Acceptance Notification (AN): It acknowledges the fact that a submitted bid conforms to the auction rules and has been accepted by the auctioneer for evaluation.
* Rejection Notification (RN): It is sent by the auctioneer and indicates that the auctioneer, before the evaluation, has rejected a submitted bid. This rejection may be caused by the expiry of the bid or the violation of auction rules.

vii. Security: We focus here on security services, common to all real-time auctions, which must be met by the communication protocol underlying auction applications. These services may be required for both sides: bidders/initiator and the auctioneer.
- On the auctioneer's side:
Identification and authentication: This would permit unauthorized posting of bids and violation of auction rules to be prevented.
Non-repudiation of bidders: This service permits to prevent a given bidder denying having sent a bid or received a TN (Omote, 2002).
- On the bidder's side:
Anonymity: This is required when bidders should not know one another's identity from the exchange of information.
Non-repudiation of the auctioneer: This service permits to prevent an auctioneer denying having received a bid (Omote, 2002).
Data integrity would allow a recipient to be sure that a received message had not been modified during transmission. Data confidentiality may also be required when submitted bids should not be accessible to unauthorized recipients (Harkavy et al., 1998).

viii. Time stamping. It might be required that exchanged messages be time stamped when sent or when received by the auctioneer. This service provides a tool for the auctioneer to manage the temporal accuracy of arrival or sending of messages (Maxemchuk and Shur, 2001,Wurman et al., 1998).

2.2 Which communication protocol for real-time auctions?

In Kaabi (Kaabi et al., 2003) we have compared the functionalities of HTTP, Internet mail, IRC and NNTP according to the defined requirements. This comparison shows that IRC is the best suited protocol with regard to the defined requirements.

IRC communications are synchronous with a push mechanism. IRC channels are defined to support group communications. With regard to fairness, IRC provides a fair distribution of messages to clients: IRC servers maintain for each client the same short period of time (2 seconds) so that all clients are served fairly by the IRC server they are connected to (IETF, 2000c). This may be considered a factor of fairness in accessing the server resources.

With regard to security, provided services are authentication and anonymity (IETF, 2000c; IETF, 2000d). The other required security services are not provided by IRC. Moreover, IRC does not provide duration of message validity nor time stamping and message-tracking.

As a result, all these missing services have to be added to the IRC functionalities in order to satisfy real-time auction requirements. In the following section, we propose the specification of a communication architecture, named AHS and based on the IRC architecture.

3. A PROPOSAL FOR COMMUNICATION ARCHITECTURE

The objective of this architecture, named AHS (Auction Handling System), is to provide the required services for real-time auction applications. It takes advantage of the previously emphasized features of the IRC. Auction applications are the end-users of the AHS architecture. This work is inspired with the ITU X435 architecture that provides suitable communication services to EDI applications by using the store and forward service provided by the X400 messaging (CCITT, 1991a; CCITT, 1991b). A similar study was done to provide communication services for EDI applications while using the Internet mail architecture (Ben Azzouz,1999; Ben Azzouz et al.,2000; Shih et al., 1997).

3.1 Description of the AHS architecture:

The AHS architecture is composed of the following functional components (see Figure 1):

The Auction Server Agent (ASA): An ASA is associated to an auction site and supports the auctioneer's activities. It may simultaneously support several auction events that have different auction protocols. An auction event may involve more than one ASA. The set of ASAs is called Auction Server System (ASS). An ASS is a distributed auctioneer.

The Buyer/Seller Agent (BSA): A BSA is a user agent that can be associated with an initiator or a bidder. A BSA is attached to one ASA. Several BSAs may be attached to the same ASA.

The Bids Store (BS): A BS provides, when required, the capacity of storing messages such as bids and PQs or TNs for further use (message-tracking requirement). The physical situation of BS is not specified. It can be situated within an ASA, a BSA or a separate entity.

3.2 The Layered Model

The AHS architecture is based on the IRC architecture (IETF, 2000a). A BSA and a BS will be implemented over an IRC client and an ASA over an IRC server.

We define a new layer, called the P-auction Layer, over the IRC layer and under the auction application layer (see Figure 1). This layer is responsible for providing the auction application with the required communication services. This layered structure has the advantage of separating between the processing (in auction applications) and communication (P-auction layer). This gives the P-auction layer flexibility and independence from the auction protocol and the technology used by the auction application

3.3 The Protocols

The following protocols are defined to specify the interactions between functional components of AHS and how the required services are provided to end-users (see Figure 1):

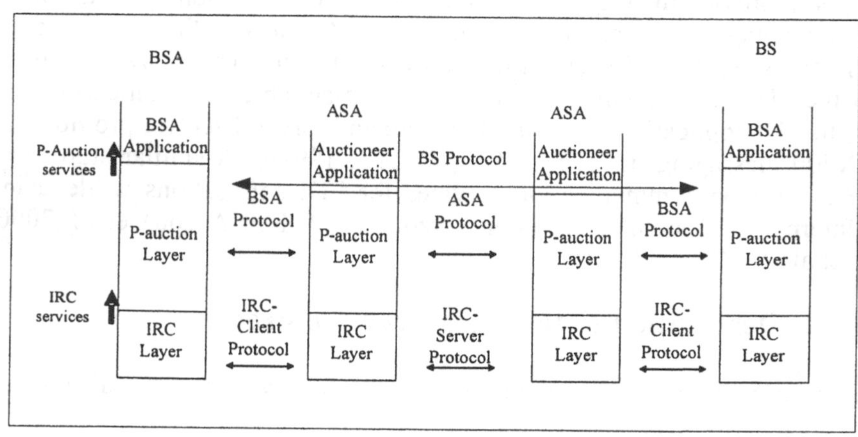

Figure 1. The AHS layered model and protocols

The BSA-protocol: This protocol specifies the interactions between a BSA and an ASA. It is encapsulated within the IRC-client protocol.

The ASA-protocol: This protocol specifies interactions between the ASAs involved in a distributed auction event. This protocol is encapsulated within the IRC-server protocol.

The BS-protocol: This protocol specifies interactions between a BSA or an ASA and a BS. In the next section we specify the BSA-protocol.

4. SPECIFICATION OF THE BSA-PROTOCOL

We first define the interactions and then, the exchanged messages.

4.1 The interactions

Interactions between a BSA and an ASA are performed according to the auction process. They are the following (Kaabi et al., 2003):

1. Bidder/initiator Registration: This is initiated by a BSA. It allows a bidder or an initiator to be registered within an ASA or to have access to an auction event managed by the ASA.
2. Setting up the auction event: This is initiated by the initiator of an auction event..
3. Access: This corresponds to the Request for participation step in the auction process. This interaction is initiated by a BSA.
4. Exit: This is initiated by a BSA that asks for leaving an auction event. It may also be initiated by an ASA to force a BSA to leave the auction event.
5. Bidding: This implies the sending of messages transporting bids and BWs and notifications between a BSA to the ASA it is attached to.
6. Price quote (PQ) announcement: This consists in the intermediary result announcement and occurs after the evaluation of submitted bids. This interaction is initiated by the ASA.
7. Transaction Notification (TN) announcement: This corresponds to the broadcasting of the final result and the sending of the request notifications. This interaction is initiated by the ASA.
8. Auction Event Closure: This is initiated by the ASA attached to the initiator's BSA. It implies the sending of a set of messages to clear the auction event.

4.2 The BSA-Protocol messages

We define three types of messages:
- Requests (.Req): These messages require a reply from the recipient.
- Responses (.Resp): These are responses to Requests.

- Indications (.Ind): These are messages that do not require replies from the recipient.

A P-auction message is composed of two parts: the Header and the Data. The Data transports the message generated by this application. The Header is processed by the P-auction layer and is structured into fields that allow the implementation of the required services.

The first field of the Header is called *code* and has a numeric value to identify each message.

Since many ASAs may constitute an auction site and many auction events may be handled by an ASA, we define two fields: *ASA-name* and *Auction-name* to identify respectively an ASA and an auction event. A field called *BSA-name* identifies a BSA.

To achieve the goal of duration of message validity and time stamping, we introduce within the Header of concerned messages respectively two fields: *Expiry-time* and *ASA-time*. The *Expiry-time* is filled by the message's sender. The *ASA-time* is filled by the ASA just before the sending, or after the reception, of time stamped messages.

To perform a control at the P-auction layer, we define two header fields that carry semantic information related either to the auction event and bids or to the BSA. These fields are the *Parameters* and the *Type*. This makes it possible to carry out a rudimentary bid validity control within this layer independently of the structure of the Data part of the message. By so doing, we save processing time in this application since a number of invalid bids are rejected at the P-auction layer (for example, a bid submitted by the seller is rejected in case of English auction).

To provide the notification service, we define a field, called *Notif-req*, in the Header of messages that may require a notification and a specific message to carry these notifications, i.e. Notif.Resp.

The other fields that compose the header are described in the following section and are classified according to BSA-ASA interactions.

5. THE ENCAPSULATION OF THE BSA-PROTOCOL IN THE IRC CLIENT PROTOCOL

5.1 Technical considerations related to IRC

These aim to use properly the interesting features of the IRC standard without modifying the source code of the IRC-server or the IRC-client protocols.

- An IRC channel is created by the ASA to which the initiator is attached and is associated to an auction event. The channel-name is the same as the auction name (*Auction-name)*.
- For security reasons, we have opted to use exclusively channels in the Invite mode. This mode disables any IRC client to join the channel unless he receives an Invite message from the channel operator (IETF, 2000b). This

way, we guarantee that the access to the channel used by an auction event is under the control of the auctioneer application.

- We define an IRC client, in addition to the IRC server within an ASA (see Figure 2). The nickname of this IRC client corresponds to the name of the ASA (*ASA-name).* This is to overcome the following limitations within the IRC specifications.

a. An IRC server has not the ability to create channels. Only clients (channel operators) have this ability (IETF, 2000b). The client ASA-name acts as a channel operator. This way, the creation of channels for auction events is controlled by the auctioneer application that asks the Client ASA-name for channel creation.
b. An IRC client cannot send textual messages to an IRC server but only to IRC clients (IETF, 2000c). BSA-protocol messages sent by a given BSA will be sent as a regular point-to-point IRC exchange between the IRC client of the BSA and the Client ASA-name.
c. An IRC server has not the ability to send point-to-point textual messages to a specific client member of a channel (IETF, 2000b). Messages sent by the auctioneer application within an ASA to specific BSAs will be sent as point-to-point exchanges between the Client ASA-name and IRC clients within these BSAs.
d. A channel operator has not the ability to close a channel. A channel is closed when the last member leaves it (IETF, 2000b). The Client ASA-name removes (Kick) all active channel members to close the auction channel.

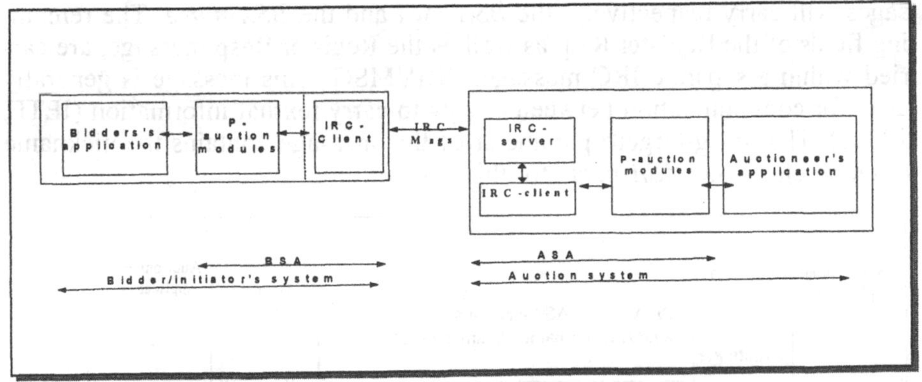

Figure 2: Design of the auctioneer's and bidders/initiators' systems

5.2 The Encapsulation

Each P-auction message will be encapsulated within an IRC message when sent and de-capsulated when received. An IRC message is composed of three fields: The Prefix, the Command and the Command parameters (IETF, 2000c). The Command parameters is composed of several fields containing information, such as <nickname>, <channel-name> and <text-to-be-

sent>. This latter transports the text sent by IRC clients. The Prefix, the Command and some Command parameters fields constitute the header of an IRC message. The encapsulation process consists of inserting a P-auction message within the Command parameters of an IRC message. This permits to leave the Prefix and the Command unchanged since they have a specific meaning for IRC protocols. The Data is carried within the <text-to-be-sent> field of the Command parameters. The Header's fields are in two categories:

- Fields that have a semantic equivalence with some fields within the Command parameters of the encapsulating IRC message. These are: BSA-name (eq. to <nickname> of the client within the BSA), BSA-pwd (eq. to <password>), Auction-name (eq. to <channel-name>) and ASA-name (eq. to <nickname> of the client ASA-name). These fields are carried by their corresponding within the Command parameters of the IRC message.
- Fields that do not have any semantic equivalence with the fields in the Command parameters of the encapsulating IRC message. These are conveyed within the < text-to-be-sent> part of the IRC message.

Table1 describes the structure of the BSA-Protocol messages as well as the syntax of their transporting IRC messages and their encapsulation within these messages.

Encapsulation during the Bidder and initiator registration: This interaction implies the sending of two messages: Register.Req and Register.Resp. A subset of the Header of the Register.Req message will be encapsulated within IRC connection registration messages. This allows using the access control facility of IRC. The Parameters of the PASS and NICK messages will carry respectively, the *BSA-pwd* and the *BSA-name*. The remaining fields of the Register.Req, as well as the Register.Resp message, are carried within a separate IRC message: PRIVMSG. This message is generally used for communication between clients to carry textual information (IETF, 2000c). The <msg-target> parameter of this message contains the nickname of target client (i.e. *ASA-name* for this case).

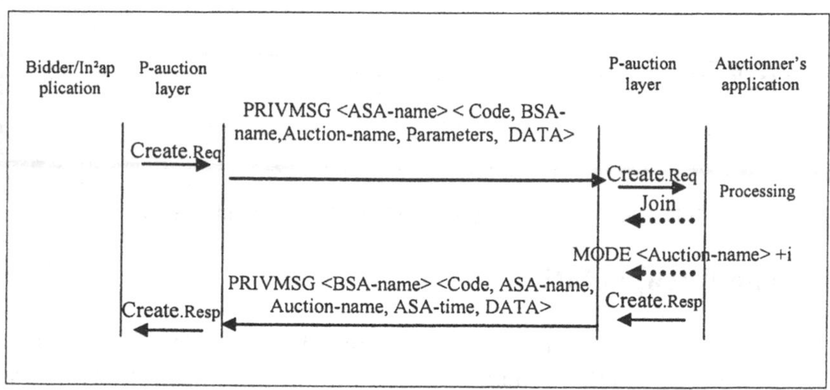

Figure 3: A successful setting up of an auction event

Encapsulation during the Setting up of the auction event: The Create.Req message is first encapsulated within a PRIVMSG (carried in the <text-to-be-sent> parameter) that is sent to the Client ASA-name. As shown in Figure 3, if the decision of the auctioneer application is an acceptance of the creation, the Client ASA-name "sends" locally to the IRC server a JOIN message to ask for the channel creation. The Create.Resp message is encapsulated within a PRIVMSG message.

Encapsulation during the Access: This interaction requires the authentication of the BSA by the auctioneer application before this latter can be invited to join the channel assigned to the auction event. The Access.Req message is encapsulated within a PRIVMSG in the <text-to-be-sent> parameter. After the auctioneer application has proceeded to the access control, the Access.Resp is carried by an INVITE IRC message (in case of acceptance) or a PRIVMSG message (in case of refusal). As a response to the INVITE, the P-auction layer of the BSA sends automatically a JOIN message.

Encapsulation during the Exit: The Exit-auction.Ind may be sent by a BSA or by an ASA. When sent by a BSA this message will be encapsulated within a PART IRC message. This IRC message is used by an IRC client to be removed from the list of active members of a channel. When sent by the ASA, the Exit-auction.Ind will be encapsulated in a KICK IRC message. This message is used by an IRC server to force the removal of an IRC client from a channel.

Encapsulation during the bidding: Bid.Req, Bw.Req and the Notif.Resp messages will be encapsulated within PRIVMSG messages.

Encapsulation during the price quote announcement: The PQ.Ind message will be encapsulated within a PRIVMSG message. To be broadcasted over the channel, the parameter <msgtarget> will contain the *Auction-name*.

Encapsulation during the transaction notification announcement: The Tr-notif.Ind will be encapsulated in a PRIVMSG message similarly to the PQ.Ind. The Tr-acc.Req and the Tr-acc.Resp will be encapsulated within a PRIVMSG message. Several Tr-acc.Req messages are sent if there are more than one winner.

Encapsulation during the auction event closure: Similarly to the Exit-auction.Ind when sent by an ASA, the Auction-closure.Ind will be encapsulated within a KICK message sent by the Client ASA-name to each active BSA. This results in the closure of the IRC channel and other resources allocated to the auction event.

5.3 Experimentation

The BSA-Protocol has been implemented in C langage and uses an IRC software irc2.10.3p3 over Linux. The IRC server has been configured (ircd.conf and config.h files) according to the installation instructions. The

IRC services are called by the use of two IRC primitives: sendto-one() for sending and m-mypriv() for receiving messages.

Table 1. BSA-Protocol messages and their encapsulation within IRC messages

P-auction message	IRC message	Encapsulation
Register.Req (Code, BSA-name, BSA-pwd, BSA-type, Notif.req, DATA)	PASS <password> NICK <nickname> USER <user-name> <mode> <hostname> <real name> PRIVMSG <msg-target><text-to-be-sent>	PASS <BSA-pwd> NICK <BSA-name> PRIVMSG <auctioneer> <Code, BSA-type, DATA>
Register.Resp (code, BSA-name, type, DATA)	PRIVMSG <msg-target><text-to-be-sent>	PRIVMSG <BSA-name> <Code, type, DATA>
Access.Req (Code, BSA-name,Auction-name, Data)	PRIVMSG <msg-target><text-to-be-sent>	PRIVMSG <ASA-name><Code,BSA-name, Auction-name, DATA>
Access.Resp (code, BSA-name, ASA-name, Type, DATA)	INVITE <nick-name><channel-name> JOIN <channel-name>{,<channel-name>}[<key>{,<key>}].	INVITE <BSA-name> <Auction-name> JOIN <Auction-name>
Exit-auction.Ind (Code, BSA-name,Auction-name, Data)	PART <chan-nel>{,<channel>} KICK <chan-nel><nickname><text-to-be-sent>	PART <Auction-name> or KICK <Auction-name><BSA-name><Data>
Bid.Req (Code, BSA-name, ASA-name, Auction-name, ASA-time, Expiry-time, Notif-req, Parameters, Data)	PRIVMSG <msg-target><text-to-be-sent>	PRIVMSG <ASA-name> <Code,BSA-name, Auction-name, ASA-time,Expiry-time, Notif-req, Parameters, DATA>
Notif.Resp (code, BSA-name, ASA-name, Auction-name, ASA-time, Expiry-time, Type, Data)	PRIVMSG <msg-target><text-to-be-sent>	PRIVMSG <BSA-name> <Code,ASA-name, Auction-name, ASA-time,Expiry-time, Type, DATA>
PQ.Ind (Code, ASA-name, Auction-name, ASA-time, Expiry-time, Data)	PRIVMSG <msg-target><text-to-be-sent>	PRIVMSG <Auction-name> <Code, ASA-name, ASA-time, Expiry-time, DATA>
Tr-notif.Ind (Code, Auction-name, ASA-name, ASA-time, Expiry-time, Data)	PRIVMSG <msg-target><text-to-be-sent>	PRIVMSG <Auction-name> <Code, ASA-name, ASA-time, Expiry-time, DATA>
Tr-acc.Req (Code, BSA-name, Auction-name, ASA-name, ASA-time, Data)	PRIVMSG <msg-target><text-to-be-sent>	PRIVMSG <BSA-name> <Code, Auction-name, ASA-name, ASA-time, Data>
Tr-acc.Resp (Code, BSA-name, ASA-name, Auction-name, ASA-time, Data)	PRIVMSG <msg-target><text-to-be-sent>	PRIVMSG <ASA-name> <Code, BSA-name, ASA-name, ASA-time, Data>
Auction-closure.Ind (Code,BSA-name,Auction-name,ASA-name,ASA-time,Data)	KICK <chan-nel><nickname><text-to-be-sent>	KICK <Auction-name> <BSA-name><Code,ASA-name,ASA-time, Data>

6. CONCLUSION

In this paper, we propose a distributed communication architecture for real-time auctions. We describe the specification of the BSA-Protocol involved between the auction site and the participants. We depict the encapsulation of this protocol in the IRC-server protocol. This specification provides the following communications services: synchronous group communications with push mode, time stamping, duration of message validity and notification. With regard to security, it provides access control, partial anonymity and non-repudiation of the auctioneer and of the winning bidders.

We are working on enhanced security for the BSA-protocol (such as signed notifications for non-repudiation, data integrity and confidentiality). The ASS architecture is under specification. Clock synchronization and fairness will be considered during this specification.

The advantages of this work lie in the fact that the AHS permit to free auction applications from communication issues and facilitates interoperability between these applications.

REFERENCES

1. Amza, C., Cecchet, E., Chanda, A., Cox, A. Elnikety, S., Gil, R., Marguerite, J., Rajamani, K., and Zwaenepoel, W. (2002). "Specification and Implementation of Dynamic Web Site Benchmarks." Proceedings of the 5th IEEE Workshop on Workload Characterization (WWC-5).
2. Asokan, N. (1998). "Fairness in Electronic Commerce." PHD dissertation, University of waterloo. http://citeseer.ist.psu.edu/4056.html.
3. Beam, C. and Segev, A. (1997). "Automated Negotiations: A survey of the State of the Art." Technical report 97-WO-1022, Haas School of Business, UC Berkeley.
4. Beam, C., Segev, A. and Shanthikumar, J.G. (1996). "Electronic Negotiation through Internet-based Auctions." CITM working paper 96-WP-1019.
5. Ben Ameur, H. (2001). "Enchères Multi-objets pour la Négociation Automatique et le Commerce Electronique." Master dissertation, Département d'Informatique, Faculté des Sciences et de Génie, Université Laval.
6. Ben Azzouz, L. (1999). "Emulation et Inter fonctionnement de messagerie EDI en environnement de messagerie interpersonnelle " Ph.D Dissertation, National School of Computer Sciences, University of Tunis II.
7. Ben Azzouz, L., Kaffel, H. and Kamoun, F. (2000). "An Edi User Agent over the Internet" Network and Information System Journal vol. 1(1), pp 483-504.
8. Ben Youssef, M., Alj, H., Vézeau, M. and Keller, R. (2001). "Combined Negotiations in E-commerce : Concepts and Architecture." Electronic Commerce Journal vol.1 No3, pp. 277-299.
9. Cardellini, V., Casalicchio, E. and Colajanni, M. (2001). "A Performance Study of Distributed Architectures for the Quality of Web Services." Proc. of The Hawaii's International Conference on System Sciences, Hawaii.

10. CCITT (1991a). Message Handling Systems: Electronic Data Interchange Messaging System, Recommendation X435.
11. CCITT (1991b). Message Handling: Electronic Data Interchange Messaging Service, recommendation F435.
12. EPRI (2001). Auction Agents for the Electric Power Industry, Research Conducted for the Electric Power Research Industry, Palo Alto, CA.
13. Harkavy, M., Tygar, J.D., Kikuchi, H. (1998). "Electronic Auctions with private bids", Proc. Third USENIX Workshop on Electronic Commerce. Boston, Massachusetts, August 31-September 3, 1998.
14. Kaabi, S., Kaffel, H. and Kamoun, F. (2001). "Evaluation of HTTP, E-mail and NNTP with regard to Negotiation Requirements in the Context of Electronic Commerce." Proc. of ICECR-4, Dallas Texas, pp. 439-447.
15. Kaabi, S., Kaffel, H. and Kamoun, F. (2003). "Specification of a Communication Protocol for Real-time Auctions." Proc. ICECR6, Dallas, Texas, pp 129-138.
16. Klein, S. (1997). "Introduction to Electronic Auctions." EM-Electronic Auctions. EM-Electronic Markets, vol..7(4), pp. 3-6.
17. Kumar, M. and Feldman, S.I. (1998a). "Business Negotiations on the Internet." Technical Report IBM Institute for Advanced Commerce, Yorktown Heights.
18. Kumar, M. and Feldman, S.I. (1998b). "Internet auctions." Proceedings of the third USENIX Workshop on Electronic Commerce, Boston, Aug 31-Sept 3, pp. 49-59.
19. Maxemchuk N.F. and Shur, D.H. (2001). "An Internet multicast system for the stock market." ACM Transactions on Computer Systems, vol.19(3), pp.384-412.
20. Omote, K. (2002). "A study on electronic auctions." submitted for Japan Advanced Institute of Science and technology in partial fulfillment of the requirements for the degree of doctor of philosophy. March 2002.
21. Panzini,F., and Shrivastava, S.K. (1999). "On the provision of replicated Internet auction Services." Proceedings of the 18th symposium on Reliable Distributed Systems, Lausanne Oct. 19, pp. 390-395.
22. Peng, C.S., Pulido, J.M., Lin, K.J. and Blough, D.M. (1998). "The design of an Internet-based Real-Time Auction System.." Proceedings of the first IEEE workshop on dependable and real-time e-commerce systems (DARE-98), Denver Jun 98, pp.70-78.
23. Shih, Jansson and Drummon. (1997). *MIME-based Secure EDI*, draft-EDIINTint-as1-04.txt.
24. Ströbel, (2000). "On auctions as the negotiation paradigm of electronic markets." EM-Electronic Markets, vol. 10, No.1,
25. Turban, E. (1997). "Auctions and Bidding on the Internet: An Assessment" EM - Electronic Auctions. EM - Electronic Markets, Vol. 7, No. 4,
26. Wurman, W.Y., Wellman, M.P., and Walsh, W.E. (1998). "The Michigan Internet AuctionBot : A Configurable Auction Server for Human and Software Agents." Proceedings of the Second Intl. Conference on Autonomous Agents 98, Minneapolis MN USA, pp. 301-308.
27. IETF (2000a). IRC-ARCH, www.ietf.org rfc2810
28. IETF (2000b). IRC-CHAN, www.ietf.org rfc2811
29. IETF (2000c). IRC-Client, www.ietf.org rfc2812
30. IETF (2000d). IRC-Server, www.ietf.org rfc2813

Part Four: Wireless Networks

AN INTERFERENCE-BASED PREVENTION MECHANISM AGAINST WEP ATTACK FOR 802.11B NETWORK

Wen-Chuan Hsieh[1], Yi-Hsien Chiu[2] and Chi-Chun Lo[3]

[1]*Shu-Te University, wch@mail.stu.edu.tw;* [2]*National Yunlin University of Science & Technology, g9223701@yuntech.edu.tw;* [3]*National Chiao-Tung University, cclo@cc.nctu.edu.tw*

Abstract: WEP has a potential vulnerability that stems from its adaptation of RC4 algorithm. As indicated by prior researches, given a sufficient collection of packets, speculation on shared key is possible by extracting IVs that matched a specific pattern. With the primary protection becomes void, there is a pressing need for new WLAN security measure. However, establishing new security protocol requires considerable time and financial resources. This research proposes an alternative solution to WEP hacking, without modification on present wireless settings, called Interference-Based Prevention Mechanism.

Key words: IEEE 802.11, Wireless Local Area Network (WLAN), RC4, Wired Equivalent Privacy (WEP)

1. INTRODUCTION

Wireless Local Area Network (WLAN) offers organizations and users a both convenient and flexible way of communication. It provides mobility, increases productivity, and lowers installation costs. However, WLAN is susceptible to attacks due to the use of radio frequency which incurs data exposure. WLAN does not have the same physical structure as LANs do, and therefore are more vulnerable to unauthorized access. While access points (AP) offers convenient and flexible way of communication, the fact that they are connected to internal network exacerbates security problem. Without additional protection, APs can as well be entries for potential attackers.

To enhance data security of wireless transmission to the level of a wired network, IEEE 802.11 standard defined WEP (Wired Equivalent Privacy), which encrypts traffics between clients and AP. However, WEP has a potential weakness stems from its adaptation of RC4 algorithm, which utilizes a plain IV (initial vector) as part of key stream computation. As indicated by prior researches[1,2,3], given a sufficient collection of packets, speculation on shared key is possible by extracting IVs that matched a specific pattern. That is, any anyone with WEP attack tools, such as AirSnort[4] and WEPCrack[5], can obtain the key in a matter of hours or days.

Obviously, WLAN suffers severe security problem and it offers merely limited privacy guarantee. Installing WLAN incurs tremendous risks since APs can as well be entries for potential attackers into internal network. With WEP, the primary WLAN protection, being compromised, the condition of wireless security is considered critical. With the primary protection becomes void, there is pressing need for new WLAN security measure.

However, security protocol requires considerable resources to upgrade legacy network for the supporting features. Instead of altering or replacing present WLAN, this research offers an alternative solution without modifying the present setting. This research proposes an Interference-Based Prevention Mechanism which is proven effective in preventing adversaries from deducing WEP key based on weak key detection.

2. THEORETICAL BACKGROUND

2.1 Wireless Security

The most prominent feature about WLAN is the absence of wires and its mobility. As compares to the traditional network, WLAN requires no complicate configuration on its physical topology. Its prestigious nature of mobility is made possible by transmitting data using radio frequency. However, as data travels through the air, it can easily be tapped by any one including unauthenticated personnel using sniffer.

Many attacks on traditional network also applied to wireless environment; for instance, DOS attack, session hijack and man-in-the-middle. Also, unauthorized clients may attempt to access WLAN without authorization. Since WLAN does not constraint users to physical connection ports, users are able to access the AP anywhere. Borisov et al.[6] conducted a detailed research on insecurity of 802.11.

As defined in IEEE 802.11b standard, WEP is applied to encrypt data so that it becomes unreadable to the intruder. Despite the effort, WEP is recently proved insecure since its key can be stolen or cracked using tools such as AirSnort. In addition, most APs are deployed with WEP setting switched off by default. APs offers MAC filter as a supplementary security feature to WEP; however, keeping track of MAC addresses list is both time consuming and inconvenient.

Because of the weaknesses in WEP security, several entities are developing stronger security technology, such as TKIP (Temporal Key Integrity Protocol) [7,8] and 802.1X[9,10]. TKIP is proposed, as part of wireless standard 802.11i, to replace WEP. 802.1X is an IEEE standard for EAP encapsulation over wired or wireless Ethernet. 802.1X is also known as EAPoL (EAP over LAN). However, the fact that majority of the legacy wireless hardware are 802.11b based requires potential adopters to either upgrade firmware or even replace the incompatible devices. The cost of such hardware and software renovation and reconfiguration is just too expensive for entities with limited budget. Therefore, for the mean time, current WLAN is considered insecure.

2.2 Wired Equivalent Privacy (WEP)

The concept of WEP is to prevent eavesdroppers by encrypting data transmitted over the WLAN from one point to another. Data encryption protects the vulnerable wireless link between clients and access points; that is, WEP does not offer end-to-end security because AP decrypts the frames before passing them to destinations that are beyond WLAN.

WEP adopts RC4 algorithm, a stream cipher, developed by RSA security. "A stream cipher operates by expanding a short key into an infinite pseudo-random key stream. The sender XORs the key stream with the plaintext to produce cipher text. The receiver has a copy of the same key, and uses it to generate identical key stream. XORing the key stream with the cipher text yields the original plaintext"[11]. In other words, RC4 is a symmetric algorithm relies on a single shared key that is used at one end to encrypt plain text into cipher text, and decrypt it at the other end[12].

Current WEP implementations support key length up to 64 bits and 128 bits; technically, the key length of both version are shorten by 24bits due to the use of plaintext Initial Vector (IV). In this research context, a key (or key combination) is a series of ASCII bytes often presented in hexadecimal; whereas a key value is one byte (8 bits) out of the total combination. Figure 1 shows a WEP encrypted frame which consists of IV(24 bits), padding(6 bits), key index(2 bits), encrypted message and Integrity Checksum Value (ICV)(32 bits). Note that the frame is transferred with the first 32 bits in plaintext and the rest of the body encrypted. This is because a sender generates IV, either incrementally or randomly, as part of inputs to encryption process. That is, the receiver must know the exact IV to decrypt the frame.

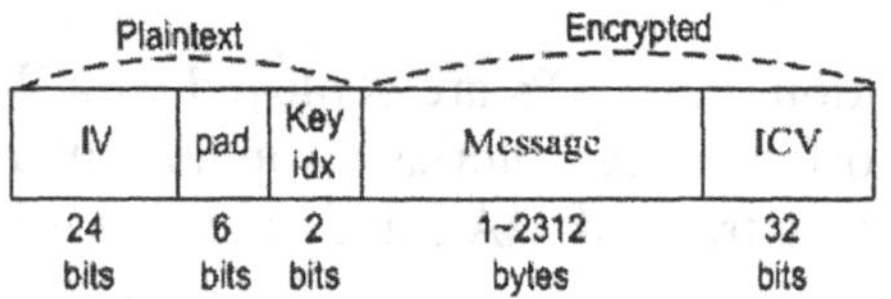

Figure 1. WEP encrypted frame format

As indicated by the length of key index in the diagram, WEP can have up to 4 (2^2) keys. However, using shared static keys can be dangerous. Therefore, the purpose of constantly changing IV is to achieve the effect as if having a greater number (2^{24}) of key combinations. This gives WEP the capability of encrypting each frame with different keys (packet key).

Figure 2 illustrates WEP encryption process which starts by generating IV and selecting a predefined key. Next, RC4 uses both IV and chosen key (k) as inputs to generate key stream. Then, plaintext message (M), along with its ICV, is combined with key stream through a bitwise XOR process, which produces ciphertext (C). Upon sending the encrypted frame, WEP appends IV in clear to the front of the frame. The encryption process can be summarized as following formula: C = (M, crc32(M)) XOR RC4(IV, k)

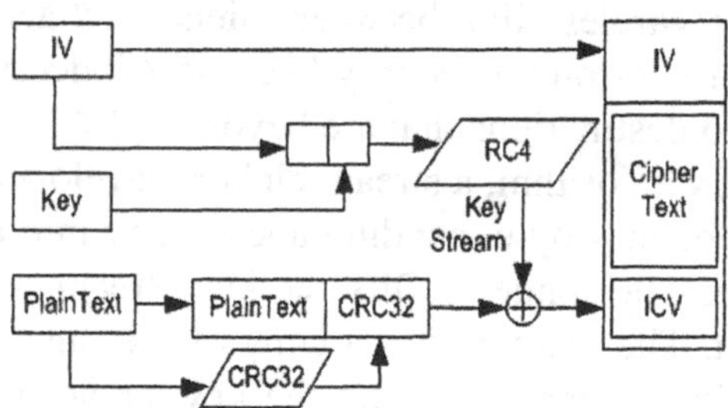

Figure 2. WEP encryption process

To decrypt, the receiving station uses the first 32bits IV and the shared key (k) as indicated by key index bits to generate the same key stream that encrypted the frame. Next, WEP XOR key stream with ciphertext (C), along with it ICV, to retrieve the plaintext (M). Note that, plaintext has ICV attached at the end. Finally, WEP computes plaintext, without ICV, CRC32 and compares the output with the ICV.

Wireless environment is prone to interference; hence, data may be lost or damaged before reaching the destination. To ensure data integrity, sender computes CRC32 against the plaintext message and inserts the output (32 bits) at the back of the message prior to encryption. The receiver ensures data integrity by matching ICV of decrypted frame with the CRC32 result done locally with the resolved message. Frames with disconfirmed check-

sum will be discarded. The decryption process can be summarized as following formula based on the encryption formula:
(M, crc32(M)) = C XOR RC4(IV, k)

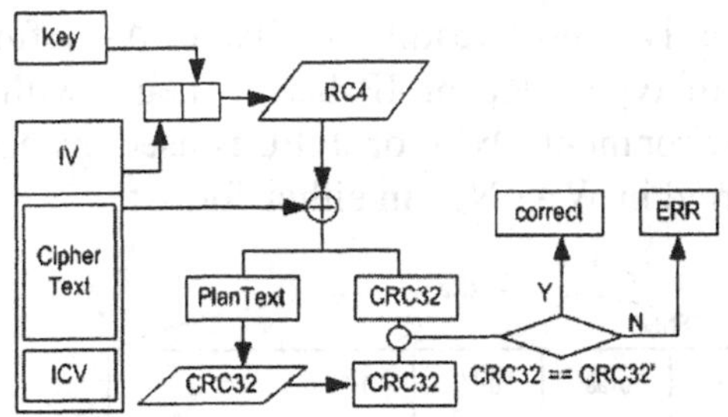

Figure 3. WEP decryption procedure

2.3 WEP Vulnerability

Though combining IV into key stream computation increases key complexity so that it appears unpredictable, the reality that IV has to be transferred in clear may divulge WEP key as first discovered in the research undertaken by Fluhrer, Martin and Shamir[1]. Specifically, frames with IV that matched (B+3, 255, X) form, where B points to the position of the key value in the combination and X can be any value between 0 and 255, may reveal key values. The probability of retrieving the right key value from the frame is 5%[13]. Given sufficient time and traffic, one is able to obtain the WEP within hours or days. For instance, an IV (4, 255, 31) may resolve the value of the K[1], where IV (7, 255, 72) may resolve the K[4] (fig 4). Often, attackers determine the key combinations by running statistic on all the potential key values computed from frames that matched such pattern.

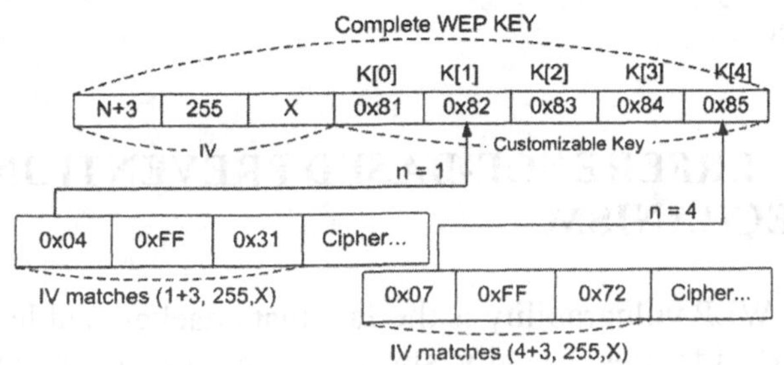

Figure 4. IV pattern resolving key combination

Part of the key value extracting concept is based on the nature of XOR. Suppose C is the result of P XOR K, then we are able to retrieve K by XOR C with P. In the case of WEP, the idea is extended and is much complicated due to RC4 algorithm; nevertheless, the fundamental idea is the same. That is, the initial step of cracking WEP key is to obtain ciphertext with its matching plaintext, which is almost readily available. As defined in 802.11 standard, any frames of type ARP or IP has to begin with 0xAA (known as SNAP). In IPX environment, 0xFF or 0xE0 is used instead. In fact, majority of the data transferred in WALN is in either format.

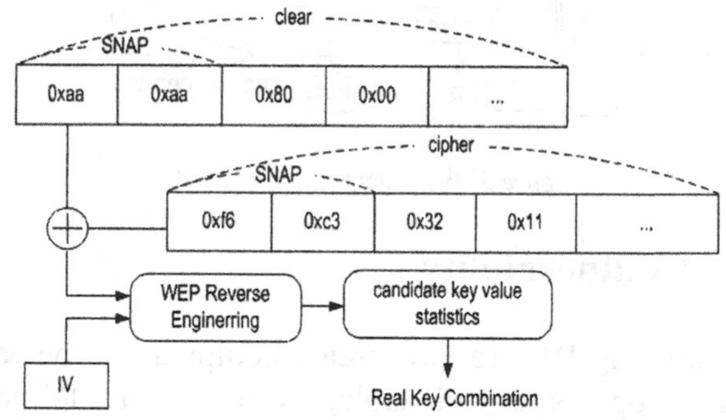

Figure 5. XOR plaintext and ciphertext to resolve key value

All in all, to crack WEP, one must first capture as much frames that matched the specified pattern as possible. Then, for each of the captured frame, XOR the first byte of the ciphertext with 0xAA to obtain the exact key stream that were used during encryption. By reverse-engineering RC4, attacker would be able to retrieve the key value (fig 5). Please refer to Fluhrer's study for detailed explanation on specific algorithms. Seth Fogie[13] has published an article which describes detailed steps of WEP cracking. Also, WEP attack implementation can be found in the research done by Stubblefield et al[2].

3. INTERFERENCE-BASED PREVENTION MECHANISM

The major WEP vulnerability is the fact that attacker is able to extract the key from gathered frames. Usually, statistics is used to assists in determining the real key values from the candidates. The real key value often has the highest occurrence among all. Therefore, it is reasonable to conclude that the resulting key is based on the amount and quality of the frames. That is, the attacker is unlikely to get the right key combination if traffic is scarce or frames reveal more false key values than that of the right ones.

Since it is impossible and unreasonable to keep WLAN traffics from increasing, we propose that the best option to prevent attacker from getting the correct key value is by poisoning the traffic with frames that are deliberately tailored to generate false result.

Based on the understanding of the frames that the attackers are interested in and the logics of detecting the key, this research devised an innovative solution called Interference-Based Prevention Mechanism (IBPM). IBPM creates interference effect by injecting spoofed frames to delude the attacker resulting in inaccurate statistic. Since injecting frames increases traffic load, hence, an effective and space-efficient method must be applied.

IBPM utilizes the same technique similar to WEP crackers. That is, IBPM monitors the traffic and keeps computing the key values. The difference is that IBPM is implemented in a client station within a WEP protected WLAN; therefore, it is assumed that IBPM station possesses the key as a legitimate user. Having the key gives it the capability of interfering network traffic in advance. Figure 6 shows IBPM generates spoofed frames whenever the speculated key value matches the real key value (we refer such event as weak-key occurrence). Consequently, the automated statistic program at the offense side takes those frames into account and increments false key values. What actually happened is that, IBPM pollutes attacker's statistic in a way that causes false key values to increase to prevent real key value becoming distinct. Since IBPM has disrupted the statistic long before it reveals the real key value, WEP is, therefore, secured. This research proposes several schemes, which are discussed under interference schemes section, to distribute spoofed frames that generates false key across all possible key values.

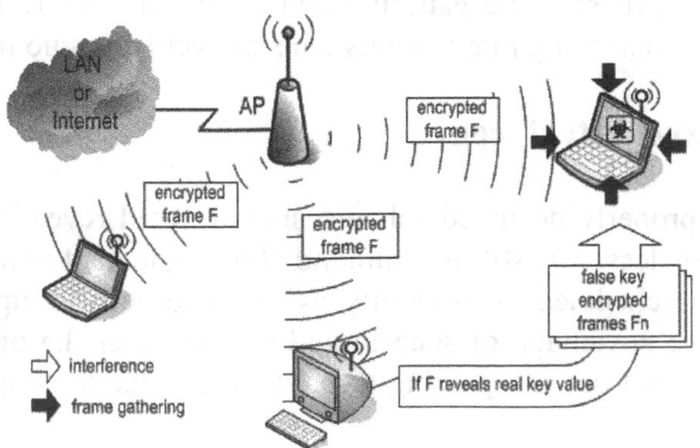

Figure 6. Interference generation

4. INTERFERENCE SCHEMES

The effect of interference is accomplished by increasing the tally of false key value whenever a weak-key is detected by IBPM. For instance, a weak-key 0xBB is detected, one may decide to increase all false-key tallies range form 0x00 to 0xFF excluding 0xBB. However, incrementing the tally arbitrarily incurs flaw that may eventually allows the attacker to discern the fixed pattern in the resulted statistic. Interference not only conceals the key, but also should prevent attackers from speculating the key based on the spoiled statistic again. Therefore, a sound interference scheme does not exhibits traceable patterns. This research proposes three schemes with each takes a different approach to poison the traffic.

4.1 Random Distribution Interference

This scheme randomly selects any amount of key values from the false set. The increment scale can also be any number. However, it is recommended to use a scale less than or equals to 3, because drastic change may ultimately cause the real key value to become the least and apparent. The scale can also be randomly assigned given a specified range.

4.2 Perfect Probability Distribution Interference

Under such scheme, every false key value is given a 50% probability of becoming candidates for increment. That is, there are roughly half of the false set members will be incremented upon each weak-key occurrence. In this case, we use a fixed increment scale of 2. This scheme is able to maintain a stable increasing rate of 1 (2 x 0.5), just like the detected weak-key, while avoiding exposing the pattern when using scale of 1. Therefore, the overall average increasing rate remains constant yet leaves no traces.

4.3 Mixed Interference

Though a properly designed scheme should avoid revealing a traceable pattern; nevertheless, it is still recommended to implement multiple schemes and have each scheduled or randomly assigned to activate upon weak key occurrence. The advantage of such mixed scheme over the others is that it prevents attackers from recognizing a fixed pattern due to adoption of a single scheme.

5. IMPLEMENTATION

5.1 System Requirements

As mentioned earlier, IBPM requires no change on the legacy network configuration and is compatible to any WEP-enabled 802.11 WLAN. The IBPM-enabled device (preferably a desktop PC) appears just like any other regular wireless clients; therefore, attackers are unlikely to realize the intension behind such deployment. As shown in IBPM system framework (fig 7), IBPM joined the WLAN as a member client station which issues bogus frames upon weak key occurrences. At the same time, the attacker, being unaware of the spoofed frames, keeps gathering the frames.

Figure 7. IBPM System Framework

IBPM involves both proactive and passive activities which include traffic sniffing and injection. Presently, this research has implemented an experimental system under Linux. Various wireless drivers are available in the open-source community[14,15,16] with each offers slightly different capabilities. This research has modified and integrated some of the drivers in achieving features to support both monitor WLAN traffics and send frames with arbitrary format, including WEP encrypted, through individual wireless network interface cards.

This research implemented IBPM using Python and C libraries under Redhat 9 Linux. The IBPM machine is equipped with two wireless network interfaces: one for sniffing frames and the other is used to inject spoofed frames whenever weak-key is detected.

5.2 Demonstration

To demonstrate the effectiveness of IBPM, two independent WEP attacks were launched against the experimental WLAN in the lab. At the end of each attack, the offender's statistical result is captured. This demonstration adopted perfect probability interference scheme. Since WEP-128 is as vulnerable as WEP-64 despite of its extended key length, therefore WEP-64 is applied just to illustrate the concept due to space limitation. AP is configured with WEP key setting as K = {76, 210, 126, 196} and 24. Figure 8 shows the result of the statistical result of the first attack without IBPM. Note that the thick line indicates the real key value.

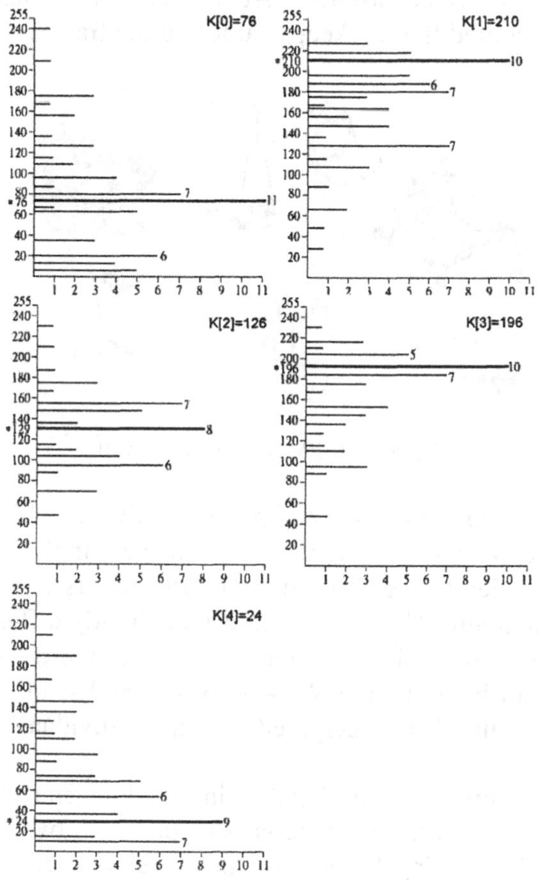

Figure 8. K[0]~K[4] statistic result without interference

Clearly, the attacker can easily points out the real key value based on the statistical result. Evidently, each of the real key value stands out prominently. In contrasting to the previous test, the result captured in the second experiment with IBPM conceals the real key values (fig 9). In addition, the overall distribution is almost random and there is no fixed pattern to follow. As for better observation, we deliberately thicken the line of the real key. In reality, the attackers will not be able to determine the real key value from such random formed statistical result.

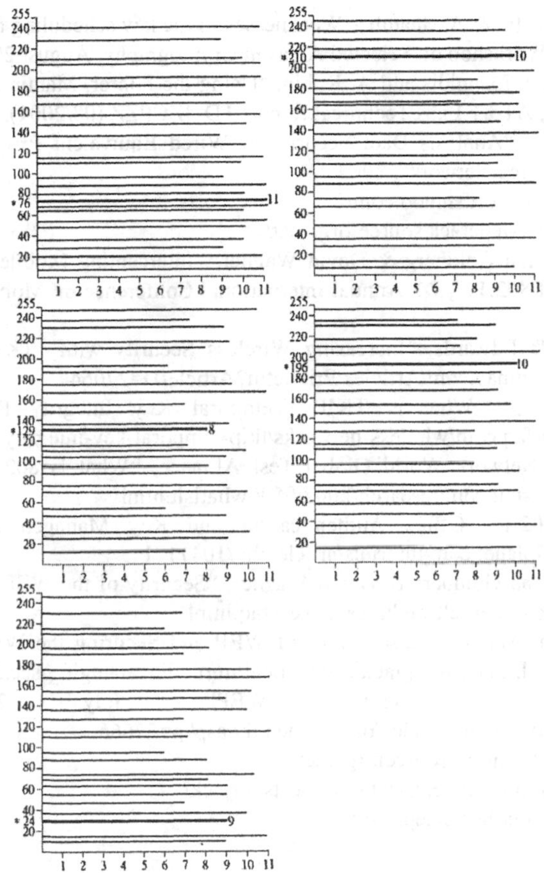

Figure 9. K[0] and K[1] statistic result with interference

6. CONCLUSION

This research covered discussion on WEP encryption, its vulnerability and basic concept on the technique applied to extract key from frames that matched the weak-key form. More importantly, we developed Interference-Based Prevention Mechanism (IBPM), which is proven to be effective in preventing attackers from speculating WEP key by means of frame gathering. Presently, two interference schemes are proposed. However, for further studies, more effort should be devoted in testing and developing of new schemes.

REFERENCES

1. S. Fluhrer, I. Mantin & A. Shamir, "Weaknesses in the key scheduling algorithm of RC4". Eighth Annual Workshop on Selected Areas in Cryptography, August 2001.
2. A. Stubblefield, J. Ioannidis and A. Rubin, "Using the Fuhrer, Mantin and Shamir Attack to Break WEP", AT&T Labs Technical Report TD-4ZCPZZ (08/2001).
3. Liao Jyun-Ruei , "Analysis Data Security in Wired Equivalent Privacy Algorithm for Wireless Local Area Network", 2002
4. AirSnort, http://airsnort.shmoo.com/
5. WEPCrack, http://wepcrack.sourceforge.net/
6. Nikita Borisov, Ian Goldberg & David Wagner, "Intercepting Mobile Communications: the insecurity of 802.11", 7th Annual International Conference on Mobile Computing and Networking
7. Mark, Joseph & Edwards, "Increasing Wireless Security with TKIP", 2002, URL: http://www.winnetmag.com/Articles/Print.cfm?ArticleID=27064
8. The TECH FAQ, "What is TKIP (Temporal Key Integrity Protocol)?", URL: http://www.tech-faq.com/wireless-networks/tkip-temporal-key-integrity-protocol.shtml
9. Joel Snyder & Network World Global Test Alliance, "What is 802.1X", 2002, URL: http://www.nwfusion.com/research/2002/0506whatisit.html
10. Jim Geier, "802.1X Offers Authentication and Key Management", 2002, URL: http://www.wi-fiplanet.com/tutorials/article.php/1041171
11. Nikita Borisov, Ian Goldberg & David Wagner, "Security of the WEP algorithm", URL: http://www.isaac.cs.berkeley.edu/isaac/wep-faq.html
12. Nuruddin Mohd. Alamgir, "Insecurities of WEP and Securing the Wireless Networks", June 2002, URL: http://www.giac.org/practical/nuruddin_alamgir_gsec.doc
13. Seth Fogie, "Cracking WEP", July 2002, URL: http://www.informit.com/articles/printerfriendly.asp?p=27666
14. SourceForge, URL: http://sourceforge.net
15. AirJack, URL: http://sourceforge.net/projects/airjack/
16. HostAP, URL: http://hostap.epitest.fi/

RESTRICTED DYNAMIC PROGRAMMING FOR BROADCAST SCHEDULING

Shuoi Wang[1,2] and Hsing-Lung Chen[3]
[1]Department of Computer Science and Information Engineering, National Taiwan University of Science and Technology, Taipei, Taiwan 106; [2]Department of Maritime Policing, Taiwan Police College, Taipei, Taiwan 116; [3]Department of Electronic Engineering, National Taiwan University of Science and Technology, Taipei, Taiwan 106

Abstract: Data broadcast has become a promising solution for information dissemination in the wireless environment due to the limited bandwidth of the channels and the power constraints of the portable devices. In this paper, a restricted dynamic programming approach which generates broadcast programs is proposed to partition data items over multiple channels near optimally. In our approach, a prediction function of the optimal average expected delay, in terms of the number of channels, the summation of the access frequencies of data items, and the ratio of the data items, is developed by employing curve fitting. Applying this function, we can find a cut point, which may be very close to the optimal cut. Thus, the search space in dynamic programming can be restricted to the interval around the found cut point. Therefore, our approach only takes $O(N\log K)$ time, where N is number of data items and K is the number of broadcast channels. Simulation results show that the solution obtained by our proposed algorithm is in fact very close to optimal one.

Key words: Data Broadcast; Data Allocation; Dynamic Programming; Multiple Channels.

1. INTRODUCTION

Recent advances in the development of portable computers and wireless communication networks make it possible for mobile clients to access data from anywhere at anytime. Broadcast-based data dissemination has become a widely accepted approach of communication in the mobile computing environment. Examples of these applications include weather forecasts, stock market quotes, and electronic newsletters. In these applications, a server periodically broadcasts a set of data items to a large community of users and clients tune in to the broadcast channel to retrieve their data of interest. Thus, the latency and the cost of data delivery are independent of the number of clients. On the contrary, an on-demand data delivery responding to a client's individual request inevitably incurs a scalability bottleneck under a heavy workload. However, in the broadcast-based system, the clients have to access data items in the broadcast channel sequentially. Some clients receive unwanted data before accessing desired data, and the corresponding response time is called the expected delay of that data item. This problem becomes worse when data access is skewed. Hence, how to allocate data items in the broadcast channel for efficient data access becomes an important issue.

Acharya[1] *et al.* propose "Broadcast Disks" architecture to minimize the average expected delay (*aed*) for the data allocation problem in a single broadcast channel. A broadcast disk involves generation of a broadcast program that schedules the data items based on their access frequencies. The broadcast is constructed by allocating data items to different "disks" of varying sizes and speeds, and then multiplexing the disks onto the same broadcast channel. This approach creates a memory hierarchy in which the fast disk contains few items and broadcasts them with high frequency while the slow disk contains more items and broadcasts them with less frequency.

In this paper, we study the data allocation problem over multiple disjoint physical channels. Such architecture has wider applicability[2,3,4]. The concept of broadcast disks can be applied to multi-channel system. That is, the disk containing data items with higher access frequencies may be distributed to a channel containing less data items such that the *aed* for those data items is reduced. The problem we study can be best understood by the illustrative example in Figure 1, where a data base contains nine items and the number of broadcast channels is three. The function of data allocation algorithm is to allocate data items into broadcast channels according to their access frequencies so as to minimize the *aed*. This is the very problem that we shall address in this paper.

Peng and Chen[5] explore the problem of generating hierarchical broadcast programs on the K broadcast channels. They develop a heuristic algorithm VF^K to minimize the *aed* of data items in the broadcast program. Although VF^K yields the *aed* close to the lower bound, it performs unstably. While the number of channels is not the power of 2, the expected delay of each channel is not balanced. Thus, the *aed* of all data items becomes worse.

For given N items with access frequencies p_r, where $1 \le r \le N$. Wong[6] shows that the lower bound of the *aed* for a periodic broadcast schedule in a single channel system is $(\sum_{r=1}^{N}\sqrt{p_r})^2/2$. Hsu[7] *et al.* extend this concept to multi-channel system, and derive that the minimal *aed* of all data items on K broadcast channels is $(\sum_{r=1}^{N}\sqrt{p_r})^2/2K$.

Yee[8] *et al.* use dynamic programming to optimally partition data items among given multiple channels. Although they determine the optimal cost, the proposed approach requires $O(KN^2)$ time and $O(KN)$ space to keep partial solutions, which makes that approach impractical in large databases.

In this paper, a prediction function which estimates the optimal *aed* of given data items on multiple channels is generated by employing curve fitting. Applying this function, a restricted dynamic programming (DP) approach is developed to allocate data items into each channel, and the resulting configuration is very close to the optimal one, but only takes $O(N\log K)$ time. The rest of paper is organized as follows. Preliminaries are given in Section 2. In Section 3, we develop a prediction function to estimate optimal *aed* by employing curve fitting. In Section 4, based on the predication function, a restricted DP approach which generates broadcast programs on multiple channels is proposed. Performance studies are presented in Section 5. This paper concludes with Section 6.

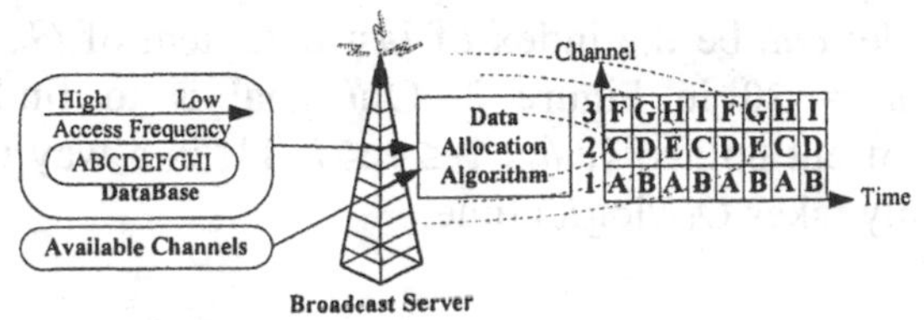

Figure 1. Data allocation problem on multiple channels.

2. PRELIMINARIES

2.1 Architectural Assumptions

This paper focuses on the wireless broadcast environment. Some assumptions should be restricted in order to make our work feasible. These assumptions include: a) A data base contains N equal-sized items, denoted as d_j, where $1 \leq j \leq N$; b) There are K equal-bandwidth physical channels for data broadcast, which can not be combined to form a single high-bandwidth one; c) Let G_i be the set of data items to be broadcast on channel i, where $1 \leq i \leq K$ and $\sum_{i=1}^{K} |G_i| = N$. The data items in each channel are sent out in a round robin manner. Each data item is broadcast only on one of these channels and a time slot is defined as the amount of time necessary to transmit an item; d) Each data item d_j has a corresponding access frequency p_j, which denotes the probability that data item d_j is requested by the clients. We assume requests are exponentially distributed, so that at each time slot the probability of a client requesting d_j is determined by p_j, where $\sum_{j=1}^{N} p_j = 1$, and; e) The mobile client can listen to multiple channels simultaneously.

2.2 Problem Statement

Our problem is to partition the data items into K groups and to allocate items in each group into an individual channel, such that the *aed* of all data is minimized. When items in set G_i are cyclically broadcast on channel i, the *expected delay* in receiving any particular data on channel i is $|G_i| / 2$. Thus, given K channels, the *aed* of all data items can be expressed as
$\sum_{i=1}^{K}\left(\frac{|G_i|}{2}\sum_{d_j \in G_i} p_{d_j}\right)$.

Theoretically, data allocation over multiple channels can be viewed as a *partition* problem for data, as shown in Figure 2. First, all items are sorted in descending order according to their access frequencies. Then, partition all data into $G_1, G_2, \ldots, G_K$ sets. Define cut_i as the cut point between G_i and G_{i+1}. For convenience, let cut_i be the index of last data item of G_i. For example, $cut_1 = 2$ and $cut_i = 20$ in Figure 2. Our goal is to find the optimal configuration set of cut points, $\{cut_i \mid 1 \leq i \leq K\text{-}1\}$, in a way that the *aed* is minimized, but only takes $O(N\log K)$ time.

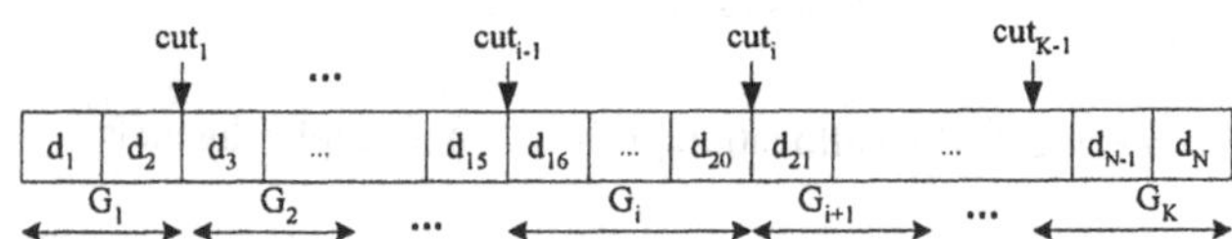

Figure 2. Partition problem for allocating N data items on K channels.

Dynamic programming[8] (DP) approach provides an optimal solution for the data allocation problem, but its time and space complexities may preclude it from practical use. Let us state the basic DP approach for the partition problem first. Data are sorted in descending order according to their access frequencies. Define $C_{i,j}$ as the *aed* of a channel containing d_i through d_j. $C_{i,j} = \frac{j-i+1}{2}\sum_{r=i}^{j} p_r$, where $1 \leq i \leq j \leq N$. Let $opt_{k,j}$ be the optimal *aed* for allocating d_1 through d_j on k channels. Given one channel, $opt_{1,j} = \frac{j}{2}\sum_{r=1}^{j} p_r$, where $1 \leq j \leq N$. The optimal *aed* for allocating N items on K channels is $opt_{K,N}$. Now, we present the basic DP algorithm for solving the partition problem in Figure 3.

In (1) as shown in Figure 3, the search space of determining the $opt_{k,j}$ is linear. Thus, by inspecting the nested loop structure of the basic DP algorithm, its running time is $O(KN^2)$. Suppose we can approximately estimate the *aed* of given data items on multiple channels, then we can find a cut point, s, which may be very close to the optimal cut. Thus the search space of the possible cut point, $\boldsymbol{r}$, can be restricted to some promising range (say seven cut points), that is, $s\text{-}3 \le \boldsymbol{r} \le s\text{+}3$, then the time to determine $opt_{k,j}$ becomes constant. Thus, the search space in DP approach can be reduced.

```
For all cut points j from 1 to N
    opt_{1,j} = (j/2) Σ_{r=1}^{j} p_r
End
For all stages k from 2 to K
    For all cut points j from k to N
        opt_{k,j} = min{opt_{k-1,r} + C_{r+1,j}}, where k-1 ≤ r ≤ j-1        (1)
    End
End
```

Figure 3. The basic dynamic programming algorithm.

2.3 Formulation of Estimated *aed* on Multiple Channels

Assume $p_1 \ge p_2 \ge \ldots \ge p_N$, and $\sum_{r=1}^{N} p_r = 1$. Denote $optimal_k^{\alpha}$ as the optimal *aed* of the last $n \le N$ data items allocated to $k \le K$ channels, where α be the summation of access frequencies of the last n data items, $\alpha = \sum_{r=N-n+1}^{N} p_r$. Thus, $optimal_K^1$ is the *aed* of all N items allocated to K channels.

Denote LB_k^{α} as the lower bound of the *aed* when the last n items are allocated to k channels. Thus, LB_1^1 is the lower bound[6] of the *aed* when all N items are allocated to a single channel, that is, $LB_1^1 = (\sum_{r=1}^{N} \sqrt{p_r})^2 / 2$.

Assume a high-bandwidth channel has bandwidth B bytes/sec, and each data item is of size D bytes. Therefore, broadcast one item on the channel will take D/B seconds. If this fast channel divides into k sub-channels, and each sub-channel has bandwidth B/k bytes/sec. Then, broadcasting each item in one of these sub-channels will take kD/B seconds. Thus, $LB_k^1 = LB_1^1 / k$. Let

$factor_k^1 = optimal_k^1 \Big/ LB_k^1 = optimal_k^1 \Big/ \frac{LB_1^1}{k} = k \times optimal_k^1 \Big/ LB_1^1$.

When the summation of access frequencies of items is not equal to 1, factor can be normalized as follows.

$$factor_k^{\alpha} = k \times \frac{\sum_{j=K-k+1}^{K} \frac{|G_j|}{2} \left(\sum_{r \in G_j} \frac{p_r}{\alpha} \right)}{\frac{1}{2} \left(\sum_{r=N-n+1}^{N} \sqrt{\frac{p_r}{\alpha}} \right)^2} = k \times \frac{\frac{1}{\alpha} \sum_{j=K-k+1}^{K} \frac{|G_j|}{2} \left(\sum_{r \in G_j} p_r \right)}{\frac{1}{2\alpha} \left(\sum_{r=N-n+1}^{N} \sqrt{p_r} \right)^2}$$

Let $LB_1^{\alpha} = (\sum_{r=N-n+1}^{N} \sqrt{p_r})^2 / 2$. Thus,

$$factor_k^{\alpha} = k \times optimal_k^{\alpha} \Big/ LB_1^{\alpha} \tag{2}$$

$$=> optimal_k^\alpha = factor_k^\alpha \times LB_1^\alpha / k \quad (3)$$

Given different data items with access frequencies and number of channels in (2), we can get different values of factor. With these values, if we can find a function to predict $factor_k^\alpha$ precisely, then we can get an estimated value very close to $optimal_k^\alpha$. Our idea is inspired by Hsu[7] *et al.* Assume data items in G_1, G_2, ..., and G_i have been allocated to the first i channels. Thus, the optimal *aed* of the unallocated data items on the (K-i) channels is $optimal_{K-i}^\alpha = factor_{K-i}^\alpha \times LB_1^\alpha / (K-i)$, where α is the summation of access frequencies of the data items on the (K-i) channels. But Hsu *et al.* use $1 \times LB_1^\alpha / (K-i)$ to estimate it.

3. PREDICTION FUNCTION OF FACTOR

To predict the value of $factor_k^\alpha$, we employ curve fitting to find a prediction function f in terms of α, k, and *ratio*, where α denotes the summation of the access frequencies of unallocated data items ($\alpha \le 1$), k denotes the number of unallocated channels ($k \le$ K), and *ratio* denotes the distribution of the data items. Intuitively, f is related to these three parameters. Simulation results confirm this intuition. Ratio[9] means (100×*ratio*)% users focus on 100×(1-*ratio*)% data items, where $0 \le ratio \le 1$.

DEFINITION: *ratio* is the summation of the access frequencies of the first 100×(1-*ratio*)% data items. In other words, $ratio = \frac{p_1 + p_2 + \dots + p_{(1-ratio)N}}{p_1 + p_2 + \dots + p_N}$

We can determine the ratio of the data items by examining whether the first N_1 data items satisfy $\sum_{j=1}^{N_1} p_j + (N_1/N) = 1$, where N_1 varies from 1 to N. Thus, $ratio = 1 - (N_1/N)$ if satisfied.

Let $est_{k,r}$ denote as the estimated *aed* of d_r through d_N allocated on k channels. Once, $f(\alpha, k, ratio)$ is found, from (3),

$$est_{k,r} = f(\alpha, k, ratio) \times LB_1^\alpha / k, \text{ where } \alpha = \sum_{q=r}^{N} p_q .$$

We call this approach as PKR estimation. It is better than Hsu *et al.*, because their $factor_k^\alpha$ are always 1.

LEMMA 1. $est_{k,r}$ can be computed in constant time.

PROOF. In initialization stage, for each data item d_j, $1 \le j \le N$, we associate a $accp_j$ with d_j, which is defined as the sum of access frequencies of data items d_1 through d_j. Initially, $accp_1 = p_1$. Other $accp_j$ can be computed by the recurrence equation $accp_j = p_j + accp_{j-1}$. Thus, $\alpha = accp_N - accp_{r-1}$. Given α, k, and *ratio*, we can compute $f(\alpha, k, ratio)$ in constant time. We also associate a acc_sqrtp_j with d_j, which is defined as $\sum_{r=1}^{j} \sqrt{p_r}$. Initially,

$acc\_sqrtp_1 = \sqrt{p_1}$. Other acc_sqrtp_i can be computed recursively by $acc\_sqrtp_j = \sqrt{p_j} + acc\_sqrtp_{j\text{-}1}$ Thus, $LB_1^{\alpha} = (\sum_{q=r}^{N}\sqrt{p_q})^2/2 = (acc\_sqrtp_N - acc\_sqrtp_{r\text{-}1})^2/2$, which also can be computed in constant time. Since $est_{k,r} = f(\alpha,k,ratio) \times LB_1^{\alpha}/k$, this lemma follows. ■

The derivation of *f*(α, *k*, *ratio*) contains the following steps.

Step A. Prediction function in terms of α.

Given data items with specific *ratio* and *k*, we can compute $optimal_k^{\alpha}$ and LB_1^{α} during the dynamic programming process. Thus, we can obtain a value of factor by (2). Let α_n be the summation of access frequencies of the last *n* data items, where $k \leq n \leq N$. For each α_n, we have a computed $factor_n$. Thus, we can gather many data points (α_n, $factor_n$). Plotting these data points and using the least-squares method for fitting data, we can get a result similar to Figure 4. The dotted line shown in Figure 4 denotes the optimal data points and the dashed line denotes the function of curve fitting we adopt, which is very close to the optimal one. The function in Figure 4 is described by

$$factor = 1 + a1\,((m_9 - ratio) \times m_{10} \times \alpha)^{a2}, \qquad (4)$$

where $a1$ and $a2$ are coefficients with specific *k* and *ratio*, $m_9 = 0.78$, and $m_{10} = 1$. Figure 4 is for the case when *k* is 3 and *ratio* is 0.80.

Step B. $a2$ in terms of $a1$.

For each input combination of *k* (e.g., 3, 4, ..., 100) and *ratio* (e.g., 0.55, 0.60, ..., 0.85), we obtain a figure similar to Figure 5 and a function has the form like (4), with different coefficients of ($a1$, $a2$). Thus, we can gather many data points ($a1_{k,ratio}$, $a2_{k,ratio}$). Plotting these data points and then employing curve fitting, we can derive a function in terms of $a1$ to denote $a2$, as shown in Figure 5. That function is

$$a2 = m_1\,(a1 - m_2) + m_3, \qquad (5)$$

where m_1, m_2, and m_3 are coefficients.

Step C. Prediction function in terms of *ratio*.

For a given *k*, we have a respective value of $a1_{k,ratio}$ for each *ratio*. Thus, we can gather many data points (*ratio*, $a1_{k,ratio}$). Plotting these data points, as the dotted line shows in Figure 6. Using curve fitting, we can get a function in terms of *ratio* to denote $a1$, as the dashed line in Figure 6. That function is

$$a1 = b1\,(ratio)^{m_8}, \qquad (6)$$

where $m_8 = 8$, $b1$ is a coefficient of a specific *k*.

Step D. Prediction function in terms of *k*.

Figure 6 is for the case when *k* is 3. For each value of *k*, we can obtain a respective coefficient of $b1_k$. Plotting these data points ($b1_k$, *k*), as the dotted line shows in Figure 7. Using curve fitting, we can obtain a function in terms of k to denote $b1$, as the dashed line in Figure 7. That function is

$$b1 = m_4\,(k + m_7)^{-m_5} + m_6. \tag{7}$$

where m_4 to m_7 are coefficients.

Step E. *factor* = $f(\alpha, k, ratio)$.

Thus, combining (4) to (7), we obtain a prediction function of factor f in terms of α, k and *ratio* with initial values of m_1 to m_{10}. For all data points <$factor_n$, α_n, k_n, $ratio_n$>, we proceed the curve fitting again to obtain a better fit. Thus, we have the final values of m_1 to m_{10} in $f(\alpha, k, ratio)$. Table 1 lists the coefficients derived in f.

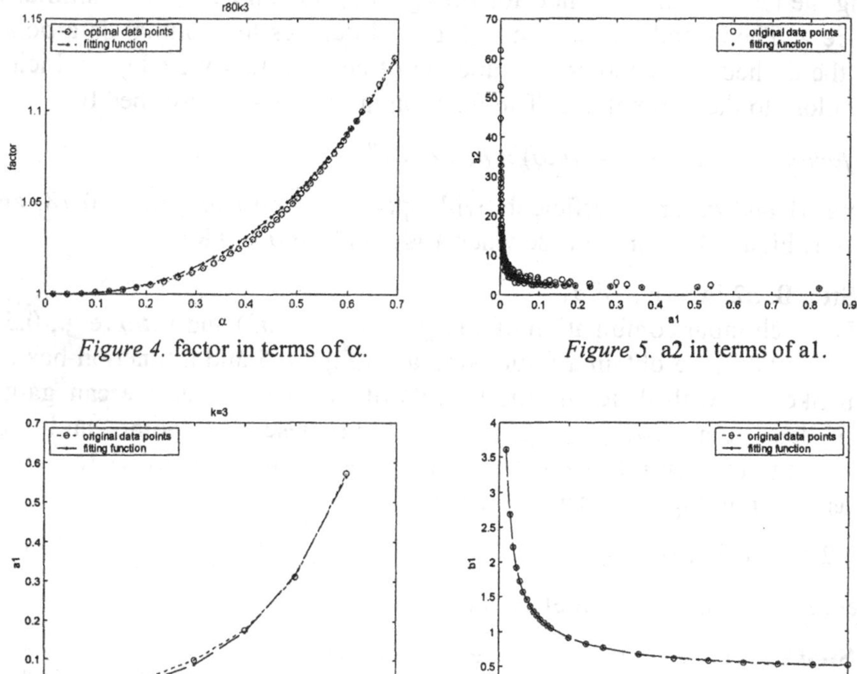

Figure 4. factor in terms of α.

Figure 5. a2 in terms of a1.

Figure 6. a1 in terms of ratio.

Figure 7. b1 in terms of k.

Table 1. Coefficients used in function f.

m_1	2.101260517	m_6	-0.12243
m_2	0.304910179	m_7	-0.19996
m_3	0.353111223	m_8	8.878037
m_4	11.72018907	m_9	28.20646
m_5	0.87676042	m_{10}	0.030049

4. A RESTRICTED DP ALGORITHM

The main idea of our proposed approach uses PKR estimation to find a cut point, which may be very close to the optimal cut. Therefore, the search space in DP can be reduced, thus, saves computations.

At stage k in the basic DP algorithm, if the cut point of $(k\text{-}1)_{th}$ partition is r, then the estimated optimal *aed* of all data items on K channels is the sum of $opt_{k-1,r}$ and $est_{K-k+1,r+1}$. Let s be a cut point such that $opt_{k-1,s} + est_{K-k+1,s+1} = \min\{opt_{k-1,r} + est_{K-k+1,r+1}\}$, where $k \le r \le N\text{-}1$. If we restrict the search space of determining $opt_{k,j}$ to the interval [s-3, s+3] in (1), then the time to determine $opt_{k,j}$ is constant. Note that, if the optimal cut point of $(k\text{-}1)_{th}$ partition is outside the interval [s-3, s+3] we predict, computed $opt_{k,j}$ is not optimal, resulting in a non-optimal solution. Now, we present our proposed Restricted DP algorithm (RDP) in Figure 8.

In Figure 8, when the minimum value of $opt_{k,j}$ is determined, the corresponding r will be stored in $lastp_{k,j}$, which can help us construct an optimal configuration. Define $selcut_i$ as the cut point of selected solution, where $1 \le i \le K\text{-}1$. Thus, $selcut_{K-1} = estcut_{K-1}$. The earlier selected cut points can be retrieved recursively by $selcut_i = lastp_{i+1,selcut_{i+1}}$, where $1 \le i \le K\text{-}2$. Therefore, the set of selected cut points can be retrieved by the backward process.

With the help of PKR estimation, the RDP algorithm can restrict the search space to some promising range. It is unlike the basic DP have to examine all possible cut points, therefore, our approach only takes $O(KN)$ time.

Some modifications in the RDP algorithm can improve the performance if some subproblems in the subproblem space need not be solved at all. As stated in Lemmas[7] 3 and 4, the number of the data items allocated to each channel possesses the hierarchical property. That is,

$$|G_{i-1}| \le |G_i| \le \frac{N - \sum_{j=1}^{i-1} |G_j|}{K-(i-1)} \le \frac{N-(i-1)}{K-(i-1)}.$$

This inequality can be used to reduce the time and space requirements in the RDP algorithm. We implement a slightly modified version of the RDP in Figure 9, called Bounded-Restricted Dynamic Programming (BRDP).

Theorem 1. The time and space complexities of the BRDP algorithm are both $O(N\log K)$

Proof. In the worst case, creating entries of the first row of the table *opt* requires N/K computations. Thus, determining $estcut_1$ takes N/K computations. Similarly, creating entries of the second row of the table *opt* requires $(N\text{-}1)/(K\text{-}1)$ computations. Then, determining $estcut_2$ spends $(N\text{-}1)/(K\text{-}1)$ computations. This process is repeatedly done until cut point $estcut_{K-1}$ is found. Thus,

$$\frac{N}{K}+\frac{N-1}{K-1}+\ldots+\frac{N-K+2}{2}\le\frac{N}{K}+\frac{N}{K-1}+\ldots+\frac{N}{1}=N\sum_{i=1}^{K}\frac{1}{i}=NH_K,$$

where $H_K = lnK + O(1)$. Therefore, The time and space complexities of the BRDP algorithm are both $O(N\log K)$. ■

```
For all cut points j from 1 to N
        opt_{1,j} = (j/2) Σ_{r=1}^{j} p_r
End
Let s be a cut point estcut_1 s.t. opt_{1,s} + est_{K-1,s+1} = min{opt_{1,r} + est_{K-1,r+1}},
where 1 ≤ r ≤ N-1.
For all stages k from 2 to K-1
    For all cut points j from max{|s-2|, k} to N
        opt_{k,j} = min{opt_{k-1,r} + C_{r+1,j}}, where max{|s-3|, k} ≤ r ≤ min{s+3, j-1}.
    End
    Let s be a cut point estcut_k such that
    opt_{k,s} + est_{K-k,s+1} = min{opt_{k,r} + est_{K-k,r+1}}, where max{|estcut_{k-1}-2|, k} ≤ r ≤ N-1.
```

Figure 8. The Restricted DP Algorithm.

```
For all cut points j from 1 to N/K
        opt_{1,j} = (j/2) Σ_{r=1}^{j} p_r
End
Let s be a cut point estcut_1 such that
opt_{1,s} + est_{K-1,s+1} = min{opt_{1,r} + est_{K-1,r+1}}, where 1 ≤ r ≤ N/K.
For all stages k from 2 to K-1
    For all cut points j from max{|s-2|, k} to (N-(k-1))/(K-(k-1))
        opt_{k,j} = min{opt_{k-1,r} + C_{r+1,j}}, where max{|s-3|, k} ≤ r ≤ min{s+3, j-1}.
    End
    Let s be a cut point estcut_k such that
    opt_{k,s} + est_{K-k,s+1} = min{opt_{k,r} + est_{K-k,r+1}},
    where max{| estcut_{k-1}-2|, k} ≤ r ≤ (N-(k-1))/(K-(k-1)).
End
```

Figure 9. The Bounded-Restricted DP Algorithm.

Table 2. Parameters used in performance evaluation.

Definition	Notation	Range
Number of data items to be broadcast	N	1000 - 5000
Number of broadcast channels	K	2 - 100
Zipf distribution parameter	*ratio*	0.55 - 0.85

Table 3. Algorithms compared in performance evaluation.

Algorithm	Notation
Optimal Solution[9]	OPT
Data-Based Algorithm[2]	HSU
PKR Approach	PKR
Bounded-Restricted Dynamic Programming Algorithm	BRDP

5. PERFORMANCE EVALUATION

Our experiments are developed by C on the computer with Intel Pentium III 1 GHz and 256MB RAM, running Windows XP. The access frequencies of broadcast items are modeled by the Zipf distribution. Let

$$p_j = \frac{j^\theta - (j-1)^\theta}{N^\theta},\ 1 \le j \le N,$$

where θ is the parameter of Zipf distribution. θ is computed[9] by

$$\theta = \frac{\log(ratio)}{\log(1-ratio)}.$$

In the Zipf distribution, the access frequencies of the data follows the *ratio*/(1-*ratio*) rule. For example, 80/20 rule means that 80 percent users are usually interested in 20 percent data items.

The simulation parameter settings for our experiments are listed in Table 2. Table 3 lists the algorithms we compared. All algorithms are implemented as described by their respective authors. Instead of using lower bound[7], PKR algorithm uses PKR estimation to estimate the *aed* of the unallocated data items on the unallocated channels.

The simulation experiments aim at studying the performance of our proposed PKR estimation approach and the BRDP algorithm, compared with another two algorithms[7,8]. Given data items with specific ratio and the number of broadcast channels K, if the set of the cut points we found is same as the set of optimal cut points, we call it *hit*. The performance metric of the algorithms is the *hit rate* (the distance to the best solution), which is the percentage of number of hits to number of experiments. For each given number of items with specific ratio, we performed 99 experiments, that is, the range of the number of channels varying from 2 to 100.

Figure 10 shows the effect of the number of the data items on the hit rate for three different approaches under different ratios. Algorithm PKR is better than Hsu's algorithm. The reason is that PKR estimation can generally predict the *aed* of all data items more precisely than the lower bound in the partition operation. Due to the fitting deviation, when the distribution is near uniform (e.g., ratio = 0.55), PKR estimation is not good enough. Perhaps there are different choices of prediction function f which would lead to a better fit. Obviously, BRDP can tolerate the estimation error. In all simulation runs under the change of skew factor ratio, number of items, and number of channels, we observe in Figure 10 the results obtained by BRDP is of high quality and is in fact very close to the optimal one. In most cases, the hit rate of BRDP is higher than 95%, and it outperforms Hsu's algorithm about 2 times higher.

6. CONCLUSIONS

We propose a restricted dynamic programming algorithm to solve the problem of allocating data items into multiple broadcast channels, and the proposed approach only takes $O(N\log K)$ time and space. In our approach, we adopt a prediction model which estimates the optimal *aed* of data items on multiple channels more precisely than lower bound. Simulation results show that the solution obtained by our algorithm is of very high quality and is very close to the optimal one.

Error! Not a valid link.Error! Not a valid link.Error! Not a valid link.Error! Not a valid link.

Figure 10. The effect of the number of data items under different ratios.

REFERENCES

1. S. Acharya, M. Franklin, S. Zdonik, and R. Alonso, Broadcast disks: data management for asymmetric communication environments, *ACM SIGMOD*, pp. 199-210, 1995.
2. C. Hsu, G. Lee, and A.L.P. Chen, Index and data allocation on multiple broadcast channels considering data access frequencies, *IEEE MDM*, pp. 87-93, 2002.
3. C. Hu and M.S. Chen, Adaptive balanced hybrid data delivery for multi-channel data broadcast, *IEEE ICC*, pp. 960-964, 2002.
4. D.A. Tran, K.A. Hua and N. Jiang, A generalized air-cache design for efficiently broadcasting on multiple channels, *ACM SAC*, 2001.
5. W.C. Peng and M.S. Chen, Dynamic generation of data broadcasting programs for a broadcast array in a mobile computing environment, *ACM CIKM*, pp. 35-45, 2000.
6. J.W. Wong, Broadcast delivery, *Proc. IEEE*, **76**(12), 1566-1577, 1988.
7. C. Hsu, G. Lee, and A.L.P. Chen, A near optimal algorithm for generating broadcast programs on multiple channels, *ACM CIKM*, pp. 303-309, 2001.
8. W.G. Yee, S.B. Navathe, E. Omiecinski and C. Jermaine, Efficient data allocation over multiple channels at broadcast servers, *IEEE Trans. Comp*, **51**(10),1231-1236, 2002.
9. D.E. Knuth, *The Art of Computer Programming*, vol. 3, Addison-Wesley, 2nd, 1981.

PERFORMANCE COMPARISON OF DISTRIBUTED FREQUENCY ASSIGNMENT ALGORITHMS FOR WIRELESS SENSOR NETWORKS

Sonia Waharte and Raouf Boutaba

University of Waterloo, School of Computer Science, 200, University Avenue West, Waterloo, ON, Canada N2L 3G1

{swaharte, rboutaba}@bbcr.uwaterloo.ca

Abstract: Minimizing energy consumption of network operations remain a major concern in wireless sensor networks due to the limited energy capacity embedded in sensor nodes. Clustering has been proposed as a potential solution to address this issue, some nodes being responsible for the data gathering of nodes located in their vicinity. However, in order to avoid inter-cluster interference, neighboring clusters must acquire different frequencies. As the specific constraints of wireless sensor networks favor a distributed approach, we analyze modified versions of distributed backtracking, distributed weak commitment and randomized algorithms with a focus on energy consumption. In this context, we find that a heuristic may achieve better results than backtracking-based algorithms.

Keywords: wireless sensor networks; frequency assignement.

1. INTRODUCTION

In exploring the possibilities of interconnecting the world but alleviated of the constraints of physical infrastructure, wireless sensor networks stand out as a promising technology. To embed sensing, communication and processing capabilities in tiny devices engendered a significant potential of applications, in fields as diverse as environment monitoring, target tracking, surveillance systems, etc.

A bulk of research activities in wireless sensor networks has focused on reducing energy consumption of network operations. By considering that the nodes at close proximity have redundant information and therefore, by limiting the number of nodes simultaneously active, significant energy savings can be achieved. Another solution consists in implementing energy-efficient network organization mechanisms such as clustering. With cluster formation, one major concern is to allow simultaneous transmissions between neighboring cells while minimizing data collisions. Hence, the problem consists in allocating different frequencies (or different codes) to neighboring clusters.

Although the frequency assignment problem has been largely addressed in the literature, new constraints pertaining to wireless sensor networks have been introduced, necessitating an evaluation from different perspectives. First, the lack of a centralized administration calls for a solution based on distributed algorithms. Second, the limited power supply embedded in sensor nodes necessitates the development of energy-efficient mechanisms. Current works in this context have emphasized on achieving optimal running time (i.e. minimizing overall delay), regardless the cost of energy consumption. However, although the evaluation of an algorithm may appear beneficial in terms of running time, it can be detrimental to applications such as environment monitoring, due to an excessive energy consumption resulting from a high processing cost or by a significant number of message exchanges.

Considering the critical importance of energy conservation in such sensor network applications, we present in this paper a bi-criteria analysis of distributed frequency allocation algorithms, based not only on the overall delay of frequency allocation, but also on the overall energy consumption during the data transmission process.

The rest of the paper is organized as follows. In Section II, we describe previous works addressing the frequency assignment problem in wireless sensor networks. An overview of the algorithms implemented in our study follows in Section III. The results of our evaluations are presented in Section IV. We conclude this paper with a description of some still unaddressed issues.

2. RELATED WORKS

Frequency assignment is a well-known NP-hard combinatorial problem, which can be formalized as an instance of the Graph Coloring problem. As this subject has been widely addressed in the literature and mainly for cellular networks, we refer the reader to previous publications for further details[6,8,10].

In wireless sensor networks, frequencies can be assigned in two different ways. The centralized approach considers that one node (e.g. the base station), at the root of the topology, can efficiently proceed with the frequency allocation process and distribute the results of the operation to the concerned nodes[2].

However, this implies that the node positions are known by the root node, which introduces significant scalability issues.

A more scalable approach consists in distributing the frequency allocation problem among the sensor nodes and in seeking for a solution locally. Yokoo and Hirayama[9] propose several algorithms for distributed constraints satisfaction problems: Asynchronous Backtracking, Asynchronous Weak Commitment[5] and Distributed Breakout Algorithm, adapted from the well-known backtracking algorithm to accommodate the constraints of distributed environments. The two first algorithms are studied in this paper. In Distributed Breakout Algorithm (DBA), after an initial setup phase during which the constraints are weighted according to some predetermined parameters, the nodes exchange information on the possible weight reduction resulting from a modification of their current frequency assignment. The node maximizing the weight reduction proceeds with the modification. As an extension, Distributed Stochastic Algorithm[1] improves over DBA by a stochastic change of the local frequency. However, these two algorithms present the drawback of requiring a global synchronization, which we believe is not suitable for sensor networks. Therefore, they will not be considered in our study.

An evaluation of algorithms developed for distributed constraints satisfaction problems in wireless networks has been conducted[7], with a focus on three specific problems: partition into coordinating cliques, distributed Hamiltonian cycle formation and conflict-free channel scheduling. For the latter problem, the authors implemented the asynchronous backtracking algorithm with a network topology composed of 25 nodes and they analyzed the number of satisfied instances according to the number of channels available with a variation of the transmission radius. However, no evaluation has been conducted on the convergence time or on the number of message exchanges.

Finally, a heuristic was proposed by Guo[3], in which each node sends its randomly chosen frequency to its two-hop neighbors. Upon reception of this information, the channel is removed from the local channel pool. The appropriateness of this algorithm for sensor networks is difficult to evaluate as no simulation has been performed. There is also no mechanism to handle the situation where the frequency pool of a node becomes empty. In this paper, we analyze a heuristic based on a similar idea, but modified and completed in order to satisfy the constraints of our specific problem.

3. ASYNCHRONOUS DISTRIBUTED ALGORITHMS

3.1 Problem Formulation

The frequency assignment problem can be stated as follows. Given a wireless network composed of n nodes and its topology graph, a frequency is assigned to each node with respect to the following characteristics:

- n nodes: $x_1..x_n$
- m frequencies: $f_1..f_m$
- 1 constraint: two adjacent nodes can not be allocated the same frequency
- each node is allocated only one frequency

We implemented three asynchronous algorithms, two of them being modified versions of the distributed backtracking and the weak commitment algorithms. The third algorithm is a heuristic, used as a comparison basis in order to evaluate the real benefit of the algorithms both in terms of convergence delay and energy consumption (represented as the number of messages sent and received by all the nodes). More details on the implementation are provided in the following subsections.

3.1.1 Distributed Backtracking Algorithm

This algorithm has been adapted from the backtracking algorithm to accommodate the constraints of distributed environments. After having informed its neighboring nodes of its frequency assignment, a node waits for any message indicating either a constraint violation or a new frequency allocation. The lowest priority node must change its value first. If no satisfactory value can be found, the node notifies a higher priority node to change its local frequency. The process is repeated until the derivation of a satisfactory solution or until the realization that no solution can be found. A node priority can be set according to its identifier, the smallest identifier having the lowest priority. Each node keeps track of the frequency assignment of all the surrounding nodes (located in the same neighborhood).

Each node is characterized by a tuple (x_i, f_i), where x_i is the identifier and f_i the frequency allocated to the node. The algorithm is described in Figure 1.

3.1.2 Asynchronous Weak-Commitment search

Static priorities introduce limitations that the Weak Commitment search tried to overcome. The approach consists in dynamically adapting nodes priorities according to their local constraints, in order to avoid an exhaustive search. The priorities are set in increasing order (the higher the value, the higher the

```
if received msg(x_j, f_j) then
    update local_table
    check_constraints
end if
```

procedure check_constraints

```
if local_table and current_value are not consistent then
    if no value in possible_values consistent with local_table then
        backtrack
    else
        select new_value from possible_values
        current_value=new_value
        send msg(identifier, new_value) to neighbors
    end if
end if
```

procedure backtrack

```
if no solution possible then
    send(no solution) with empty value;
else
    sendmsg(identifier, x_i, f_i)) where identifier refers to the local node and x_i is the agent
    with the lowest priority
    remove (x_i, f_i) from local_table
end if
```

Figure 1. Distributed Backtracking Algorithm

priority). Each time a node modifies its frequency (and thus consumes energy to inform its neighbors of its new frequency allocation), it increases its priority by one. Compared to an identifier-based priority assignment, this mechanism guarantees fairness (based on the number of frequency modifications). Further extensions are envisioned by setting priorities based on the energy level or on a node willingness to participate in the frequency allocation process.

A description of the algorithm is given in Figure 2.

3.1.3 Heuristic Algorithm

In order to analyze the efficiency of the previous algorithms, we implemented an algorithm based on a random frequency assignment. It allowed us to relax our system from the constraint of node identification which can generate significant overhead. The algorithm is based on the following principle: the nodes randomly choose a frequency among a predefined set, after a random waiting period. In the meanwhile, if a node receives a frequency assignment notification from a neighboring node, it updates its local table and randomly chooses a frequency in the remaining frequencies pool. If a node receives a

procedure check_constraints

```
if local_table and current_value are not consistent then
   if no value in possible_values consistent with local_table then
      backtrack
   else
      select new_value from possible_values while minimizing the number of constraints
      violation with lower priority agent
      current_value=new_value
      send msg(identifier, new_value) to neighbors
   end if
end if
```

procedure backtrack

```
if no solution possible then
   send(no solution) with empty value;
else
   send msg(identifier, x_i, f_i)) where x_i is the agent with the lowest priority
   remove (x_i, f_i) from the local table
   current priority++
   send msg (x_i, f_i, current_priority) to neighbors
end if
```

Figure 2. Distributed Weak Commitment Algorithm

conflicting frequency notification after it has already chosen a frequency, the receiver node has to change its frequency and has to inform its neighbors of its new frequency. If no frequency is available, the node re-initializes its local table (i.e. it removes all values previously registered from the table), picks up a random frequency and informs its neighbors of its new value. Once no local constraint is violated, the algorithm stops. A sketch of the algorithm is given in Figure 3.

4. EVALUATION

Whereas previous studies mainly evaluated the performance of distributed algorithms based on the execution time, we believe that, in sensor networks, energy consumption is a more critical criterion. Although technological advances have been achieved in the domain of energy supplies, a sensor node's lifetime still remains constrained. It thus becomes necessary to focus not only on the execution time but also on the energy consumption during the network operations.

In the following section, we first analyze the impact of the network size on the execution time of the algorithms and the energy consumption (in terms of

```
Initialization
wait random time
if received msg(f_j) then
    update local_table
    check_constraints
else
    choose random frequency among possible values
end if

procedure check_constraints
if local_freq and value in local_table are not consistent then
    if no value in possible_values consistent with local_table then
        reinitialize local_table
        choose new_fr and send msg(new_fr) to neighbors
    else
        select new_fr from possible_values
        local_freq=new_fr
        send msg(local_freq) to neighbors
    end if
end if
```

Figure 3. Heuristic Algorithm

number of messages sent and received). Then, we study the influence of the size of the frequency group.

4.1 Assumptions

In our study, we made the following assumptions:

- The propagation delay is considered negligible compared to the transmission delay. This hypothesis is justified if we consider that the transmission radius is around 20 meters and that the propagation speed is the speed of light. The propagation delay is then in the order of the nanosecond.
- The propagation distance can not exceed 20 meters.
- For the algorithms needing node identification (distributed backtracking and weak commitment), we assume that the node identifiers are set before the network deployment and are unique.
- All the packets are successfully delivered.

4.2 Simulation Results

The following results represent an average over 50 simulation runs, with test topologies uniformly generated at random.

The energy consumption is evaluated according to the radio propagation model described by Heinzelman et al.[4], where the energy to transmit a packet E_{Tx} and to receive a packet E_{Rx} can be computed such as:

$$E_{Tx} = lE_{elec} + l\epsilon d^2$$

$$E_{Rx} = lE_{elec}$$

where $E_{elec} = 50\text{nJ/bit}$, l is the packet size, $\epsilon = 100\text{pJ/bit/m}^2$ and d is the transmission distance.

In the simulations, we assume a 25-byte packet size, with a 2-byte node identifier and 5 bits reserved for the frequency assignment.

Figure 4 depicts the impact of the network size on the execution time of the studied algorithms. The Backtracking algorithm and the Weak Commitment algorithm significantly outperform the Heuristic algorithm, and achieve an improvement in term of execution time around five times higher than the Heuristic algorithm for a topology of 100 nodes. As expected, the Weak Commitment algorithm also performs on average better than the Backtracking algorithm, especially when the size of the network increases.

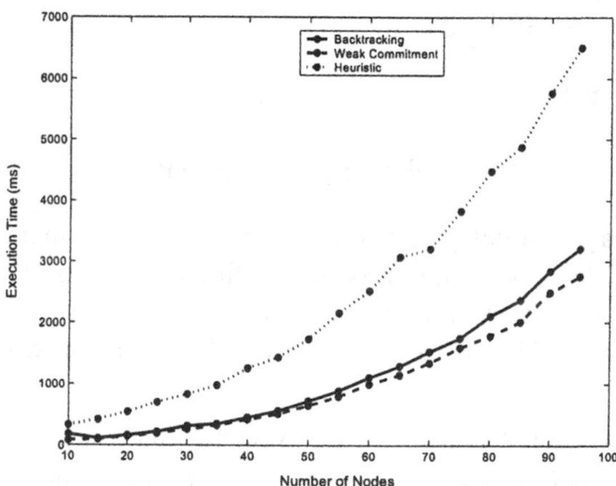

Figure 4. Impact of the network size on the execution time

However, from an energy perspective, the Heuristic algorithm achieves excellent results compared to the two other algorithms (Figure 5). The difference with the backtracking-based algorithms becomes more significant with the increase of the network size.

For an average of four neighbors per node, the increase of the number of frequencies in the frequency pool has little impact on the relative performance

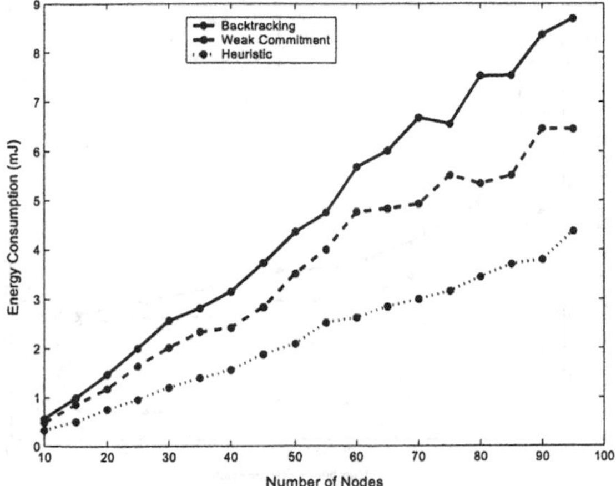

Figure 5. Impact of the network size on the energy consumption

of the algorithms (Figure 6), the Weak Commitment algorithm still performing the best.

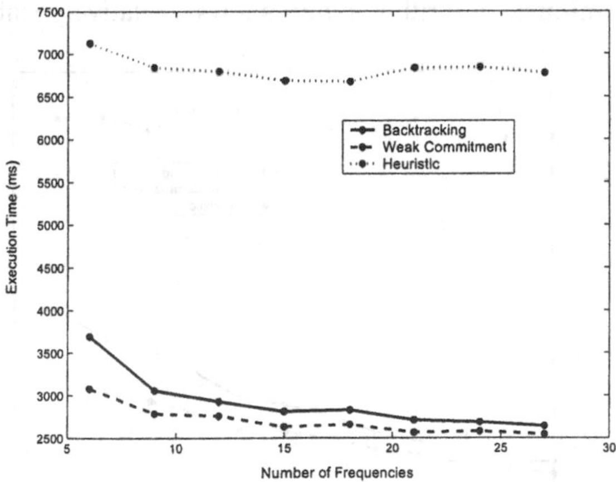

Figure 6. Influence of the number of frequencies on the execution time

When the number of frequencies increases, the energy performance of the Heuristic algorithm remains excellent, independently of the number of frequencies (Figure 7). The exponential decrease of the energy consumption can be explained by the fact that the probability that two nodes choose the same frequency progressively decreases when the size of the frequency pool increases.

The increase of the network density does not have any impact on the performance of the Heuristic (Figure 8). Even if its performance is relatively poor

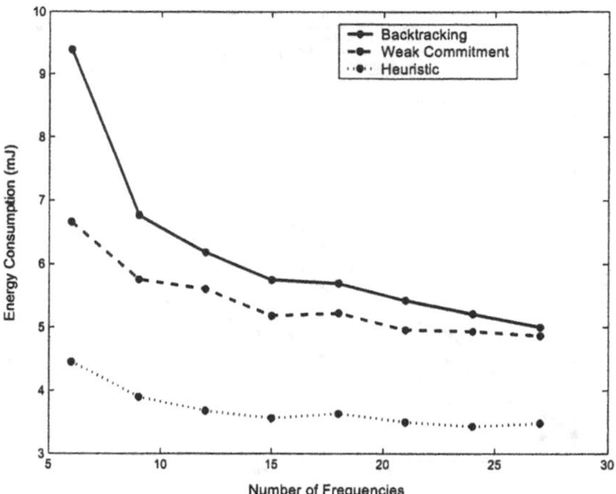

Figure 7. Influence of the number of frequencies on the energy consumption

compared to the Weak Commitment and Backtracking algorithms, it presents a more stable behavior than the other two algorithms. The performance of the backtracking algorithm deteriorates with the increase of the network density. The Weak Commitment algorithm in turn shows a relatively stable behavior.

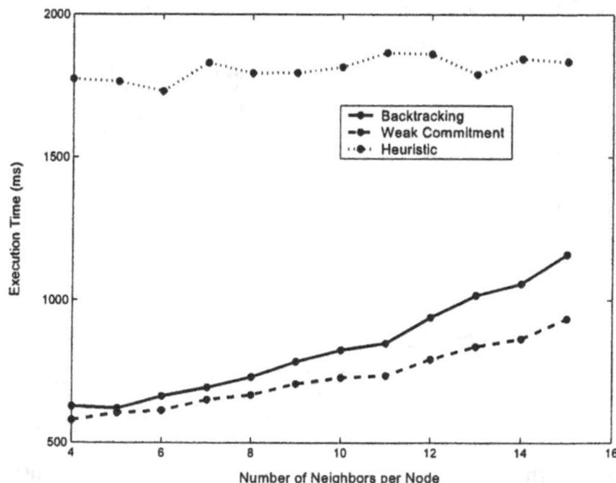

Figure 8. Influence of the network density on the execution time for a network topology of 50 nodes and 20 frequencies available

When the network density increases, the Heuristic algorithm achieves the best performance in terms of energy consumption (Figure 9).

The results obtained in the simulations are summarized in Table 1. The algorithms which perform best for each category of tests are pointed out.

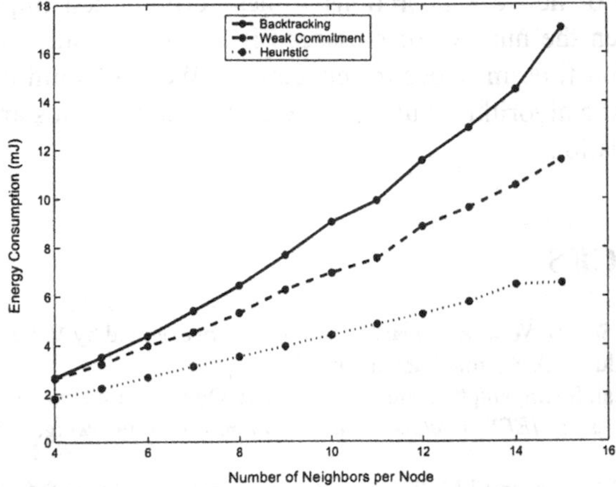

Figure 9. Influence of the network density on the energy consumption for a network topology of 50 nodes and 20 frequencies available

Table 1. Best performing algorithms according to the tested criteria

		Weak Commitment	*Heuristic*
Scalability	Execution Time	X	
	Energy		X
Frequencies	Execution Time	X	
	Energy		X
Network Density	Execution Time	X	
	Energy		X

5. CONCLUSION AND FUTURE WORK

In the context of Wireless Sensor Networks, the frequency allocation problem does not yield a unique solution. From an execution time perspective, the Weak Commitment algorithm performs the best whereas the Heuristic algorithm achieves the best results in term of energy consumption. A tradeoff between energy conservation and speed of convergence seems to be necessary. A fast running time may be critical for some applications whereas for some other applications a longer lifetime may suffice.

To determine which algorithm would be preferable, one solution would be to extend the experiments by introducing some criteria which have been ignored so far. Indeed, as stated previously, the transmission time has been neglected but may have an influence for low bandwidth implementations (in the order of several kb/s). Moreover, packet loss should be introduced for a more realistic representation of network operations. These criteria can impact the results especially when the number of messages increases. A shift of performance of the execution time may occur between the Weak Commitment algorithm and the Heuristic algorithm, but tests and further simulations are needed for a complete validation.

REFERENCES

1. *Distributed Sensor Networks, a multiagent perspective*, edited by V. Lesser, C.L. Ortiz, Jr and M. Tambe (Kluwer Academic Publications, 2003).

2. V. Annamalai, S.K.S. Gupta, , and L. Schwiebert, On tree-based convergecasting in wireless sensor networks, in: *IEEE Wireless Communications and Networking*, Vol.3, 1942 – 1947 (2003).

3. C. Guo, L.C. Zhong, and J.M. Rabaey, Low power distributed mac for ad hoc sensor radio networks, in *IEEE Global Telecommunications Conference*, Vol. 5, 2944 – 2948 (2001).

4. W.B. Heinzelman, A.P. Chandrakasan, and H. Balakrishnan, An application-specific protocol architecture for wireless microsensor networks, in *IEEE Transactions on Wireless Communications*, Vol. 1, 660670 (2002).

5. K. Hirayama and M. Yokoo, The effect of nogood learning in distributed constraint satisfaction, in *20th IEEE International Conference on Distributed Computing Systems*, 169 – 177 (2000).

6. I. Katzela and M. Naghshineh, Channel assignment schemes for cellular mobile telecommunication systems: a comprehensive survey, *IEEE Personal Communications*, 3(3), 10 – 31 (1996).

7. B. Krishnamachari, S. Wicker, R. Bejar, and C. Fernandez, On the complexity of distributed selfconfiguration in wireless networks, *Telecommunication Systems, Special Issue on Wireless Networks and Mobile Computing*, 169 – 177 (2000).

8. M. Sengoku, H. Tamura, S. Shinoda, and T. Abe, Graph and network theory and cellular mobile communications, in *IEEE International Symposium on Circuits and Systems*, Vol. 4, 2208 – 2211 (1993).

9. M. Yokoo and K. Hirayama, Algorithms for distributed constraint satisfaction: A review, *Autonomous Agents and Multi-Agent Systems*, 3(2), 185 – 207 (2000).

10. D. Youngs, Frequency assignment for cellular radio networks, in *Fifth IEEE Conference on Telecommunications*, 179 – 183 (1995).

FAST HANDOFF SUPPORT IN AN IP-EVOLVED UMTS ARCHITECTURE

Lila Dimopoulou, Georgios Leoleis, Iakovos S. Venieris
School of Electrical and Computer Engineering, National Technical University of Athens, 9 Heroon Polytechniou str, 157 73 Athens, Greece

Abstract: IP technology will play a key role in beyond 3G systems, which face the great challenge of integration in order to provide seamless service to users anywhere and anytime. Apart from its natural role as a unifier, IP also comprises the main drive for network evolution towards all-IP network infrastructures. In this regard, we exploit IP as an enabler for the evolution of the UMTS packet-switched core network, eliminating its duality at user and transport level. We focus on mobility management in the core network, which is handled by pure IP mechanisms (Mobile IPv6, MIPv6), and on the support of fast handoff across UMTS access networks by means of the IETF's Fast MIPv6 proposal. Emphasis is put on identifying the proper interaction points between the Fast MIPv6 operation and the UMTS-specific Serving Radio Network Subsystem (SRNS) relocation procedure in order to provide a seamless handoff service to the user while not compromising the network's performance and scalability.

Key words: Fast Handoff; Fast Mobile IPv6; Beyond 3G System; SRNS Relocation.

1. INTRODUCTION

IP technology has evidently played a key role in cellular mobile systems, such as the UMTS, where it has been adopted as the transport means of the packet switched (PS) core network, as an effort to support packet-based services in an efficient and cost effective manner. Although its impact has been primarily economic, considering the homogenization it achieves on the network resulting in the latter's easier maintenance and operation, it has become an imperative need from technological standpoint in future mobile networks – beyond 3rd Generation (3G) networks – facing the great challenge of integration[1] in order to provide seamless service to users anywhere and anytime.

Such integration involves all networks that simply co-exist nowadays, namely 3G cellular systems, Wireless Local Area Networks (WLANs), mobile ad-hoc networks, Personal Area Networks etc. World-wide roaming between various network technologies, continuously growing in heterogeneity, necessitates their seamless integration. IP is naturally the best choice to serve the internetworking and unification of these diverse technologies. However, apart from its role as a unifier, IP technology will also be the main drive for network evolution towards all-IP core networks. By that we mean that cellular infrastructures are directed to adopting IP as the basis for networking in the core network, for the latter to become access technology agnostic, and to allow a more efficient and straightforward integration with diverse access technologies.

In this paper, we address the IP technology as an enabler for network evolution, on the basis of the UMTS infrastructure. We propose a target architecture, where the UMTS Terrestrial Radio Access Network (UTRAN) forms the user's access means towards an all-IP core network. The duality of IP in application (user) and transport level[2], as it is the case in the UMTS PS core network, - jeopardizing network efficiency and performance - is eliminated, and a single IP layer is employed for all functionalities. Our focus has been brought on mobility management in the core network where pure IP-based solutions, and in particular Mobile IPv6[3] –MIPv6–, are adopted. Our contribution however concerns the application of the IETF's Fast MIPv6 mechanism[4] at the borders of the core network for achieving seamless handoffs across UMTS access networks. This necessitates an in depth study of the UMTS procedures[5] executed during handover for rendering their seamless interworking with IP handoff procedures feasible.

The rest of the paper is structured as follows. Section 2 presents the background in Mobile IPv6, IPv6 address autoconfiguration and Fast Handoff issues with the aim to identify the main contributions of handoff delay. In section 3, the target architecture is introduced and the Fast Handoff procedure in the context of the proposed environment is described in great detail. Certain issues regarding the handoff procedure and enhancements to it are examined in section 4 while section 5 concludes our paper.

2. BACKGROUND

2.1 Mobile IPv6

Mobile IPv6[3] comprises the IETF solution for handling the mobility of hosts in IPv6 networks. It extends the basic IPv6 functionality by means of header extensions rather than being built on top of it, as it is the case with MIPv4. Its fundamental principle is that a mobile host should use two IP addresses, a permanent address – the *home address*, assigned to the host and acting as its global identifier – and a temporary address – the *care-of address* (CoA), providing the host's actual location –. A mobile host (MH), while attached to its home network, is able to receive packets destined to its home address, and being forwarded by means of conventional IP routing. When the host crosses the boundaries of its serving network, movement detection is performed in order to identify its new point of attachment and further acquire a new CoA (nCoA). Once configured with a CoA, the MH needs to send a Binding Update (BU) message to its Home Agent (HA) to register this 'temporary' address. This CoA is obtained through IPv6 address autoconfiguration mechanisms; however, the time needed for autoconfiguration and for the Binding Management to complete sets the MIP operation inefficient for fast intra-domain movements.

According to the typical Mobile IP operation, the correspondent host (CH) addresses the MH at the latter's home address, and consequently does not need to implement the specific IPv6 extensions, which actually form MIPv6. In the opposite case – when the CHs are augmented with the MIPv6 functionality – then route optimization can be used for the direct delivery of packets to the MH without the intervention of the HA. The CHs are able to associate the MH's home address with a CoA – via BUs transmitted by the MH to CHs. However data packets will not be encapsulated for delivery to the MH, as is the case in MIPv4, but instead an IPv6 Routing header will be used for this purpose. These packets have as destination address the MH's CoA. The 'home address' information, required to preserve transparency to upper layers and ensure session continuity, is included in the routing header. In the reverse direction, packets have as source address the host's CoA while the home address is included in the newly defined *home address* destination option.

2.2 Handoff View of IPv6 Address Autoconfiguration

There are two mechanisms defined for the allocation of an IPv6 address to a node: the stateless and the stateful autoconfiguration. The stateful mechanism requires a Dynamic Host Configuration Protocol server to perform the address assignment to the IPv6 node. On the other hand, the stateless autoconfiguration procedure does not need any external entity involved in the address autoconfiguration (apart from the entity functioning as the first hop IP router, referred to as the access router –AR–). The stateless mechanism[6] allows a host to generate its own addresses using a combination of locally available information and information advertised by ARs. The latter advertise prefixes that identify the subnet(s) associated with a link, while hosts generate an 'interface identifier' that uniquely identifies an interface on each subnet. A global address is formed by combining the two. The formation of an address must be followed by the Duplicate Address Detection (DAD) procedure in order to avoid address duplication on links. Since it is the interface identifier that guarantees the uniqueness of the address –all hosts on the link use the same advertised prefix–, it suffices to perform DAD on the link local address. The latter is formed by appending the interface identifier to the well-known link-local prefix (FE80::) and only allows for IP-connectivity with nodes located at the same link. Global addresses formed from the same interface identifier need not be tested for uniqueness. In brief, the address autoconfiguration is composed of the following steps:

1. The host generates a link-local address for its interface on a link
2. It then performs DAD to verify the uniqueness of this address, i.e. verify the uniqueness of the interface identifier on the link
3. It uses the prefix(es) advertised by routers for forming a global address. DAD is not needed if the same interface identifier as in the link-local address is used.

When a host handoffs to a new subnet, it needs to be configured with a new global address, which is topologically correct, for being able to receive data on the new link. As it is expected, DAD needs to be performed for this address so as to verify its uniqueness on the link. As before, DAD can be once executed on the host's link-local address, given that the newly formed global address uses the same interface identifier. The basic deficiency coming along with DAD execution is that it adds delay to the handover. For the IP connectivity to be regained, after the establishment of link-level connectivity, some additional time is needed. In particular, during DAD, the host transmits a *Neighbor Solicitation* for the tentative link-local address and waits for *RetransTimer* milliseconds[7] till it considers the address unique. More precisely, the exact number of times the *Neighbor Solicitation* is (re)transmitted and the period between consecutive solicitations is link-specific and may be set by system management. DAD only fails if in the mean time, the host receives a *Neighbor Advertisement* for the same address, meaning that another host is using the being questioned address or if another host is in the progress of performing DAD for the same address and has also transmitted a *Neighbor Solicitation.* From the above it is deduced that at least a link-wide round-trip is needed for performing DAD while 1.5 - 2 round-trips are required in total for the whole autoconfiguration procedure if the router discovery (step 3) is performed in the sequence.

2.2.1 UMTS links

Let us examine how the IPv6 stateless address autoconfiguration mechanism is supported in UMTS[8], as it is recommended by the IETF IPv6 WG[9], and which of its inherent delays are eliminated. We should first present the

UMTS architecture in an IP centric view and understand how UMTS links are viewed from the IP layer. According to UMTS terminology, the Packet Data Protocol (PDP) context defines a link between the cellular host and the GGSN (Gateway GPRS Support Node), over which packets are transferred. An IP address is initially assigned to a primary PDP context while zero or more secondary PDP contexts use the same IP address. Thus, all PDP contexts using the same IP address define a link (a point-to-point link, referred to as UMTS link). A host may have activated more than one primary PDP contexts, i.e. may have more than one links to the GGSN(s).

Over the UMTS links, the host should configure a link-local address for on-link communication with the GGSN, and global address(es) for communication with other hosts. What is most important here is that the GGSN assigns the interface identifier to the host for forming its link-local address (corresponding to a UMTS link). The GGSN only has to ensure that there will be no collision between its own link-local address and the one of the cellular host, i.e. between its own interface identifier and the one assigned to the host. As a consequence, the host does not need to perform DAD for this address. Moreover, the cellular host must form a global address, based on the prefix(es) advertised by the GGSN. However here, the GGSN assigns a *prefix* that is *unique within its scope* to each primary PDP context. The uniqueness of the prefix suffices to guarantee the uniqueness of the MH's global address. This approach has been chosen taking into account that hosts may use multiple identifiers (apart from the one assigned by the GGSN) for forming global addresses, including randomly generated identifiers (e.g. for privacy purposes). This avoids the necessity to perform DAD for every address built by the MH. To sum up, the way address autoconfiguration is performed in UMTS eliminates the need for DAD messages over the air interface and therefore removes this factor of delay. The afore-described concepts are summarized in Table 1.

Table 1. IPv6 autoconfiguration concepts

	Shared Links	UMTS Links (Point-to-Point)
Interface Identifier	Generated by the host	Assigned by GGSN
Link-local address uniqueness	Guaranteed, DAD is performed	Guaranteed, GGSN is the only neighbor on the link with a different Interface ID
Prefix	Assigned by the Router to all hosts on the link	Unique prefix assigned to host by GGSN
Global address uniqueness	Guaranteed if using the Interface ID of link-local address	Guaranteed due to Unique Prefix
DAD	Needed for the link-local address and for global addresses using other Interface IDs	Not needed

2.3 MIPv6 Enhancements towards Fast Handoffs

As obviated earlier, MIPv6 presents some deficiencies due to the inherent delays introduced by address autoconfiguration and binding management. Fast MIPv6[4] (FMIPv6) comes to address the following problem: how to allow a mobile host to send packets as soon as it detects a new subnet link, and how to deliver packets to a mobile host as soon as its attachment is detected by the new access router. In other words, FMIP's primary aim is to eliminate the two factors of delay in address autoconfiguration. It achieves this by informing the mobile host, prior to its movement, of the new AR's advertised prefix, IP address and link layer address. The mobile host is already configured with the new address at the time it attaches to the new link. Supposing that the uniqueness of the address is guaranteed, the host can start sending packets in the uplink direction, setting the new address as the source address of these packets. In the downlink direction, a factor of delay is yet introduced before the new AR (nAR) can start delivering packets to the host. The nAR typically starts the Neighbor Discovery operation as soon as it receives packets for a host, in order to detect its presence and resolve its link layer address. This operation results in considerable delay that may last multiple seconds. In order to circumvent this delay, the FMIPv6 procedure requires from a MH to announce its attachment through a *Fast Neighbor Advertisement* (FNA) message that allows nAR to consider it reachable.

FMIPv6 is also essential for de-correlating the packet reception and transmission capability of the host from the time needed for the Binding Updates to HA and CHs to complete. This is required for two reasons:

1. The MH cannot start sending packets to CHs it communicates with, setting as source address the new CoA, prior to sending a BU to them, since the CHs will drop these packets.

2. The MH will not be able to receive packets from CHs at its new address, till the CHs update their caches for the host.

These two problems are basically addressed by setting up a bidirectional tunnel between the old AR and the MH at its nCoA. The tunnel remains active until the MH completes the Binding Update with its communicating hosts. To CHs, the mobile host is located at the old subnet; the old path is temporarily extended with the branch *old AR – nCoA of host* for allowing communication to continue during the IP handoff transition period. The full path is reestablished when the Binding Update procedure completes.

In brief, the operation of the protocol is as follows: the host sends a *Router Solicitation for Proxy* (RtSolPr) message to its AR so as to obtain information – e.g. prefix – related to available access points. The AR serving the user responds with a *Proxy Router Advertisement* (PrRtAdv) containing the requested information and thus allowing the mobile host to perform address autoconfiguration as if it had already migrated to a new link. The host, after formulating a prospective new CoA, sends a *Fast Binding Update* (FBU) to its AR for requesting the tunneling of packets addressed to its old CoA (oCoA) towards its nCoA. The AR serving the host (referred to as old AR, oAR) exchanges *Handover Initiate* (HI) and *Handover Acknowledge* (HAck) messages with the nAR for initiating the process of the MH's handover, while possibly validating the nCoA formed by the host. The oAR responds to the MH with a *Fast Binding Acknowledge* (FBack) message on both links (old and new) and starts the tunneling of arriving data. The MH, as soon as it attaches on the new link, transmits an FNA for informing the nAR of its presence. Packets from this point on can be delivered to the MH.

3. TARGET ARCHITECTURE

Before presenting our reference architecture, we shall elaborate on the requirements that the architecture should meet. We aim at a fully homogenized, access-agnostic core network (CN) solution that will cover the majority of access technologies under the IP suite umbrella. As a first step, we shall use the advanced radio access technologies offered by UTRAN. Moreover, the proposed solution should leave the UTRAN intact, that is, the core network should only view the standard Iu interface[10] and utilize its standardized capabilities. Further, the architecture, being IP-centric, should support all functionalities implemented by legacy UMTS protocols. Our focus however is placed on mobility management. Last, Fast Handoff should not be an access-dependent capability. Instead, it should be supported by the access-agnostic core network for handling the movement of the users across any access network.

Based on these requirements, we propose a target architecture, being evolved from the legacy UMTS network, with a view to bring the IP layer closer to the access network while maintaining performance, efficiency and scalability. To this aim, we integrate both GPRS Support Nodes (GSNs) of the UMTS PS core network, to form a single node, named as UMTS Access Router (UAR), and situated at the border of the RAN and the CN of the cellular infrastructure. In this way, the UMTS PS core network is replaced with a fully IP-compatible backbone where the UAR acts as the first-hop IP router. Traditional UMTS mechanisms in the CN, such as mobility management, being based until now on the heavy GPRS Tunneling Protocol (GTP) operation[5], are handled by IP-oriented ones. Mobile IPv6 is adopted as the ultimate mobile IP solution and hence, the user's movement across UARs will trigger the Binding Update procedure with the HA and/or CHs and not the re-establishment of GTP tunnels. However, the basic strength of the proposed architecture lies in that UARs run Fast MIPv6 for enabling the fast handoffs of Quality-of-Service (QoS) stringent sessions across RANs that do not allow for soft handoff[1]. This IP-based fast handoff capability also renders the UMTS Gn interface[2] among UARs redundant, since location management functionalities are now covered by FMIPv6.

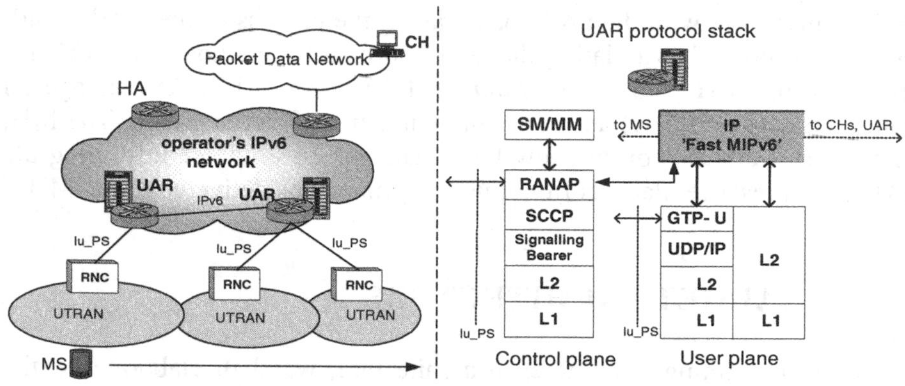

Figure 1. Target Architecture and UAR protocol stack

Fig. 1 depicts the target architecture; as it is shown, the UAR is mediated among UTRAN and the IP-based core network while trying to hide the peculiarities of the former towards the latter (by terminating legacy UMTS Non-Access-Stratum protocols for Session and Mobility Management – SM, MM). Moreover, no change in the UMTS protocol stack over the communication links with UTRAN (Iu-PS interface) has been effected, in both control and user planes. Fast MIP runs over the UTRAN user plane towards the Mobile Stations (MSs) while it also interoperates with RANAP (Radio Access Network Application Protocol), as it will be described below, for efficiently handling the MS's handoff across UARs.

3.1 Fast Handoff Procedure

In this section, we provide the full details of interoperation between UMTS and IP mechanisms during the handoff procedure. Our main objective has been to find the appropriate points of interaction (triggers) between the two layers – UMTS and IP – for enhancing overall handoff performance. The SRNS (Serving Radio Network Subsystem) relocation[5], as performed in UMTS when the user is being served by a new RNC (Radio Network Con-

[1] *Soft handoff* – macro-diversity – enables a seamless type of handover at layer 2. This is not applicable to inter-technology handoffs or even intra-technology ones, such as UTRAN-UTRAN handoffs, where the RANs are not interconnected (Iur interface not available).

[2] The Gn interface lies between two SGSNs and is implemented by GTP. It mainly carries out location management functionality.

troller), is the basis of the Fast Handoff procedure introduced, while interoperation with the FMIP protocol additionally takes place.

The handoff scheme, as depicted in Fig. 2, is triggered by UTRAN functions – sbased on measurements between the MS and the RNC serving the user. The SRNS relocation procedure starts and is composed of two main phases; a) the reservation of resources on the new link –between the target RNC and target UAR–, b) the target RNC takes over the serving RNC (SRNC) role and starts delivering data to the MS. According to the UMTS standardized procedure, a temporary GTP tunnel is used between the concerned RNCs for the forwarding of data following the old path until the new one – GGSN to target RNC – is updated. Although this forwarding capability is a built-in functionality of SRNS relocation, we choose instead to use the IP tunnel –used in FMIP– between the involved UARs.

Looking at the procedure in Fig. 2, the *Relocation Required* message comprises the L2 trigger at oUAR for transmitting a *PrRtAdv* to the MS. For the sake of simplicity, we will assume that the MS has activated one primary PDP context with possibly multiple secondary ones and therefore has been assigned one prefix for forming its global address(es). Before transmitting the Advertisement, the oUAR performs a target RNC – nUAR resolution for being able to fill the advertisement with the relevant information for the nUAR. At this point, we will assume that the oUAR knows the new prefix to be assigned to the MS for the corresponding PDP context, and we explain later how this is performed. The advertisement reveals to the MS information, such as the prefix used for address autoconfiguration and the nUAR's IP address. The message is sent unsolicited, in which case it acts as a network-initiated handover trigger, and not in response to a *Router Solicitation for Proxy* message from the MS where it would be required that the MS be aware of an identification of its attachment point (e.g. RNC).

The MS, in response to the PrRtAdv, formulates a new CoA based on the advertised prefix and sends a Fast Binding Update message to the oUAR. The source address is set to the oCoA while the alternate care-of address option is set to the nCoA. This message declares to the oUAR that it should forward data addressed to the oCoA towards the nCoA. The oUAR needs however to ensure that the new address can be used; note that address – prefix– assignment is performed by nUAR. Moreover, the oUAR will only start the forwarding if it is assured that resources have been reserved in the new path. Optionally, it may start buffering data, for example, in the case of lossless PDPs, in parallel to forwarding the data to the source RNC. The procedure continues with a *Handover Initiate* to the nUAR. This message requests from nUAR to verify the validity of the host's nCoA, and in the negative case the assignment of a new one. The nUAR may also create a host route entry for oCoA in case the nCoA cannot be accepted or assigned.

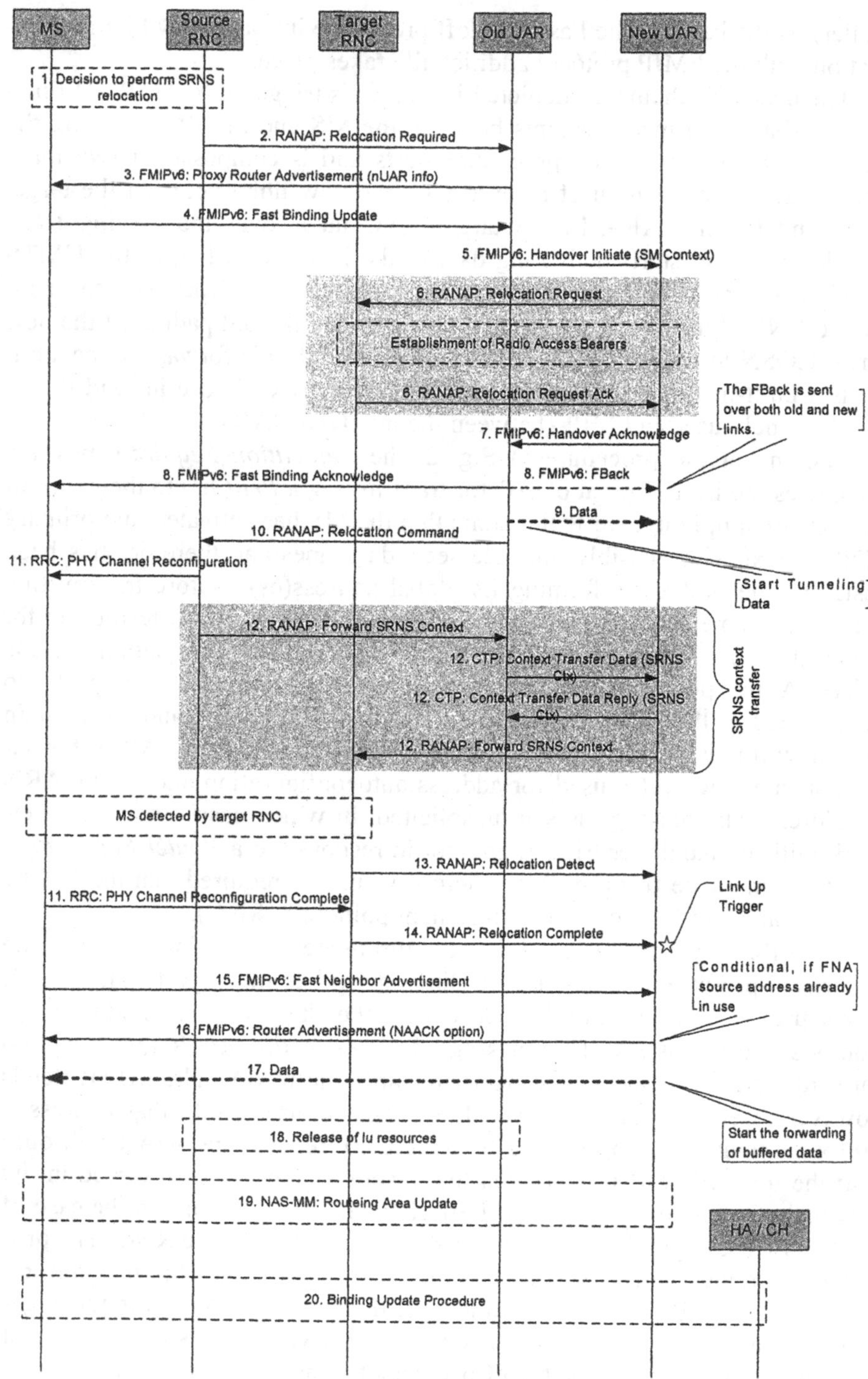

Figure 2. Fast Handoff scheme

The nUAR will only respond to HI after it relocates the Iu bearers. Note that the context information for these bearers is carried in the HI message, by means of a new option defined for this purpose. This mainly involves Session Management parameters for the MS's PDP contexts, such as QoS related information etc. The nUAR instructs the reservation of resources towards the new RNC by means of the *Relocation Request* message. In parallel, it checks whether the nCoA is valid and if not, it assigns a new address to the host – i.e., it only modifies the prefix –. After receiving the confirmation for the relocation of Iu bearers –*Relocation Request Acknowledge*–, it can respond to the oUAR with a *Handover Acknowledge* indicating that it accepts handover and that the forwarding of data can start. The new CoA is also included in the message if the nUAR does not accept the one assigned by the oUAR. Note here that the nUAR has to update the PDP contexts with the new address assigned to the host and also creates a proxy neighbor entry for this address and starts defending it.

The oUAR, after receiving the HAck, responds to the MS with a *FBack* where it indicates the nCoA – carried in the *alternate care-of address* option – to be used, if needed. This message is sent both over the old and new link. In the former case, it has as destination the oCoA while in the latter case the nCoA is used. Typically, this will be the first packet buffered at the nUAR for the MS's nCoA. The oUAR at this point instructs the source RNC to start the execution of the relocation – *Relocation Command* –, meaning that the source RNC from this moment will no longer be able to deliver packets to the host. In the meantime, the oUAR after receiving the *HAck* may also start the forwarding of data to the nCoA. After receiving the *Relocation Command*, the source RNC will instruct the MS to reconfigure itself and set the target RNC as its serving one. This is performed by means of a Radio Resource Control (RRC) message sent to the MS; the latter informs accordingly the target RNC, with an RRC message, that it has been reconfigured, which means that theoretically the target RNC is capable of delivering data to the MS. In parallel to the physical channel reconfiguration, the involved RNCs also exchange SRNS contexts for the supported Radio Access Bearers (RABs). This context information will allow the target RNC to support delivery order or lossless service for RABs, when this has been requested.

When the target RNC detects the MS, it notifies accordingly the nUAR – *Relocation Detect* – and starts the SRNC operation. This means that it may start processing upstream packets coming from the MS and forward them towards the nUAR. Moreover, it can start forwarding downstream packets, if available, towards the MS, given that the latter has reconfigured itself. The *Relocation Complete* comprises the link-up L2 trigger to the nUAR, which is informed of the MS's arrival on the new link. However, the forwarding of buffered packet towards the new RNC can not immediately start upon receipt of this message. Recall that the nUAR buffers packets addressed to nCoA after the request to do so by oUAR via the HI message.

At this point, we should examine when exactly packet forwarding to the MS is performed. If the MS has received the FBack message over the old link, then it has already updated its PDP context with the nCoA. In this way, packets buffered at the nUAR and addressed at the nCoA can be delivered over this PDP context. This is communicated to the nUAR by means of an FNA message, with the expected nCoA as source address, transmitted by the MS as soon as it is reconfigured. If a different address is identified by the nUAR (e.g. when FBack is not received over the old link and a non-valid address was assigned by oUAR), then the latter responds with a Router Advertisement with the *Neighbor Advertisement Acknowledge* option for indicating an alternate prefix to the MS. The MS subsequently forms a new CoA and modifies the respective PDP context. The nUAR may now start the forwarding of packets and also updates its neighbor cache. As expected, the MS will receive the FBack message on the new link. At this point, the MS receives packets addressed to its oCoA and being encapsulated by the oUAR towards its nCoA. In the reverse direction, packets have as source address the oCoA for not interrupting communication with the CHs, and are reverse tunneled towards the oUAR. Next, the MH may proceed with the MIPv6 Binding Update procedure with its HA and CHs.

3.2 Further Issues

3.2.1 Interface Identifier and Prefix Assignment

We shall analyze how the assignment of interface identifiers and prefixes may be performed by UARs in the context of an operator's IP domain. Each UAR is configured with an interface identifier which is used for its link-local address. Moreover, the UAR has to assign a unique interface identifier to each MS, for the latter to form its link-local address. The only requirement is that the UAR's identifier must not collide with the one assigned to the MS. Taking into consideration that the MS might roam to a new UAR within the IP domain, we propose that each UAR knows the Interface Ids belonging to all UARs and assigns Interface Ids to MSs that do not collide with the ones assigned to UARs. When the MS camps away from its serving UAR, it can continue using the same link-local address with the new UAR.

As for the Prefixes advertised by PrRtAdvs, we present an example of how they can be calculated by UARs, trying to avoid collisions with already assigned ones. Assuming that the operator's network, comprised of four UARs, has been assigned an n-bit prefix, then the operator assigns 2^{64-n-2} prefixes –links– to each UAR. Prefixes are identified as belonging to each UAR by means of the two bits *64-n-1* and *64-n-2*. When an UAR proxies an advertisement from a neighboring router, it simply sets these 2 bits for the advertised prefix to belong to the router being proxied. Certainly, collisions may arise, if this prefix is already assigned to a MS being served by the neighboring router. However, we can avoid collisions if neighboring ARs are configured to start assigning different prefixes to MSs – e.g. one starts from the low order bits and the other from the high order bits –. In areas where roaming between UARs is a frequent event, the operator may even resort to configuring UARs to set non-overlapping bits in their prefixes – i.e. one sets only the low bits while the other sets only the high bits.

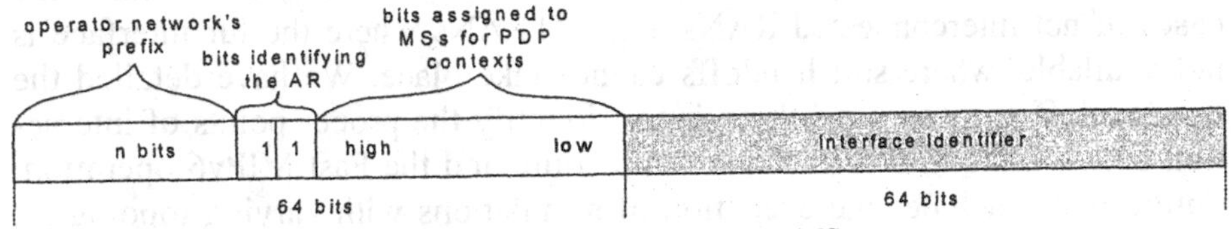

Figure 3. Prefix Assignment to MSs

3.2.2 Context Transfer

During the SRNS relocation procedure in UMTS, context transfer is performed at two levels of the architecture; between SGSNs –MM, PDP contexts– and between RNCs –SRNS contexts–. We attempt to identify which of these contexts are still meaningful within the new architecture. MM and PDP contexts are kept unchanged and are transferred between UARs by means of an extension to the HI message, as indicated earlier. As for the SRNS contexts, their main parameters are the GTP-U sequence numbers (SNs) next to be transmitted in both directions and the PDCP[11] (Packet Data Convergence Protocol) SNs to be used for data transmission from and to the MS. This information refers to each Radio Access Bearer –corresponding to a PDP– activated for the MS. Note that GTP-U packet sequencing is used when 'delivery order' is requested. In this case, the GGSN and RNC GTP-U protocol entities have to maintain the sequence of GTP-U packets transmitted in both directions. In our architecture, this requirement is relaxed due to the fact that *both* GTP-U entities are initialized during handoff – in UMTS, the GGSN GTP-U entity continues running – and therefore there is no problem in starting sequencing from scratch[3]. The PDCP SNs, on the other hand, are used when lossless PDCP service is requested. The target RNC needs this information in order to synchronize with the MS for the next PDCP packet expected in the uplink and downlink direction for each lossless radio bearer. Since the concerned RANs are not interconnected, this context transfer follows a path via the involved UARs. As an option, we can apply the IETF Context Transfer Protocol at the inter-UAR interface for transferring SRNS context.

[3] Delivery order is not required for IPv6 PDPs and it is not of our concern in the context of the proposed fast handoff procedure.

4. CONCLUSIONS

We have presented an evolved UMTS architecture, where the GSNs have been integrated to form one node, the UMTS Access Router, and the PS core network has been replaced by a fully IP-compatible backbone. Mobility in the core network is handled by Mobile IPv6, which has replaced the heavy GTP operation, while fast handoff across UARs is also supported by means of the IETF Fast MIPv6 protocol. This feature is particularly important in the cases of not interconnected RANs, e.g. UTRANs where the Iur interface is not available, where soft handoffs cannot take place. We have detailed the fast handoff procedure, while trying to identify the proper points of interaction between the SRNS relocation procedure and the Fast MIPv6 operation. Future work includes the execution of simulations with varying topologies, number of hosts, user moving patterns and active sessions regarding their QoS needs in order to evaluate the protocol's performance from both user and network perspectives.

REFERENCES

1. Jun-Zhao Sun et al., "Features in Future: 4G Visions From a Technical Perspective", in proceedings of IEEE Globecom'01, vol. 6, 2001.
2. Fabio M. Chiussi et al., "Mobility Management in Third Generation All-IP Networks", IEEE Communications Magazine, vol. 40, no. 9, September 2002, pp. 124-135.
3. D. Johnson et al., "Mobility Support in IPv6", Internet Draft, draft-ietf-mobileip-ipv6-21.txt, August 2003.
4. Rajeev Koodli et al., "Fast Handovers for Mobile IPv6", Internet Draft, draft-ietf-mipshop-fast-mipv6-01.txt, January 2004.
5. 3GPP TS 23.060 v6.3.0, "General Packet Radio Service (GPRS); Service description; Stage 2", December 2003.
6. S. Thomson et al., "IPv6 Stateless Address Autoconfiguration", Internet Draft, draft-ietf-ipv6-rfc2462bis-00.txt, February 2004.
7. T. Narten et al., "Neighbor Discovery for IP Version 6 (IPv6)", RFC 2461, December 1998.
8. 3GPP TS 29.061 v5.8.0, "Interworking between the Public Land Mobile Network (PLMN) supporting packet based services and Packet Data Networks (PDN)", December 2003.
9. M. Wasserman et al., "Recommendations for IPv6 in Third Generation Partnership Project (3GPP) Standards", RFC 3314, September 2002.
10. 3GPP TS 25.413 v6.1.0, "UTRAN Iu interface RANAP signalling", March 2004.
11. 3GPP TS 25.323 v6.0.0, "Packet Data Convergence Protocol (PDCP) Specification", December 2003.

Part Five: Intelligent Networks

STORAGE CAPACITY ALLOCATION ALGORITHMS FOR HIERARCHICAL CONTENT DISTRIBUTION*

Nikolaos Laoutaris, Vassilios Zissimopoulos and Ioannis Stavrakakis
Dep. of Informatics and Telecommunications, University of Athens, 15784 Athens, Greece
{laoutaris,vassilis,istavrak}@di.uoa.gr

Abstract: The addition of storage capacity in network nodes for the caching or replication of popular data objects results in reduced end-user delay, reduced network traffic, and improved scalability. The problem of allocating an available storage budget to the nodes of a hierarchical content distribution system is formulated; optimal algorithms, as well as fast/efficient heuristics, are developed for its solution. An innovative aspect of the presented approach is that it combines all relevant subproblems, concerning node locations, node sizes, and object placement, and solves them jointly in a single optimization step. The developed algorithms may be utilized in content distribution networks that employ either replication or caching/replacement.

Keywords: content distribution; web caching; storage allocation; heuristic algorithms.

1. INTRODUCTION

Recent efforts to improve the service that is offered to the ever increasing internet population strive to supplement the traditional bandwidth-centric internet with a rather non-traditional network resource – storage. Storage capacity (or memory) is employed to bring valuable information in close proximity to the end-users. The benefits of this tactic are quite diverse: end-users experience smaller delays, the load imposed on the network and on web-servers is reduced, the scalability of the entire content provisioning/distribution chain in the internet is improved. In most cases the engagement of the memory has been done in an ad hoc manner. Such an uncoordinated deployment, can seriously impair the effectiveness of the new resource.

This paper attempts to answer the question of how to allocate a given storage capacity budget to the nodes of a generic hierarchical content distribution system. Such a system can materialize as any one of the following: a hier-

*This work and its dissemination efforts have been supported in part by the IST Program of the European Union under contract IST-2001-32686 (Broadway).

archical cache comprising cooperating proxies from different organizations; a content distribution network offering hosting services or leasing storage to others that may implement hosting services on top of it; a dedicated electronic media system that has a hierarchical structure (e.g., video on demand distribution). The dimensioning of web caches and content distribution nodes has received a rather limited attention as compared to other related issues such as replacement policies Cao and Irani, 1997; Fan et al., 2000, proxy placement algorithms Krishnan et al., 2000; Li et al., 1999; Qiu et al., 2001; Cronin et al., 2002, object placement algorithms Korupolu et al., 1999; Kangasharju et al., 2002, and request redirection mechanisms Pan et al., 2003. In fact, the only published paper on the dimensioning of web proxies that we are aware of is due to Kelly and Reeves, 2001, whereas the majority of related works in the field have disregarded storage dimensioning issues by assuming the existence of infinite storage capacity Rodriguez et al., 2001; Gadde et al., 2002.

The limited attention paid to this problem probably owes to the fact that the rapidly decreasing cost of storage combined with the small size of typical web objects (html pages, images), make *infinitely large caches* for web objects realizable in practice, thus potentially obliterating the need for storage allocation algorithms. Although we support that storage allocation algorithms are marginally useful when considering typical web objects – which have a median size of just 4KB – we feel that recent changes in the internet traffic mix prompt for the development of such algorithms. A recent large scale characterization of http traffic from Saroiu et al., 2002 has shown that more than 75% of internet traffic is generated by P2P applications that employ the http protocol, such as KaZaa and Gnutella. The median object size of these P2P systems is 4MB which represents a thousand-fold increase over the 4KB median size of typical web objects. Furthermore, the access to these objects is highly repetitive and skewed towards the most popular ones thus making them highly amenable to caching. Such objects can exhaust the capacity of a cache or a CDN node, even under a low price of storage thus eliminating the assumption of infinitely large caches.

2. OUR APPROACH TOWARDS STORAGE CAPACITY ALLOCATION

The current work addresses the problem of allocating a storage resource differently than previous attempts, taking into consideration related resource allocation subproblems that affect it. Previous attempts have broken the problem of designing a content distribution network into a number of subproblems consisted of: (1) deciding where to install proxies (and possibly their number too); (2) deciding how much storage capacity to allocate to each installed proxy; (3) deciding on which objects to place in each proxy. Solving each one of the problems independently (by assuming a given solution for the others) is bound to lead to a suboptimal solution, due to the dependencies among them. For instance, a different storage allocation may be obtained by assuming different object placement policies and vice versa. The dependencies among the subproblems are not neglected under the current approach and, thus, an optimal solution for all the subproblems is concurrently derived, guaranteeing optimal overall performance.

Our methodology can be used for the optimization of existing systems (e.g. to re-organize more effectively the allocation of storage in a hierarchical cache) but hopefully will be the approach to be followed in developing future systems where the memory resource will be utilized dynamically and on-demand. The current work makes the following contributions towards the above mentioned uses:

- Introduces the idea of provisioning memory using a very small granule as an alternative to/extension of known paradigms (mirror placement, proxy placement) and models the problem with an integer linear program. The derived solution provides for a joint optimization of storage capacity allocation and object placement and can be exploited in systems that perform replication, as well as in those that perform caching.
- Develops fast efficient heuristic algorithms that approximate closely the optimal performance in various common scenarios but can execute very fast, as required by self organizing systems (and as opposed to planning/dimentioning processes that can employ slow algorithms). Moreover these algorithms may be executed incrementally when the available storage changes, thus obliterating the need for re-optimization from scratch.

The work focuses on hierarchical topologies. There are several reasons for this: (1) many information distribution systems have an inherent hierarchical structure owing to administrative and/or scalability reasons (examples include hierarchical web caching Wessels and Claffy, 1998, hierarchical data storage in Grid computing Ranganathan and Foster, 2001, hierarchical peer-to-peer networks Garces-Erice et al., 2003); (2) although the internet is not a perfect tree as it contains multiple routes and cycles, parts of it are trees (due to the actual physical structure, or as a consequence of routing rules) and what's more, overlay networks on top of it have no reason not to take the form of a tree if this is called for; (3) it is known that once good algorithms exist for a tree they may be applied appropriately to handle general graph topologies Bartal, 1996.

3. THE STORAGE CAPACITY ALLOCATION PROBLEM

3.1 Problem statement

The storage capacity allocation problem is defined here as that of the distribution of an available storage capacity budget to the nodes of a hierarchical content distribution system, given known access costs and client demand patterns. The proposed algorithms allocate storage units that may contain any object from a set of distinct objects (thus this is a multi-commodity problem as opposed to the single commodity k-median) and employ objective functions that are representative of the exact content of each node. Additionally, it is not assumed that a node can hold the entire set of available objects; in fact, this set need not contain objects from a single web-server, but it potentially includes object from different web-servers outside the hierarchy. As compared to works that study the object placement problem Kangasharju et al., 2002; Korupolu et al., 1999 where each proxy has a known capacity, the current approach adds *an additional degree of freedom* by performing the dimensioning of the proxies along with the object placement. Note that this may lead to a significant improvement in performance because even an optimal object placement policy will perform poorly if poor storage capacity allocation decisions have preceded (e.g., a large amount of storage has been allocated to proxies that receive only a few requests, whereas proxies that service a lot of requests have been allocated a limited storage capacity).

The input to the problem consists of the following: a set of N distinct unit-sized objects, $\mathcal{O}$; an available storage capacity budget of S storage units; a set of m clients, $\mathcal{J}$, each client j having a distinct request rate λ_j and a distinct object demand distribution $p_j : \mathcal{O} \rightarrow [0,1]$; a tree graph T with a node set of n nodes, $\mathcal{V}$, and a distance function $d_{j,v} : \mathcal{J} \times \mathcal{V} \rightarrow R^+$ associated with the jth leaf node and node v; this distance captures the cost paid when client j retrieves an object from node v. Each client is co-located with a leaf node and represents a local user population (with size proportional to λ_j). A client issues a request for an object and this request must be serviced by either an ancestor node that holds the requested object or by the origin server. In any case, a client always receives a given object from the same unique node. The storage capacity allocation problem amounts to identifying a set $\mathcal{A} \subseteq \mathbb{A}$ with no more than S elements (node-object pairs) (v, k), $v \in \mathcal{V}$, $k \in \mathcal{O}$; $\mathbb{A}$ is the set that contains all node-object pairs. $\mathcal{A}$ must be chosen so as to minimize the following expression of cost:

$$\min_{\mathcal{A} \subseteq \mathbb{A}:|\mathcal{A}| \leq S} \sum_{j \in \mathcal{J}} \lambda_j \sum_{k \in \mathcal{O}} p_j(k) \cdot d_j^{min}(k), \tag{1}$$

where $d_j^{min}(k) = \min\{d_{j,os}, d_{j,v}\} : v \in ancestors(j), (v,k) \in \mathcal{A}$; $d_{j,os}$ is the distance between the jth client (co-located with the jth leaf node) and the origin server, while $d_{j,v}$ is the distance between the jth leaf node and an ancestor node v. This cost models only "read" operations from clients. Adding "write" (update) operations from content creators is possible but as stated in Rabinovich, 1998 the frequency of writes is negligible compared to the frequency of reads and, thus, it does not seriously affect the placement decisions.

The output of the storage capacity allocation problem prescribes where in the network to place storage, how much of it, and which objects to store, so as to achieve a minimal cost (in terms of fetch distance) subject to a single storage constraint. This solution can be implemented directly in a real world content distribution system that performs replication of content. Notice that the exact specification of objects for a node also produces the storage capacity that must be allocated to this node. Thus, an alternative strategy is to disregard the exact object placement plan and just use the derived per-node capacity allocation in order to dimension the nodes of a hierarchical cache that operates under a dynamic caching/replacement algorithm (e.g., LRU, LFU and their variants). Recently there has been concern that current hierarchical caches are not appropriately dimensioned Williamson, 2002 (e.g., too much storage has been given to underutilized upper level caches). Thus, the produced results can be utilized by systems that employ replication as well as by those that employ caching.

3.2 Integer linear programming formulation of an optimal solution

In this section the storage capacity allocation problem is modeled with an integer linear program (ILP). Let $X_{j,v}(k)$ denote a binary integer variable which is equal to one if client j gets object k from node v where v is an ancestor of client j (including the co-located jth leaf node, excluding the origin server), and zero otherwise. Also let $\delta_v(k)$ denote an auxiliary binary integer variable which is equal to one if object k is placed at the ancestor node v, and zero otherwise. The two types of variables are related as follows:

$$\delta_v(k) = \begin{cases} 1 & \text{if} \quad \sum_{j \in leaves(v)} X_{j,v}(k) > 0 \\ 0 & \text{otherwise} \end{cases} \tag{2}$$

Equation (2) expresses the obvious requirement that an object must be placed at a node if some clients are to access it from that node. The following ILP gives an optimal solution to the storage capacity allocation problem.

Maximize:

$$z = \sum_{j \in \mathcal{J}} \lambda_j \sum_{k \in \mathcal{O}} p_j(k) \cdot \sum_{v \in ancestors(j)} (d_{j,os} - d_{j,v}) \cdot X_{j,v}(k) \tag{3}$$

Subject to:

$$\sum_{v \in ancestors(j)} X_{j,v}(k) \leq 1 \qquad j \in \mathcal{J}, k \in \mathcal{O} \tag{4}$$

$$\sum_{j \in leaves(v)} X_{j,v}(k) \leq U \cdot \delta_v(k) \qquad v \in \mathcal{V}, k \in \mathcal{O}, U \geq |\mathcal{J}| \tag{5}$$

$$\sum_{v \in \mathcal{V}} \sum_{k \in \mathcal{O}} \delta_v(k) \leq S \tag{6}$$

$$X_{j,v}(k), \delta_v(k) \quad \text{binary decision variables} \quad v \in \mathcal{V}, j \in \mathcal{J}, k \in \mathcal{O}$$

The maximization of (3) is equivalent to the minimization of (1) (the two objectives differ by sign and a constant). Notice that only the $X_{j,v}(k)$'s contribute to the objective function and the $\delta_v(k)$'s do not.

In the sequel, the above mentioned ILP will only be employed for the purpose of obtaining a bound on the performance of an optimal storage capacity allocation. Such a bound is derived by considering the LP-relaxation of the ILP (removing the requirement that the decision variables assume integer values in the solution) which can be derived rapidly by a linear programming solver.

3.3 Complexity of an optimal solution

The ILP formulation of Sect. 3.2 is generally NP-hard thus cannot be used for practical purposes. In Laoutaris et al., 2004a we have shown that the optimal solution to the problem discussed can be obtained in $O(\max\{n^4N, n^2N^2\})$. From a theoretical point of view, this result is attractive as it involves small powers of the input. In practice, however, such a result might be difficult to apply due to the quadratic dependence of complexity on the number of distinct objects N, which may assume very big values. For this reason, this work is primarily focused on the development of efficient heuristic algorithms. The developed algorithms are easy to implement, incur lower complexity, provide for a close approximation of the optimal in all our scenarios, and lend themselves to incremental use (such need arising when the available storage changes dynamically).

3.4 The Greedy heuristic

The Greedy heuristic begins with an empty hierarchy and enters a loop placing one object in each iteration, thus, in exactly S iterations all the storage capacity is allocated. Objects are placed in a globally greedy fashion according to the gain that is produced with each placement and past placement decisions are not subject to future placement decisions in subsequent iterations. The gain of an object at a certain node depends on the location of the node, the popularity of the object, the aggregate request rate for this object from all clients on the leaves of the subtree rooted at the selected node, and on prior iterations that have placed the same object elsewhere in the tree. In the first iteration the algorithm selects an node-object pair (v_1, k_1) that yields a maximum gain and places k_1 at v_1. Subsequent decisions place an object k in node v when the $gain_v(k)$ is maximum among all (v, k) pairs that have not been selected yet; $gain_v(k)$ is defined as:

$$gain_v(k) = \sum_{\substack{j \in leaves(v) \\ k \notin path(j,v)}} (d_{j,par_v(k)} - d_{j,v}) \cdot p_j(k) \cdot \lambda_j \qquad (7)$$

The parenthesized quantity in (7) is the distance reduction that is achieved by client j when fetching object k from node v instead from node $par_v(k)$, i.e., v's closest parent that caches k; initially it is the origin server that is the closest parent for all objects and all nodes but this changes as additional copies get replicated elsewhere in the tree.

Greedy is presented in detail in Table 1. Lines 1-7 describe the initialization of the algorithm. For each node v the gain of placing object k in v is computed and these values are inserted in a max-heap Cormen et al., 2001 data structure

$g(v, \cdot)$ (n max-heaps, one for its node v); the max-heaps are used so as to allow locating the most valuable object for each node in $O(1)$ (this does not require sorting the N objects). In the S iterations of the algorithm the following three steps are executed: (1) (v^*, k^*), the node-object pair that produces the maximum gain among the set of node-object pairs that have not been placed, $\mathcal{P}$, is selected, removed from $\mathcal{P}$, and the max-heap $g(v^*, \cdot)$ is re-organized (lines 9-10); (2) for each ancestor u of v^* up to $par_v(k)$ that does not hold k^*, the (potential) gain incurred if k^* is selected for u at a later iteration is updated and the corresponding max-heap is re-organized (lines 11-14) – the update of the potential gain $g(u, k^*)$ is necessary because the clients belonging to the subtree below v^* will not be fetching k^* from u but from v^* or its subtree, thus effectively reducing its previous potential gain at u; (3) for each descendant u of v^* that does not hold k^* and has v^* as closest parent with k^*, the potential gain incurred if k^* is selected for u at a later iteration is updated and the corresponding max-heap is re-organized (lines 15-18) – the update is in this case necessary because v^* becomes now the closest ancestor with k^* for some of its descendants, thus effectively reducing the previously computed gain for them that was based on a more distant parent. Notice that the various affected max-heaps need to be re-organized since one of their elements changes value; this must be done so as to maintain the max-heap property (i.e., have the maximum value accessible in $O(1)$ from the root of the max-heap).

A straightforward evaluation of the gain function $gain_v(k)$ requires $O(n)$ complexity. Since there are n nodes, evaluating $gain_v(k)$ for a given k for all nodes v would require $O(n^2)$ complexity, if each evaluation were to be carried out independently. Such an independent operation would involve however much overhead due to unnecessary repetitious work. See that the evaluation of the gain function depends on knowing the request rate that goes into the node and the closest parent that stores the object. To obtain the request rate for object k at node v, it suffices to know the corresponding rates at its children, and then sum the rates that go into children that do not cache k; knowing these rates, makes re-examining the entire subtree of v down to the leaf level redundant. Similar observation can be made regarding the identification of the closest parent. We make use of these observations in order to be able to evaluate $gain_v(k)$ for a particular pair (v, k) in $O(1)$. This allows calculating the gain for placing k in each of the n nodes in $O(n)$ instead of $O(n^2)$. In Laoutaris et al., 2004b we show how this can be achieved by first pre-computing information pertaining to request rates and closest parents and then using it to calculate up to n gain function for a given object in just $O(n)$. Since the gain function is $O(1)$ following the pre-computation step (occurring once at the beginning of each iteration), the complexity of each iteration of the Greedy algorithm depends on the number of nodes that are involved in the iteration and the update of the corresponding data structures.

The initial creation of the n max-heaps can be done in $O(nN)$ (each max-heap containing N values). At the beginning of each iteration there is the pre-processing step to get $rate_v(k)$ and $par_v(k)$ in $O(n)$ as explained in the appendix of Laoutaris et al., 2004b. The first step of each iteration requires that the highest value in all n max-heaps be selected. Finding the largest value in a max-heap requires $O(1)$ time thus the largest value in all n max-heaps can

```
for each v ∈ V
   for each k ∈ O
      g(v,k) = gain_v(k)
       insert g(v,k) in max-heap g(v,·)
   end for
end for
i = 1, P = {(v,k) : v ∈ V, k ∈ O}
while i ≤ S
   select (v*,k*) ∈ P : g(v*,k*) ≥ g(v,k)∀(v,k) ∈ P
    P = P − {(v*,k*)}, re-organize max-heap g(v*,·)
   for each u ∈ path(v*, par_v*(k*)) not caching k*
      g(u,k*) = gain_u(k*)
      re-organize max-heap g(u,·)
   end for
   for each u ∈ subtree(v*) not caching k*
                with par_u(k*) = v*
      g(u,k*) = gain_u(k*)
      re-organize max-heap g(u,·)
   end for
   i = i + 1
end while
```

Table 1. The Greedy algorithm.

be identified in $O(n)$ by a simple linear search or in $O(\log n)$ if an additional max-heap is maintained, containing the highest value from each of the n max-heaps $g(v,\cdot)$ (the latter might be unnecessary since n is typically rather small). The second step requires in the worst case the update of $L-1$ ancestors that do not cache k^*, L being the height of the tree. The function $gain_v(k^*)$ is re-evaluated for these nodes (each evaluation in $O(1)$) and the corresponding max-heap is re-organized in order to maintain the heap property; this can be done in $O(\log N)$ for a heap with N objects. The third step requires updating the descendants of v^* that are affected by storing k^* at v^*; this can be done in $O(n \log N)$. As a result the S iterations of the algorithm require $O(S \cdot (n + L \log N + n \log N))$, which simplifies to $O(S \cdot n \log N)$ by noting that L is at most n. Thus the overall complexity of Greedy (initialization + iterations) is $O(\max\{nN, Sn \log N\})$ which is linear in either N or S.

A salient feature of Greedy is that it can be executed incrementally, i.e., if the available storage budget changes from S to S' (e.g., because more storage has become available) and the user access patterns have not changed significantly then no re-optimization from scratch is required; it suffices to continue Greedy from its last iteration and add (or remove) $|S' - S|$ objects. This can present a significant advantage when the algorithm must be executed frequently for different S.

3.5 The improved Greedy heuristic (iGreedy)

In the previous Greedy algorithm one can make the following simple observation. Since clients are located at the leaves of the tree, if an object is placed at all children of a node u, then it is meaningless to also store it in u since no request will reach it there. This situation leads to the "waste" of storage units in "barren" objects. The Greedy algorithm often introduces barren objects as a result of its greedy mode of operation; an object is at some point placed at the father u while at that time not all children store it but with subsequent iterations it is also placed at all children thus rendering barren the copy at the father. This situation is not an occasional one but it is repeated quite frequently, resulting in wasting a substantial amount of the storage budget. The situation may be resolved by executing an additional check when placing an object k^* at a node v^*. The improved algorithm checks all peer nodes of v^* (at the same level, belonging to the same father) and if it finds that all store k^* then it also check whether their father u also stores it. In such a case it removes it from u freeing one storage unit; the resulting algorithm is called improved Greedy (iGreedy). The additional step of iGreedy is given in Table 2 and is executed between lines 10 and 11 of the basic Greedy algorithm.

iGreedy performs slightly more processing as compared to Greedy due to the following two additional actions: (1) in each iteration a maximum of Q peers need to be examined against k^*, Q denoting the maximum node degree of the tree; (2) each eviction of a barren object increases the number of iterations by one by freeing one storage unit which will have to be allocated in a subsequent iteration. Searching the Q peers does not affect the asymptotic per-iteration complexity of Greedy which is $O(n \log N)$. The increase in the number of iteration has a somewhat larger impact on the required processing. The following proposition establishes an exact upper bound on the number of iterations performed by iGreedy (the proof is included in a longer version of this article Laoutaris et al., 2004b).

PROPOSITION 1 *The maximum number of iterations performed by iGreedy cannot exceed* $T(S) = 2 \cdot S - 1$.

Thus in the worst case iGreedy will perform $2 \cdot S - 1$ iterations, with each iteration incurring the same complexity as with the basic Greedy. This means that the asymptotic complexity of iGreedy is identical to that of Greedy.

3.6 Numerical results under iGreedy

In this section the presented numerical results attempt to accomplish the following: (1) demonstrate the effectiveness of iGreedy in approximating the optimal performance; (2) present possible applications of the developed algorithms. When not stated otherwise, the clients are assumed to be sharing a common Zipf-like demand distribution p_j over $\mathcal{O}$ with a typical skewness parameter $a = 0.9$ and equal request rates $\lambda_j = 1, \forall j \in \mathcal{J}$. A Zipf-like distribution is a power-law dictating that the ith most popular object is requested with a probability C/i^a, where $C = (\sum_{j=1}^{N} \frac{1}{j^a})^{-1}$. The skewness parameter a captures the degree of concentration of requests; values approaching 1 mean that few distinct objects receive the vast majority of requests, while small values indicate progressively uniform popularity. The Zipf-like distribution is generally recognized as a good model for characterizing the popularity of various types of measured workloads, such as web objects Breslau et al., 1999 and multimedia clips Chesire et al., 2001. The popularity of P2P Saroiu et al.,

```
10.1: allpeers=1
10.2: for each v ∈ peers(v*)
10.3:     if k* not cached in v
10.4:         allpeers = 0
10.5:         break
10.6:     end if
10.7: end for
10.8: if k* cached in u=father(v*)
                 and allpeers=1
10.9:     remove k* from u
                 and set i = i − 1
10.10:end if
```

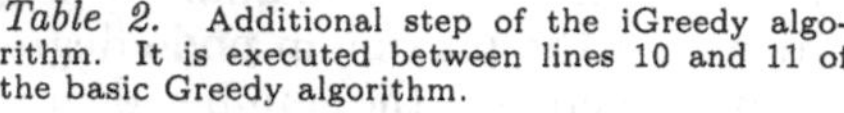

Table 2. Additional step of the iGreedy algorithm. It is executed between lines 10 and 11 of the basic Greedy algorithm.

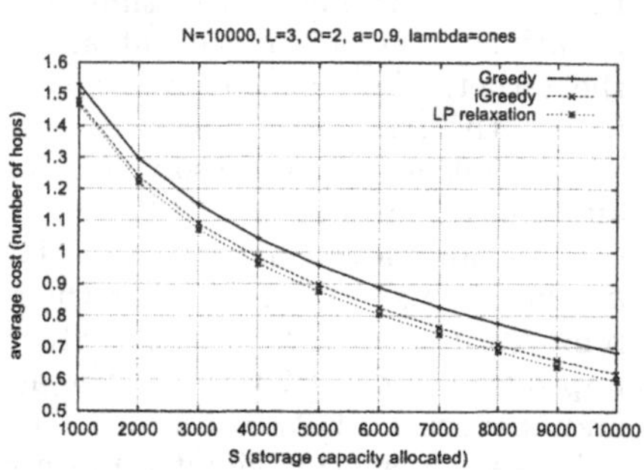

Figure 1. The average cost of Greedy, iGreedy and LP-relaxation.

2002 and CDN content has also been shown to be quite skewed towards the most popular documents, thus approaching a Zipf-like behavior.

As far as the topology of the experiments is concerned, regular Q-ary trees are used in all examples. Regular Q-ary trees are commonly used for the derivation of numerical results for algorithms operating on trees Rodriguez et al., 2001. The entire set of parameters (demand and topology) for each experiment is indicated in the title of the corresponding graph. The distance function $d_{j,v}$ capture the number of hops between client j (co-located with the jth leaf thus $d_{j,j} = 0$) and node v. The distance of the origin server is $d_{j,os} = L$ for an L level hierarchy.

Quality of the approximation. Figure 1 shows the average cost per request for Greedy and iGreedy (expressed in number of hops to reach an object). The performance of the heuristic algorithms is plotted against the bound of the corresponding optimal performance obtained from the LP-relaxation of the ILP of Sect.3.2. The x-axis indicates the number of available storage units in the hierarchy (S) with each storage unit being able to host a single object. From the graph it may be seen that iGreedy is no more than 3% away from the optimal while Greedy may deviate as much as 14% in the presented results. The performance gap between the two owes to the waste of a significant amount of storage in barren objects under Greedy.

The effects of skewness and non-homogeneous demand. The following two figures focus on the vertical allocation of storage under iGreedy. Figure 2 shows the effect of the skewness of the demand distribution on the per-level allocation of storage. Highly skewed distributions (the skewness parameter a approaching 1) increase the amount of storage that is allocated to the leaves (level-1), while less skewed distributions allocate more storage to the root (level-3). This effect is explained as follows. Under a highly skewed distribution a small number of popular objects attracts the majority of requests

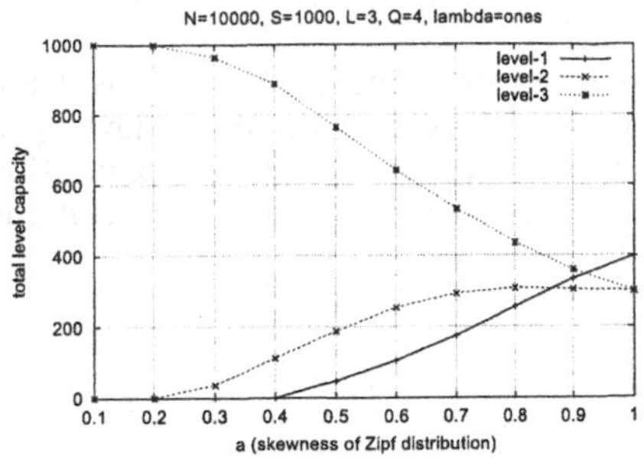

Figure 2. The effect of skewness of popularity on the per-level allocation of storage under iGreedy.

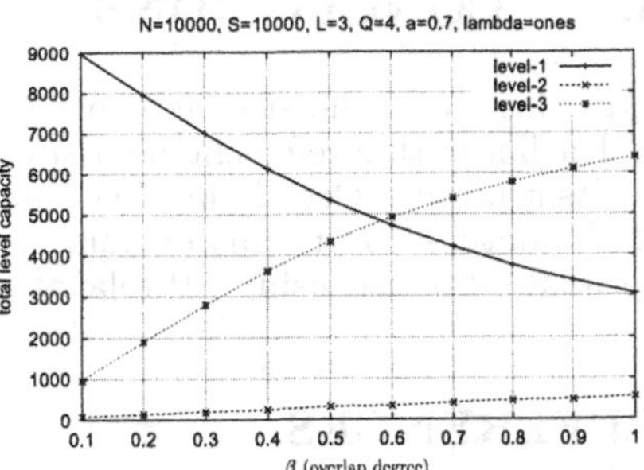

Figure 3. The effect of non-homogeneous demand on the per-level allocation of storage under iGreedy.

and, thus, these objects are intensively replicated at the lower levels, leading to the allocation of most of the storage to the lower levels. When the distribution tends to be "flat" it is better to limit the number of replicas per object and instead increase the number of distinct objects that can be replicated. This is achieved by sharing objects, i.e., by placing them higher in the hierarchy that leads to the allocation of more storage to the higher levels.

Figure 3 illustrates the effect of the degree of homogeneity in the access patterns of different clients. Two clients are non-homogeneous if they employ different demand distributions. In the presented results each client j references N objects; βN objects are common to all clients while the remaining $(1-\beta)N$ are only referenced by client j. A Zipf-like distribution is created for each client by randomly choosing an object from its reference set and assigning it the next higher value from a Zipf-like distribution and then repeating the same action until all objects have been assigned probabilities. The parameter β will be referred to as the *overlap degree*; values of β approaching 1 mean that most objects are common to all clients (although each client may request a common object with a potentially different probability) while small values of β mean that each client references a potentially different set of objects. Figure 3 shows that the root level (level-3) of a hierarchical system is assigned more storage when there is a substantial amount of overlap in client reference patterns. Otherwise most of the storage goes to the lower levels. This behavior is explained as follows. Storage is effectively utilized at the upper levels when each placed object receives an aggregate request stream from several clients. Such an aggregation may only exist when a substantial amount of objects are common to all clients; otherwise it is better to allocate all the storage to the lower levels – thus sacrificing the (ineffective) aggregation effect – and instead reduce the distance between clients and objects.

4. CONCLUSIONS

In this paper the storage capacity allocation problem has been considered and a linear time efficient heuristic algorithm, iGreedy, has been developed for its solution. iGreedy has been shown to provide for a good approximation of the optimal by means of numerical comparison against the bound of the optimal (obtained using LP-relaxations).

REFERENCES

Bartal, Y. (1996). On approximating arbitrary metrics by tree metrics. In *Proceedings of the 37th Annual IEEE Symposium on Foundations of Computer Science (IEEE FOCS).*

Breslau, Lee, Cao, Pei, Fan, Li, Philips, Graham, and Shenker, Scott (1999). Web caching and Zipf-like distributions: Evidence and implications. In *Proceedings of the Conference on Computer Communications (IEEE Infocom)*, New York.

Cao, Pei and Irani, Sandy (1997). Cost-aware WWW proxy caching algorithms. In *Proceedings of the USENIX Symposium on Internet Technologies and Systems*, pages 193–206.

Chesire, Maureen, Wolman, Alec, Voelker, Geoffrey M., and Levy, Henry M. (2001). Measurement and analysis of a streaming-media workload. In *Proceedings of USITS.*

Cormen, Thomas H., Leiserson, Charles E., Rivest, Ronald L., and Stein, Clifford (2001). *Introduction to Algorithms, 2nd Edition.* MIT Press, Cambridge, Massachusetts.

Cronin, Eric, Jamin, Sugih, Jin, Cheng, Kurc, Anthony R., Raz, Danny, and Shavitt, Yuval (2002). Constraint mirror placement on the internet. *IEEE Journal on Selected Areas in Communications*, 20(7).

Fan, Li, Cao, Pei, Almeida, Jussara, and Broder, Andrei Z. (2000). Summary cache: a scalable wide-area web cache sharing protocol. *IEEE/ACM Transactions on Networking*, 8(3):281–293.

Gadde, Syam, Chase, Jeff, and Rabinovich, Michael (2002). Web caching and content distribution: A view from the interior. *Computer Communications*, 24(2).

Garces-Erice, Luis, Biersack, Ernst W., Ross, Keith W., Felber, Pascal A., , and Urvoy-Keller, Guillaume (2003). Hierarchical P2P systems. In *Proceedings of ACM/IFIP International Conference on Parallel and Distributed Computing (Euro-Par)*, Klagenfurt, Austria.

Kangasharju, Jussi, Roberts, James, and Ross, Keith W. (2002). Object replication strategies in content distribution networks. *Computer Communications*, (4):376–383.

Kelly, T. and Reeves, D. (2001). Optimal web cache sizing: scalable methods for exact solutions. *Computer Communications*, 24(2):163–173.

Korupolu, Madhukar R., Plaxton, C. Greg, and Rajaraman, Rajmohan (1999). Placement algorithms for hierarchical cooperative caching. In *Proceedings of the 10th Annual Symposium on Discrete Algorithms (ACM-SIAM SODA)*, pages 586 – 595.

Krishnan, P., Raz, Danny, and Shavit, Yuval (2000). The cache location problem. *IEEE/ACM Transactions on Networking*, 8(5):568–581.

Laoutaris, Nikolaos, Zissimopoulos, Vassilios, and Stavrakakis, Ioannis (2004a). Joint object placement and node dimensioning for internet content distribution. *Information Processing Letters*, 89(6):273–279.

Laoutaris, Nikolaos, Zissimopoulos, Vassilios, and Stavrakakis, Ioannis (2004b). On the optimization of storage capacity allocation for content distribution. *Computer Networks.* [submitted].

Li, Bo, Golin, Mordecai J., Italiano, Giuseppe F., Deng, Xin, and Sohraby, Kazem (1999). On the optimal placement of web proxies in the internet. In *Proceedings of the Conference on Computer Communications (IEEE Infocom)*, New York.

Pan, Jianping, Hou, Y. Thomas, and Li, Bo (2003). An overview DNS-based server selection in content distribution networks. *Computer Networks*, 43(6).

Qiu, Lili, Padmanabhan, Venkata, and Voelker, Geoffrey (2001). On the placement of web server replicas. In *Proceedings of the Conference on Computer Communications (IEEE Infocom)*, Anchorage, Alaska.

Rabinovich, Michael (1998). Issues in web content replication. *Data Engineering Bulletin (invited paper)*, 21(4).

Ranganathan, K. and Foster, I. (2001). Identifying dynamic replication strategies for a high performance data grid. In *Proceedings of the International Workshop on Grid Computing*, Denver, Colorado.

Rodriguez, Pablo, Spanner, Christian, and Biersack, Ernst W. (2001). Analysis of web caching architectures: Hierarchical and distributed caching. *IEEE/ACM Transactions on Networking*, 9(4).

Saroiu, Stefan, Gummadi, Krishna P., Dunn, Richard J., Gribble, Steven D., and Levy, Henry M. (2002). An analysis of internet content delivery systems. In *Proceedings of the 5th Symposium on Operating Systems Design and Implementation (OSDI 2002)*.

Wessels, Duane and Claffy, K. (1998). ICP and the Squid web cache. *IEEE Journal on Selected Areas in Communications*, 16(3).

Williamson, Carey (2002). On filter effects in web caching hierarchies. *ACM Transactions on Internet Technology*, 2(1).

AN INFERENCE ALGORITHM FOR PROBABILISTIC FAULT MANAGEMENT IN DISTRIBUTED SYSTEMS

Jianguo Ding[1,2], Bernd Krämer[1], Yingcai Bai[2] and Hansheng Chen[3]
[1]*FernUniversität Hagen, D-58084, Germany;* [2]*Shanghai Jiao Tong University, Shanghai 200030, P. R. China, Jianguo.Ding@sjtu.edu.cn;* [3]*East-china Institute of Computer Technology, Shanghai 200233, P. R. China*

Abstract: With the proliferation of novel paradigms in distributed systems, including service-oriented computing, ubiquitous computing or self-organizing systems, an efficient distributed management system needs to work effectively even in face of incomplete management information, uncertain situations and dynamic changes. In this paper, Bayesian networks are proposed to model dependencies between managed objects in distributed systems management. Based on probabilistic backward inference mechanisms the so-called Strongest Dependency Route (SDR) algorithm is used to compute the set of most probable faults that may have caused an error or failure.

Keywords: fault management; uncertainty; Bayesian network; backward inference.

1. INTRODUCTION

As distributed systems grow in size, heterogeneity, pervasiveness, and complexity of applications and network services, their effective management becomes more important and more difficult. Individual hardware defects or software errors or combinations of such defects and errors in different system components may cause the degradation of services of other (remote) components in the network or even their complete failure due to functional dependencies between managed objects. Hence an effective distributed fault detection mechanism is needed to support rapid decision making in distributed systems management and allow for partial automation of fault correction. In the past decade, a great deal of research effort has been focused on improving a management system in fault detection and diagnosis.

Rule-based methods were proposed for fault detection[11,6]. Finite State Machines (FSMs) were used to model fault propagation behaviour and duration[4,27]. Coding-based methods[18,26] and Case-based methods[17] were used for fault identification and isolation. However most of these solutions are unable to deal with the incomplete and imprecise management information effectively. Probabilistic reasoning is another effective approach for fault detection in distributed systems management[5,9,24].

On the practical side, most of the current commercial management software, such as IBM Tivoli, HP OpenView, or Cisco serial network management software still lack facilities for exact fault localization, or the automatic execution of appropriate fault recovery actions. A typical metric for on-line fault identification is 95% fault location accuracy and 5% faults can not be located and recovered in due time[22]. Hence for large distributed systems including thousands of managed components it may be rather time-consuming and difficult to resolve the problems in a short time by exhaustive search in locating the root causes of a failure.

In this paper we apply Bayesian networks (BNs) to model dependencies among managed objects and provide efficient methods to locate the root causes of failure situations in the presence of imprecise management information. Our ultimate goal is to automate part of the daily management business. A Strongest Dependence Route (SDR) algorithm for backward-inference in BNs is presented. The SDR algorithm will allow users to trace the strongest dependency route from some malicious effect to its causes, so that the most probable causes are investigated first. The algorithm also provides a dependency ranking of a particular effect's causes.

2. BAYESIAN NETWORKS FOR DISTRIBUTED SYSTEMS MANAGEMENT

2.1 BNs model for distributed systems management

Bayesian networks, also known as Bayesian belief networks, belief networks, causal networks or probabilistic networks, are effective means to model probabilistic knowledge by representing cause-and-effect relationships among key entities of a managed system. BNs can be used to generate useful predictions about future faults and decisions even in the presence of uncertain or incomplete information. BNs have been applied to problems in medical diagnosis[20,26], map learning[1] and language understanding[2]. BNs use DAGs (Directed Acyclic Graphs) with probability labels to represent probabilistic knowledge. BNs can be defined as a triplet (V, L, P), where V is a set of variables (nodes of the DAG), L is the set of causal links among the variables (the directed arcs between nodes of the DAG), *P* is a set of probability distributions defined by: $P=\{p(v|\pi(v))|v \in V\}$; $\pi(v)$ denotes the parents of node *v*. The DAG is commonly referred to as the dependence structure of a BN.

In BNs, the information included in one node depends on the information of its predecessor nodes. The former denotes an effect node; the latter represents its causes. This dependency relationship is denoted by a probability distribution in the interval [0, 1]. An important advantage of BNs is the avoidance of building huge joint probability distribution tables that include permutations of all the nodes in the network. Rather, for an effect node, only the states of its immediate predecessor need to be considered.

Figure 1 shows a particular detail of the campus network of the FernUniversität in Hagen.

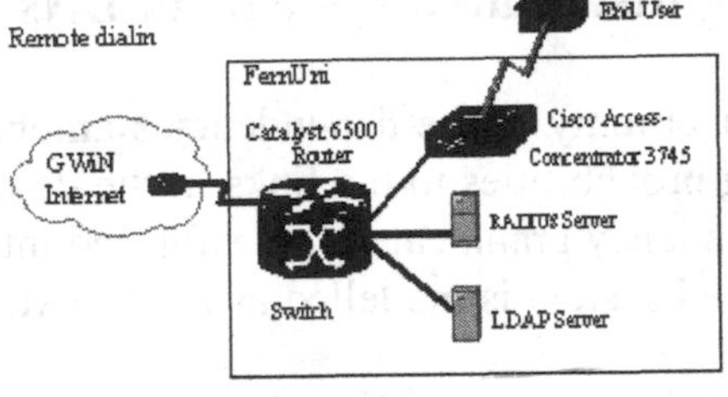

Figure 1. Example of distributed system

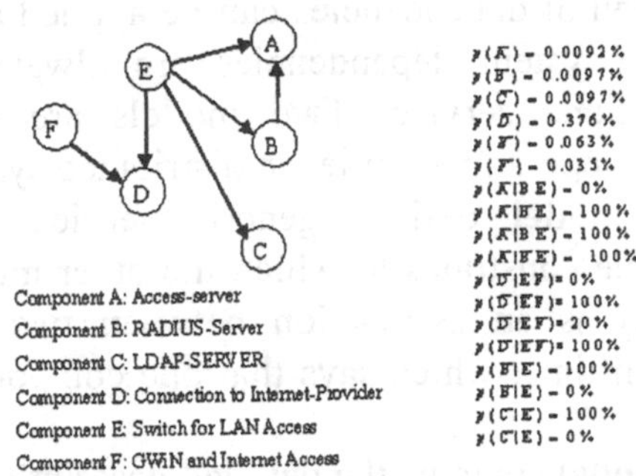

Figure 2. Example of Bayesian Network for Figure 1

When only the connection service for end users is considered, Figure 2 illustrates the associated BN. The arrows in the BN denote the dependency from causes to effects. The weights of the links denote the probability of dependency between the objects. In this example, the annotation $p(\overline{D}|\overline{E}F)=100\%$ denotes the probability of the non-availability of component D is 100% when component F is in order but component E is not. Other annotations can be read similarly.

Due to the dense knowledge representation and precise calculations of BNs, BNs can represent large amount of interconnected and causally linked data as they occur in distributed systems. Generally speaking:

(1) BNs can represent knowledge in depth by modeling the relationship between the causes and effects among network components and network services.

(2) They can provide guidance in diagnosis. Calculations over a BN can determine both the precedence of detected effects and the network part that needs further investigation in order to provide a finer grained diagnosis.

(3) They have the capability of handling uncertain and incomplete information due to their grounding in probability theory.

2.2 Mapping distributed systems to BNs

We represent uncertainty in the dependencies among distributed system entities by assigning probabilities to the links in the dependency or causality graph[13,15]. This dependency graph can be transformed into a BN[10].

When a distributed system is modelled as a BN, two important processes need to be resolved:

(1)Ascertain the dependency relationship between managed entities.

When one entity requires a service performed by another entity in order to execute its function, this relationship between the two entities is called a dependency. The notion of dependencies can be applied at various levels of granularity. The inter-system dependencies are always confined to the components of the same service. Two models are useful to get the dependency between cooperating entities in distributed systems[12].

The functional model defines generic service dependencies and establishes the principle constrains to which the other models are bound. A functional dependency is an association between two entities, typically captured first at design time, which says that one component requires some services from another.

The structural model contains the detailed descriptions of software and hardware components that realize the service. A structural dependency contains detailed information and is typically captured first at deployment or installation time.

(2)Obtain the measurement of the dependency.

When BNs are used to model distributed systems, BNs represent causes and effects between observable symptoms and the unobserved problems, so that when a set of evidences is observed the most likely causes can be determined by inference technologies.

Single-cause and multi-cause are two kinds of general assumptions to consider the dependencies between managed entities in distributed systems management. A non-root node may have one or several parents (causal nodes). Single-cause means any of the causes must lead to the effect. While multi-cause means that one effect is generated only when more than one cause happens simultaneity. Management information statistics are the main source to get the dependencies between the managed objects in distributed systems. The empirical knowledge of experts and experiments are useful to determine the dependency.

Some researchers have performed useful work to discover dependencies from the application view in distributed systems[8,12,7].

3. BACKWARD INFERENCE IN BAYESIAN NETWORKS

3.1 Inference in Bayesian networks

The most common approach towards reasoning with uncertain information about dependencies in distributed systems is backward inference, which traces the causes from effects. We define E as the set of effects (evidences) which we can observe, and C as the set of causes.

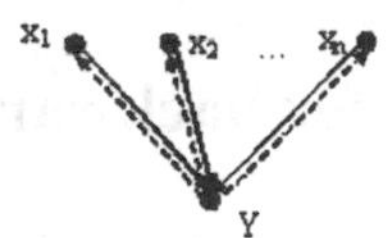

Figure 3. Basic model for backward inference in Bayesian networks

Before discussing the complex backward inference in BNs, a simple model will be examined. In BNs, one node may have one or several parents (if it is not a root node), and we denote the dependency between parents and their child by a JPD (Joint Probability Distribution).

Figure 3 shows the basic model for backward inference in BNs. Let $X=(x_1, x_2, \ldots, x_n)$ be the set of causes. According to the definition of BNs, the following variables are known: $p(x_1), p(x_2), \ldots, p(x_n), p(Y|x_1, x_2, \ldots, x_n)=p(Y|X)$. Here $x_1, x_2, \ldots, x_n$ are mutually independent, so

$$p(X) = p(x_1, x_2, \ldots, x_n) = \prod_{i=1}^{n} p(x_i) \tag{1}$$

$$p(Y) = \sum_X [p(Y|X)p(X)] = \sum_X \left[p(Y|X) \prod_{i=1}^{n} p(x_i) \right] \tag{2}$$

by Bayes' theorem, $$p(X \mid Y) = \frac{p(Y|X)p(X)}{p(Y)} = \frac{p(Y|X) \prod_{i=1}^{n} p(x_i)}{\sum_X \left[p(Y|X) \prod_{i=1}^{n} p(x_i) \right]} \tag{3}$$

which computes to $$p(x_i|Y) = \sum_{X \backslash x_i} p(X|Y) \tag{4}$$

In Eq. (4), $X \backslash x_i = X - \{x_i\}$. According to the Eqs. (1)-(4), the individual conditional probability $p(x_i|Y)$ can be achieved from the JPD $p(Y|X)$, $X=(x_1, x_2, \ldots, x_n)$. The backward dependency can be obtained from Eq. (4). The dashed arrowed lines in figure 3 denote the backward inference from effect Y to individual cause x_i, $i \in [1,2, \ldots, n]$.

In Figure 2, when a fault is detected in component D, then based on Eqs. (1)-(4), we obtain $p(\bar{F} \mid \bar{D}) = 67.6\%, p(\bar{E} \mid \bar{D}) = 32.4\%$. This can be interpreted as follows: when component D is not available, the probability of a fault in component F is 67.6% and the probability of a fault in component E is 32.4%. Here only the fault related to connection service is considered.

3.2 SDR algorithm for backward inference

In distributed system management, the key factors that are related to the defect in the system should be identified. The Strongest Dependency Route (SDR) algorithm is proposed to resolve these tasks based on probabilistic inference.

Before we describe the SDR, the definition of strongest cause and strongest dependency route are given as follows:

Definition 3.2.1 *In a BN let C be the set of causes, E be the set of effects. For* $e_i \in E$, C_i *be the set of causes based on effect* e_i, *iff* $p(c_k|e_i) = Max[p(c_j|e_i), c_j \in C]$, *then* c_k *is the strongest cause for effect* e_i.

Definition 3.2.2 *In a BN, let C be the set of causes, E be the set of effects, let R be the set of routes from effect* $e_i \in E$ *to its cause* $c_j \in C$, $R=(R_1, R_2, \ldots, R_m)$. *Let* M_k *be the set of transition nodes between* e_i *and* c_j *in route* $R_k \in R$. *Iff* $p(c_j|M_k, e_i) = Max[p(c_j|M_t, e_i), t=(1, 2, \ldots, m)]$ *(here* $p(c_j|M_t, e_i)$ *can be derived from* $p(M_t|e_i) * p(c_j|M_t)$*), then* R_k *is the strongest route between* e_i *and* c_j.

The detailed description of the SDR algorithm is described as follows:

3.2.1 Pruning of the BNs

Generally speaking, multiple effects (symptoms) may be observed at a moment, so $E_k = \{e_1, e_2, ..., e_k\}$ is defined as initial effects. In the operation of pruning, every step just integrates current nodes' parents into BN' and omits their brother nodes, because their brother nodes are independent with each other. To achieve this, a prune operation is defined as follows:

Algorithm 3.2.1

Prune $(BN=(V, L, P),\ E_k \subset E,\ E_k = \{e_1, e_2, ..., e_k\})$
    new $BN' = (V', L', P')$;
    $V' = E_k$;   *//add* E_k *to* V',
    $L' = \Phi$;   *//*Φ *denotes empty set.*
    for $e_i \in E_k$ $(i=1,..,k)$
        $v_i = e_i$,
        while $v_i \neq NIL$ ***do***
            $V' = V' \cup \{\pi(v_i)\}$,   *// add vertex* $\pi(v_i)$ *to* V'
            $v_i \leftarrow \pi(v_i)$,
            $L' = L' + <\pi(v_i), v_i>$; *// add edge* $<\pi(v_i), v_i>$ *to* L'
    return BN';

The pruned graph is composed of the effect node E_k and its entire ancestor.

3.2.2 Strongest Dependency Route (SDR) trace algorithm

After the pruning algorithm has been applied to a BN, a simplified sub-BN is obtained. Between every cause and effect, there may be more than one dependency routes. The questions now are: which route is the strongest dependency route and among all causes, which is the strongest cause? The SDR algorithm use product calculation to measure the serial strongest dependencies between effect nodes and causal nodes. Suppose $E_k \subset E$, $E_k = \{e_1, e_2, ..., e_k\}$. If $k=1$, the graph will degenerate to a single-effect model.

Algorithm 3.2.2 (SDR):

Input: V: the set of nodes (variables) in BNs; L: the set of links in the BN; P: the dependency probability distribution for every node in BN; $E_k = \{e_1, e_2, ..., e_k\}$: the set of initial effect nodes in BN.

Output: T: a spanning tree of the BN, rooted on vertex E_k, and a vertex-labelling gives the probability of the strongest dependency from e_i to each vertex.

Variables: *depend[v]*: the probability of the strongest dependency between v and all its descendants; $p(v|u)$: the probability can be calculated from JPD of $p(u|\pi(u))$ based on Eqs. (1)-(4), v is the parent of u; $\varphi(l)$: the temporal variable which records the strongest dependency between nodes.

Initialize *the SDR tree T as* E_k; // E_k *is added as root nodes of T*
Write *label 1 on* e_i // $e_i \in E_k$
While *SDR tree T does not yet span the BN*
  For *each frontier edge l in BN*
    Let *u be the labelled endpoint of edge l,*
    Let *v be the unlabeled endpoint of edge l (v is one parent of u),*
    Set $\varphi(l)=depend[u]*p(v|u)$;
  Let *l be a frontier edge for T that has the maximum* φ*-value;*
  Add *edge l (and vertex v) to tree T;*
  $depend[v]=\varphi(l)$;
  Write *label depend[v] on vertex v;*
Return *SDR tree T and its vertex labels;*

The result of the SDR algorithm is a spanning tree T. Every cause code $c_j \in C$ is labeled with $depend[c_j]=p(c_j|M_k,e_i)$, $e_i \in E_k$, M_k is the transition nodes between e_i and c_j in route $R_k \in R$.

3.2.3 Proof of the SDR algorithm

Now we prove the correctness of SDR algorithm. Algorithm 3.2.2 gives a way to identify the strongest route from effect e_i ($e_i \in E_k$) to c_j ($c_j \in C$). If the route $<e_i, u_1, u_2, \cdots, u_n, c_j>$ is the strongest dependency route, $<e_i, \delta_1, \delta_2, \ldots, \delta_m, c_j>$ is any route from e_i to c_j. Then

$$p(u_1|e_i)*p(u_2|u_1)*\ldots*p(c_j|u_n) \geq p(\delta_1|e_i)*p(\delta_2|\delta_1)*\ldots*p(c_j|\delta_m) \quad (5)$$

Define $weight(u,\pi(u)) = -lg(p(\pi(u)|u))$, Eq. (5) is transferred to:

$$weight(e_i,u_1) + weight(u_1,u_2) + \ldots + weight(u_n,c_j)$$
$$\leq weight(e_i,\delta_1) + weight(\delta_1,\delta_2) + \ldots + weight(\delta_m,c_j) \quad (6)$$

Lemma: when a vertex u is added to spanning tree T, $d[u] = weight(e_i,u) = -lg(depend[u])$.

$0<depend[\delta_j] \leq 1$, so $d[\delta_j] \geq 0$. (Note $depend[\delta_j] \neq 0$, or else exists empty link between δ_j *and its children.*)

Proof: suppose to the contrary that at some point the SDR algorithm first attempts to add a vertex u to T for which $d[u] \neq weight(e_i, u)$.

Consider the situation just prior to the insertion of u. See Figure 4. Consider the true strongest dependency route from e_i to u. Because $e_i \in T$, and $u \in V \backslash T$, at some point this route must first take a jump out of T. Let (x, y) be the edge taken by the path, where $x \in T$, and $y \in V \backslash T$. We now prove that $d[y] = weight(e_i, y)$.

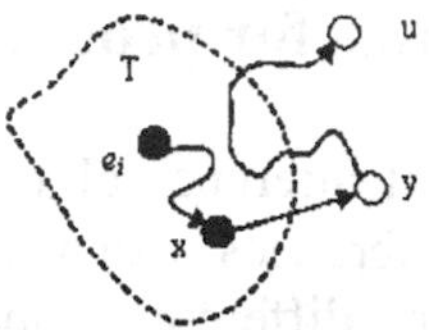

Figure 4. Proof of SDR algorithm

We have computed x, so $d[y] \le d[x] + weight(x, y)$ (7)

Since x was added to T earlier, by hypothesis, $d[x]=weight(e_i, x)$ (8)

Since $<e_i, \ldots, x, y>$ is sub-path of a strongest dependency route, by Eq.(8),

$$weight(e_i,y)= weight(e_i, x) + weight(x, y) = d[x] + weight(x, y) \quad (9)$$

By Eq. (7) and Eq. (9), we get $d[y] \le weight(e_i, y)$.
Hence $d[y] = weight(e_i, y)$.
Now note that since y appears midway on the route from e_i to u, and all subsequent edges are positive, we have $weight(e_i, y)<weight(e_i, u)$, and thus $d[y] = weight(e_i, y)< weight(e_i, u) \le d[u]$.
Thus y would have been added to T before u, in contradiction to our assumption that u is the next vertex to be added to T.
Since the calculation is correct for every effect node. It is also true that for multiple effect nodes in tracing the strongest dependency route. At the end of the algorithm, all vertices are in T, thus all dependency (weight) estimates are correct.

3.2.4 Complexity analysis of the SDR algorithm

To determine the complexity of SDR algorithm, we observe that every link (edge) in BN is only calculated one time, so the size of the links in BN is consistent with the complexity. It is known in a complete directed graph that the number of edges is $n(n-1)/2=(n^2-n)/2$, where n is the size of the nodes in the pruned spanning tree of BN. Normally a BN is an incomplete directed graph. So the calculation time of SDR is less than $(n^2-n)/2$. The complexity of SDR is $O(n^2)$.
According to the SDR algorithm, the strongest routes between effects and causes can be obtained by Depth-First search in the spanning tree.

Meanwhile the values of the labels in the cause nodes generate a dependency ranking of causes based on the effects. This dependency sequence is useful reference for fault diagnosis and related maintenance operations.

3.3 Related algorithms for probabilistic inference

There exist various types of inference algorithms For BNs. They can be classified into two types of inferences: exact inference[18,21] and approximate inference[19]. Each class offers different properties and works better on different classes of problems. This situation is true for almost all computational problems and probabilistic inference using general BNs has been shown to be NP-hard by Cooper[3].

Pearl's algorithm, the most popular inference algorithm in BNs, can not be extended easily to apply to acyclic multiply connected digraphs in general[23,14]. Another popular exact BN inference algorithm is the clique-tree algorithm[18]. It transforms a multiply connected network into a clique tree by clustering the triangulated moral graph of the underlying undirected graph first, and then performs message propagation over the clique tree. But it is difficult to record the internal nodes and the dependency routes between particular effect nodes and causes. In distributed systems management, the states of internal nodes and the key route, which connect the effects and causes, are important for management decisions. Moreover, the sequence of localization for potential faults is very useful for reference to systems managers. For system performance management, the identification of related key factors is also important. Few algorithms give satisfactory resolution for this case.

Compared to other algorithms, the SDR algorithm belongs into the class of exact inferences and it provides an efficient method to trace the strongest dependency routes from effects to causes and to track the dependency sequences of the causes. Moreover it can treat multiple connected networks modelled as DAGs.

4. CONCLUSIONS AND FUTURE WORK

In distributed systems of realistic size and complexity, managers have to live with unstable, uncertain and incomplete information. It is reasonable to use BNs to represent the knowledge about managed objects and their dependencies and apply probabilistic reasoning to determine the causes of failures or errors. Bayesian inference is a popular mechanism underlying probabilistic reasoning systems. The SDR algorithm introduced in this paper presents an efficient method to trace the causes of effects. This is useful for systems diagnosis and fault location and further can be used to improve performance management.

Most distributed systems, however, dynamically update their structures, topologies and their dependency relationships between management objects. Due to the special requirements in distributed systems and network management, the following topics need to be considered for future work:

First, we need to accommodate sustainable changes and maintain a healthy management system based on learning strategies that allows us to modify the cause-effect structure and also the dependencies between the nodes of a BNs correspondingly

Secondly, if the status of concerned managed objects is predictable, it is also possible to get a prospective view of the whole system. So an effective prediction strategy, which takes into account the dynamic changes in distributed systems, is important. This is related to discrete nonlinear time series analysis. Nonlinear regression theory[25] is useful to capture the trend of changes and give reasonable predictions of individual components and the whole system. Further the state of particular causal managed component can be predicted from future possible states of the effect components based on the inference rules in BNs. The future performance of the system can also be evaluated.

REFERENCES

1. K. Basye, T. Dean, J. S. Vitter. Coping with Uncertainty in Map Learning. *Machine Learning* 29(1): 65-88, 1997.
2. E. Charniak, R. P. Goldman. A Semantics for Probabilistic Quantifier-Free First-Order Languages, with Particular Application to Story Understanding. *Proceedings of IJCAI-89*, pp1074-1079, Morgan-Kaufmann, 1989.
3. G. Cooper. Computational complexity of probabilistic inference using Bayesian belief networks. *Artificial Intelligence*, 42:393-405, 1990.
4. C. Chao, D. Yang, A. Liu. A LAN fault diagnosis system. *Computer Communications*. Vol24, pp1439-1451, 2001.
5. R. H. Deng, A. A. Lazar, and W. Wang. A probabilistic Approach to Fault Diagnosis in Linear Lightwave Networks. *IEEE Journal on Selected Areas in Communications*, Vol. 11, no. 9, pp. 1438-1448, December 1993.
6. M. Frontini, J. Griffin, and S. Towers. A knowledge-based system for fault localization in wide area networks. *In Integrated Network Management*, II. Amsterdam: North-Holland, 519-530, 1990.
7. J. Gao, G. Kar, P. Kermani. Approaches to Building Self Healing Systems using Dependency Analysis. *IEEE/IFIP NOMS'04*, April, 2004.
8. M. Gupta, A. Neogi, M. K. Agarwal and G. Kar. Discovering Dynamic Dependencies in Enterprise Environments for Problem Determination. *DSOM 2003, LNCS 2867*, ISBN 3-540-20314-1, pp221-233, 2003,
9. C. S. Hood and C. Ji. Proactive network-fault detection. *IEEE Transactions on Reliability*, 46(3):333-341, September 1997.

10. D. Heckerman and M. P. Wellman. Bayesian networks. *Communications of the ACM*, 38(3):27-30, Mar. 1995.
11. C. A. Joseph, A. Sherzer, K. Muralidhar. Knowledge based fault management for OSI networks. *Proceedings of 3th International conference on Industrial and engineering applications of artificial intelligence and expert systems*, pp.61-69, June 1990.
12. A. Keller, U. Blumenthal, G. Kar. Classification and Computation of Dependencies for Distributed Management. *Proceedings of 5th IEEE Symposium on Computers and Communications*. Antibes-Juan-les-Pins, France, July 2000.
13. I. Katzela and M. Schwarz. Schemes for fault identification in communication networks. *IEEE Transactions on Networking*, 3(6): 733-764, 1995.
14. F. L. Koch, and C. B. Westphall. Decentralized network management using distributed artificial intelligence. *Journal of Network and Systems Management*, Vol. 9, No. 4, Dec. 2001.
15. S. Klinger , S. Yemini , Y. Yemini , D. Ohsie , S. Stolfo. A coding approach to event correlation. *Proceedings of the fourth international symposium on Integrated network management IV*, p.266-277, January 1995.
16. C. Lo, S. H. Chen, B. Lin. Coding-based schemes for fault identification in communication networks. *Int. Journal of Network Management* 10(3): 157-164, 2000.
17. L. Lewis. A case-based reasoning approach to the resolution of faults in communication networks. *In Integrated Network Management*, III, 671-682. Elsevier Science Publishers B.V., Amsterdam, 1993.
18. S. L. Lauritzen and D. J. Spiegelhalter. Local computations with probabilities on graphical structures and their application to expert systems. *Journal of the Royal Statistical Society*, Series B 50:157-224, 1988.
19. R. M. Neal, Probabilistic inference using Markov chain Monte Carlo methods. *Tech. Rep. CRG-TR93-1*, University of Toronto, Department of Computer Science, 1993.
20. D. Nikovski. Constructing Bayesian networks for medical diagnosis from incomplete and partially correct statistics. *IEEE Transactions on Knowledge and Data Engineering*, Vol. 12, No. 4, pp. 509 - 516, July 2000.
21. J. Pearl. A constraint-propagation approach to probabilistic reasoning. *Uncertainty in Artificial Intelligence*. North-Holland, Amsterdam, pp357-369, 1986.
22. The International Engineering Consortium. Highly available embedded computer platforms become reality. http://www.iec.org/online/tutorials/acrobat/ha_embed.pdf.
23. H. J. Suermondt and G. F. Cooper. Probabilistic inference in multiply connected belief network using loop cutsets. *International Journal of Approximate Reasoning*, vol. 4, 1990, pp. 283-306.
24. M. Steinder and A. S. Sethi. Non-deterministic diagnosis of end-to-end service failures in a multi-layer communication system. *In Proc. of ICCCN*, Scottsdale, AR, 2001. pp. 374–379.
25. A. S. Weigend, and N. A. Gershenfeld. Time Series Prediction. *Addison-Wesley*, 1994.
26. W. Wiegerinck, H.J. Kappen, E.W.M.T ter Braak, W.J.P.P ter Burg, M.J. Nijman. Approximate inference for medical diagnosis. *Pattern Recognition Letters*, 20:1231-1239, 1999.
27. C. Wang, M. Schwartz. Fault detection with multiple observers. *IEEE/ACM transactions on Networking*. 1993; Vol 1: pp48-55
28. S. A. Yemini, S. Kliger, E. Mozes, Y. Yemini, and D. Ohsie. High speed and robust event correlation. *IEEE Communications Magazine*, 34(5):82-90, 1996.

NEW PROTOCOL FOR GROUPING DATA USING ACTIVE NETWORK

A. Moreno[1], B. Curto[2] and V. Moreno[2]
Computer Science
University of Salamanca (Spain)
[1] *amoreno@usal.es*
[2] *control@abedul.usal.es*

Abstract: Active networks provide an ideal support for the incorporation of intelligent behaviors into networks. This ones make possible to introduce a more general functionality, that supports the dynamic modification of the behavior of switching networks. By means of this approach, users can dynamically insert code into nodes, thus adding new capabilities.

In this paper we propose a protocol to improve network services, implemented from the point of view of active networks. The specific purpose of the protocol is to reduce the network traffic, thus increasing the ratio of the volume of transmitted data to the number of frames that pass through the network. More precisely, we intend to merge data coming from the same node or from different nodes in intermediate points of the path toward the receiver, so that they arrive as closely grouped as possible. This would reduce the congestion in networks that support a great volume of traffic, like those that are used in industrial environments, which can also suffer from a low bandwidth.

For the implementation of this protocol we use ANTS (*Active Node Transport System*), a freeware tools for the construction of active networks developed by MIT(*Massachusetts Institute of Technology*). ANTS makes use of one of the most innovative and daring approaches for injecting programs into active nodes, by means of a mobile code called *capsules*.

Keywords: Active networks, network protocols, mobile code, data grouping.

INTRODUCTION

A current tendency in distributed control systems is the use of general purpose networks, like *Ethernet*, to communicate different processing elements.

This can produce poor network performance in segments that support high flows of data coming from multiple devices that are linked to it. This situation appears in industrial environments, particularly in data acquisition distributed systems.

A possible solution is the development of new protocols that allow us to improve network services by looking for alternatives that adapt to the specific necessities of each moment. However, the introduction of a new protocol is a slow and difficult task. This is due, first of all to the fact that it is necessary to carry out a standardization process in order to guarantee the system interoperability. Then, once a protocol has been accepted, its development presents many difficulties since there is no automatic form to include the new protocol with those already in existence. And one must add to this, the problem of compatibility with existing versions. This means one must take into account the times needed to solve the previous problems, thus making viable solutions to appear very slow in the evolution of network services. In this sense, it is foreseen that the deployment of IPv6[1] will take no less than approximately fifteen years to be completed.

Thus, *active networks* seem to be a perfect solution to tackle the aforementioned problems[2; 3]. The Active Network Program is a DARPA-sponsored research program born during the years 1994 and 1995. Users can program the network by loading their own programs for the realization of specific tasks. The main idea in active networks is to standardize a communication model, instead of individual communication protocols.

There exist two approaches for the construction of active networks: one of them is *discrete* and the other is *integrated.* In the first one, we separate the procedure to inject programs into active nodes from the processing of the packets that cross the node. Therefore, there will be a mechanism for the users to include their programs into active nodes, and there will be a mark on the packets to indicate which is the program that will process them. A different approach is to include the code with which packets will be processed inside the very same packets. This way, packets (called *capsules*) will contain both data and programs[3].

At the moment, one important research field in active networks[2] is centered in the creation of tools that facilitate their construction to a greater extent. One of them, ANTS (*Active Node Transport System*), has been developed by MIT and is freely distributable[4]. It was the first tool to use the capsules pattern. More recently, MIT has designed PAN(*Practical Active Networks*)[5], which is based on the ANTS architecture, but whose objective is to be more efficient than its predecessor, both in execution and in the code mobility system. Utah University is also working on a system based on ANTS, known as Janos[6], that improves node resource administration and tries to make a clearer and surer separation among the different user programs that will be executed in one node. The University of Pennsylvania in collaboration with Bell Switchware [7], which has taken the discrete approach as its focus, makes special emphasis on node security. Georgia Tech is creating an operating system for nodes, called Bowman[8], and an execution environment called CANE (*Composable Active Network Elements*)[9], also with a discrete approach.

Our work, based on the integrated approach, tries to find a definition for protocols that intend to solve the overload problems or network congestion that appears in industrial environments, typically in data acquisition distributed systems. The proposed solution contemplates the grouping of data generated by different sources that head toward a common point. We try to increase the ratio of useful information to the overhead of these protocols. This leads to a reduction of the traffic that goes through the network, thus producing a better response time. One must also take into account the possibility that the time necessary to carry out a grouping could be detrimental when trying to have data arrive without undue delay. As we shall see, we have thought about the introduction of a timeout procedure that guarantees as much as possible that packets are sent in a reasonable interval of time.

The remainder of this paper is organized as follows: in second and third sections we describe the motivation and the goals that have guided the realization of this work. In the next section we detail the specifications of the grouping protocol. In the fifth section we describe the implementation details, and afterward we show a study case. The seventh and last section closes the work and we present our main conclusions.

1. MOTIVATION

Let us consider a communication scheme in which a group of sender nodes distributed in the network must transmit a large amount of data that are being produced, and a receiving entity must operate taking those data as input. This situation can be found in a distributed data acquisition system (Figure 1), where several nodes capture data coming from different sensors. These nodes send the data they have picked up to a receiver, all within some established time constraint.

If each node sends its data one by one and data arrive separately, coming from different nodes, the network will experience an unnecessary overload. This problem is specially severe in the segment connected to the receiver, since it collects all incoming traffic.

If we use this approach, each data is encapsulated inside a frame when crossing the network, and we can have the paradoxical situation that the information of the protocol can actually take up more space than data being sent.

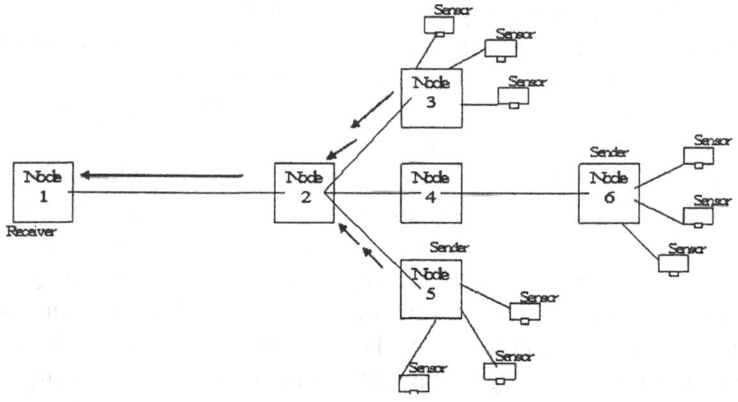

Figure 1. Data transmission without group

A possible solution to the problem would be that senders themselves grouped their data before emitting them. However, it can be interesting not to overload to the originators with grouping tasks. Then this solution would not be feasible. In any case, this scheme does not solve one of the problems, since data coming from different nodes will still reach the receiver separately.

2. GOALS

The main goal of this work is to propose a solution to improve the ratio between the amount of transmitted information and the protocol-related overhead in environments with heavy traffic. In this way we try to enhance the performance of low-bandwidth links, thus enhancing their response time.

The solution is based on an active network system that makes it possible to generate new network protocols that adapt to the specifications of different services. The solution of the problem we have described will be a protocol in which data are grouped (Figure 2) as they traverse the network toward their target. The points where data can be grouped would be the active nodes, in such a way that data can reach the receivers in as close a group as possible.

The new data frame maintains the same size, with a bigger amount of useful information. The grouping of data sent from the same node would be no longer be the responsibility of sender, but rather it could be delegated to the closest node in the network. The protocol could be also built in such a way that the sender could configure the emission, thus indicating the node that must do the grouping, its size, the maximum wait time for data grouping, etc, all this in an attempt to enhance the time response of the network.

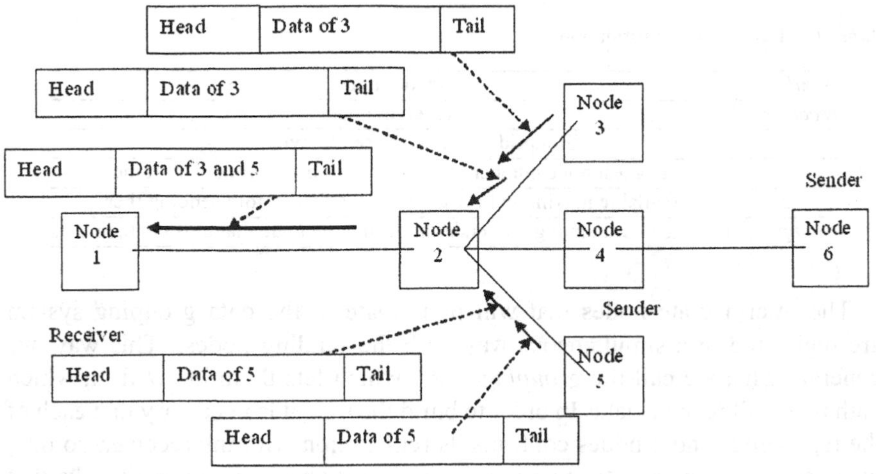

Figure 2. Data transmission with group

3. THE PROPOSED GROUPING PROTOCOL

The data grouping protocol makes it possible to join data coming from one or different sources in best possible way. In order to carry out this task, it performs three important functions:

- Configuration of the grouping system. The receiving node has the capability to configure the system so that an efficient grouping of data is done.
- Grouping of data coming from the same originator. This task is that of building groups with data that come from the same node. It can be the responsibility of the same sending node, or that of any neighbor.
- Grouping of data coming from different senders. The nodes that receive data coming from different sources should encapsulate them all in one frame in an orderly way. This task will be carried out by nodes placed at the intersections.

Configuration of the grouping system

First, the receiver must select the nodes from which it requires data. All these senders will be part of a *merge data group*. The receiver will send a set of parameters (Table 1) to each data source. This process shall be called *registration*.

Table 1. Registration information

sender	Sender node
receiver	Receiver node
group	Group-id to which the sender will belong
units	Highest node number of data that can be grouped in a frame
time	Available maximum time to group the data before sending them
jump	Distance limit among the sender node and the node that clusters their data

The intermediate nodes that will participate in the data grouping system are registered in a simultaneous way with the sending nodes. This way, we generate what we call the *groupings tree*, which lets them know from which path they will receive data. In order to build the tree, it is necessary that each of the registered sender nodes confirms its registration with the receiver, so that, when a node receives confirmation, it can know about the network links that will belong to the tree branch and store them in the *registration array* (RM).

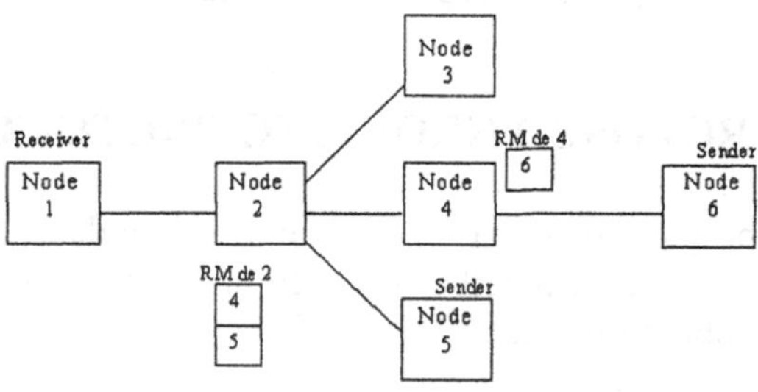

Figure 3. Registration operation

In order to show the registration process we refer to the network topology that we have shown previously. Node 1 sends a registration frame to nodes 5 and 6, and with their answers the tree is built (Figure 3). Nodes 2 and 4 have generated a *registration array* (RM) that serves to locate the tree branches. Node 3 is excluded because it has not been registered.

Grouping data from the same sender node

Nodes that have been registered as data sources send their data in an individual way toward the node where they will group. This will be configured by the *jump* parameter or in an mandatory way a intersection node (a node whose *registration array* has more than one element). Frames with individual data are named *unit capsules* and the place where the units will group shall be called *grouping point.* At this point, the unit capsule itinerary is finished and this will be the place where data are stored in the cache together with the data from other unit capsule that come from the same source. The first capsule of the grouping should start a timer whose timeout value is specified in the configuration.

When the unit capsule that completes the group arrives, that is to say, when the limit of elements (*units*) fixed in the configuration phase is reached, the whole group is stored in a *grouping capsule* (described in the next section), and this capsule is transmitted to the receiver. The cache is cleared and the timer is stopped. If time runs out before a group is completed, the *grouping capsule* is created with whatever elements are available at the moment; then the frame is marked as urgent and sent, and the cache is cleared.

If we had configured in the previous example the *jump* parameter with a value greater than two, then the unit capsule would cluster in the intersection node 2, since they are not able to group beyond an intersection (Figure 4a). If the value of *jump* is 1, then the node 6 capsules would cluster in 4 and those of the node 5 would cluster in 2. They would only leave the *grouping point* when the time runs out, or when the group is completed (Figure 4b).

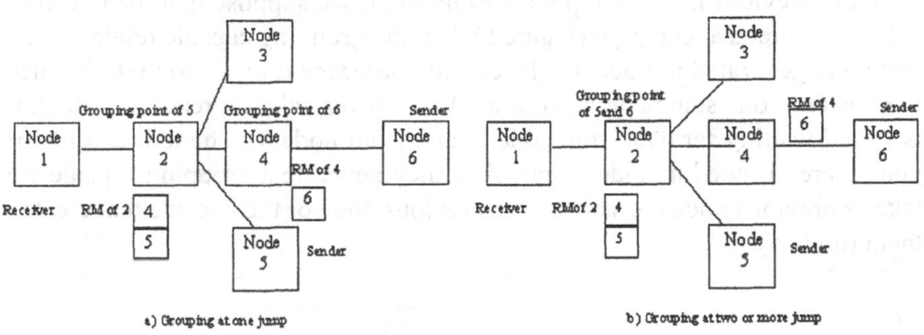

Figure 4. Grouping

Grouping data from different senders

The second phase of the protocol is the grouping of data that come from different senders. In this phase, data are no longer grouped according to source, but depending on their tree branches. It is necessary to ensure that no information about the original data sender is lost.

As in the previous case, when the first datum to be grouped arrives, a timer is started. Data are not sent to the following node unless one of these four cases happens:

- Data of all the tree branches have been stored, that is to say, a complete grouping has been made. In that case a group is formed with all the data and they are sent to the next node of the tree. Afterward, the timer is stopped.

- Data arrive from some of the branches and there are data remaining in the area corresponding to that branch. A group is formed with the data, and they are sent to the next node. The timer is stopped. The data that have just arrived are stored in the proper place. Since they are the first data of the next *grouping capsule*, the timer is started again.

- Data have arrived marked as urgent. In that case, data are sent as urgent data and the timer is stopped.

- Time has run out. A group is formed and it is sent as urgent.

In the previous network topology example, if we suppose that the unit capsule are grouped at one jump(Figure 5) then the grouping capsule related to the node 6 is generated in node 4. The capsule proceeds directly to node 2, since there is only one source branch (node 4). However, when it reaches node 2, it is stored waiting for data from node 5, and from node 4. The unit capsule of node 5 are grouped in node 2, but when they generate a grouping capsule we take as previous node the same as the previous node of the capsule that created them (node 5).

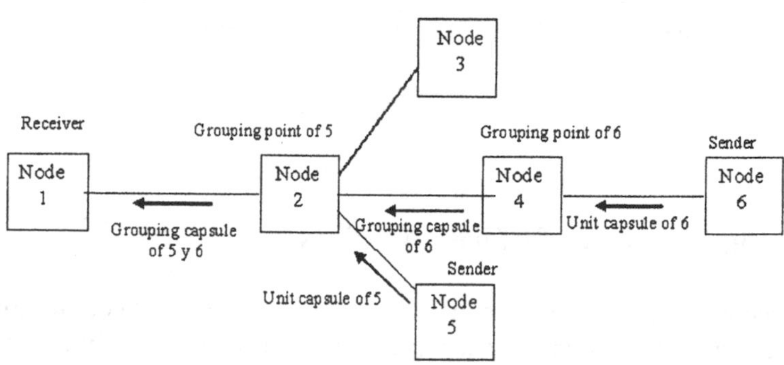

Figure 5. Final scheme of the grouping

In the previous description we have assumed that only one group had been generated. However, the protocol supports the generation of various groups in a direct way.

Table 2. Classes and methods of ANTS

Class	Methods
Application	send, receive, node
Capsule	evaluate, length, register, serialize, deserialize
Node	address, get, put, routerfornode, delivertoapp
Protocol	startProtocolDefn, startGroupDefn, addCapsule, endProtocolDefn, endGroupDefn

4. IMPLEMENTATION

In order to implement the protocol[10], we have selected ANTS as a tool for the construction of active networks, because it is written in a general-purpose language (JAVA), as opposed to others that define their own language. Thus we achieve a double goal: the portability of the applications and the possibility of modifying some aspects of the tool, since we can access the source code.

ANTS

ANTS provides a programming model that allows to define protocols, a code distribution system to load new protocols into the network and an environment to execute them. It is based on the capsule-oriented network model, in which the frame can take a reference to some executable code in each node, so that if the node does not have code it can make a request to previous nodes.

In this model, users adapt the network to their necessities by means of the definition of protocols based on ANTS classes (Table 2). Thus, to develop your own protocol, an derivation of the abstract class *Protocol* is created. It is necessary to identify the protocol data units (PDU) that will be inserted in the network and the different processing routines. Each type of PDU and its routines are specified in a derivation of the abstract class *Capsule*. A new application, an entity that makes use of the network to send and receive capsules, must be developed by means of class derived from abstract class *Application*. Active nodes constitute the environment in which a capsule is executed: they receive it, load it and they execute its routines, and they plan its transmission. Thus, an instance of the *Node* class represents the environment of local execution of ANTS.

The protocol

As we have mentioned, in order to use ANTS to codify the protocol it is necessary to generate the classes that will implement the protocol, the capsules and the applications that use them.

Hence, we have defined a class called *FussionProtocol* that registers the different kinds of capsules that compose the protocol: *FussionRegCap*, *FussionUnitCap* y *FussionCapsule*. The first one defines both the registration of sender nodes and the registration acknowledgment. *FussionUnitCap* transports the data toward the *Grouping Points* and it also implements the algorithm that groups individual data. The capsule *FusionCapsule* will be received by the reception node and it will contain grouped data. It will be created by a *FussionUnitCap* and it implements the algorithm to fuse data groups.

Applications constitute the programs that are going to use an active network protocol. They inherit the properties from the *Application* class and in our implementation three classes have been defined: *ReceptionApp*, *EmissionApp* and *MFussionApp*. *ReceptionApp* allows us to register emitter applications, in such a way that the group of fusion data and its parameters can be configured by sending a capsule *FussionRegCap*. It also performs the reception of fused data that arrive from the *FussionCapsule*. The class *EmissionApp* represents the data emitter that makes use of protocol *FussionProtocol*; it receives registrations from a receptor and it sends automatically an acknowledge.

Finally, the *MFusionApp* is executed at every node of the network to monitor their activities. It reads data from the cache when it receives a *FussionUnitCap* capsule or a *FussionRegCap* capsule, and it shows the *GoupingTree* for the node before the last actualization.

The tree branches are filled at the actual moment and the senders nodes that groups their data at this node.

The protocol requires a timing system to control data arrivals. The ANTS nodes do not provide any timing service. Nevertheless, ANTS has a extension system of functionalities that in this work has allowed us to implement a module that offers this timing service. Timing is done by means of a *thread* that will wait for the indicated time period, and then will send a finish signal. In order to perform this task, it has been necessary to create a class named *WaitTime*, that performs a timing task, and a interface, *TimInt*, that must be implemented by every class that need to be timed.

5. CASE OF STUDY

In order to show the effectivity and the performance of the grouping protocol we have built, we shall test with a network topology like the one shown in Figure 6.

The receiving application that is executed at node 20.20.20.1 registers the following emitter nodes at group 0: 20.20.20.3, 20.20.20.4 and 20.20.20.5. This means that the grouping task will be performed at the previous node

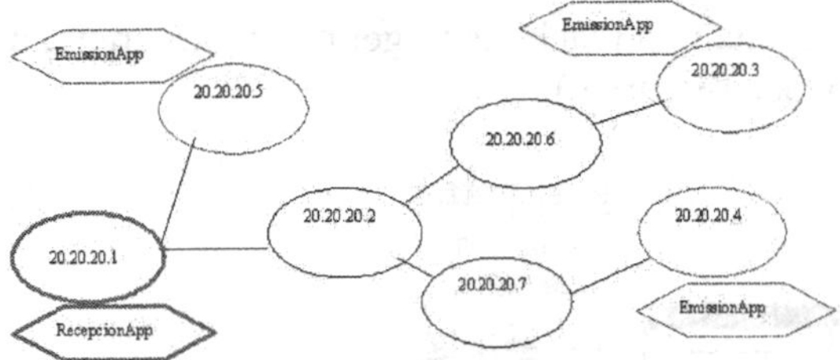

Figure 6. Active Network Topology

(*Jump*=1), where three data will be grouped (*Unit*=3) and that the waiting time will be 10 seconds (*Time*=10).

A register capsule of type *FusionRegCap* is sent to every node and their applications reply with a *CapRegFusion* as an acknowledgment (Figure 7).

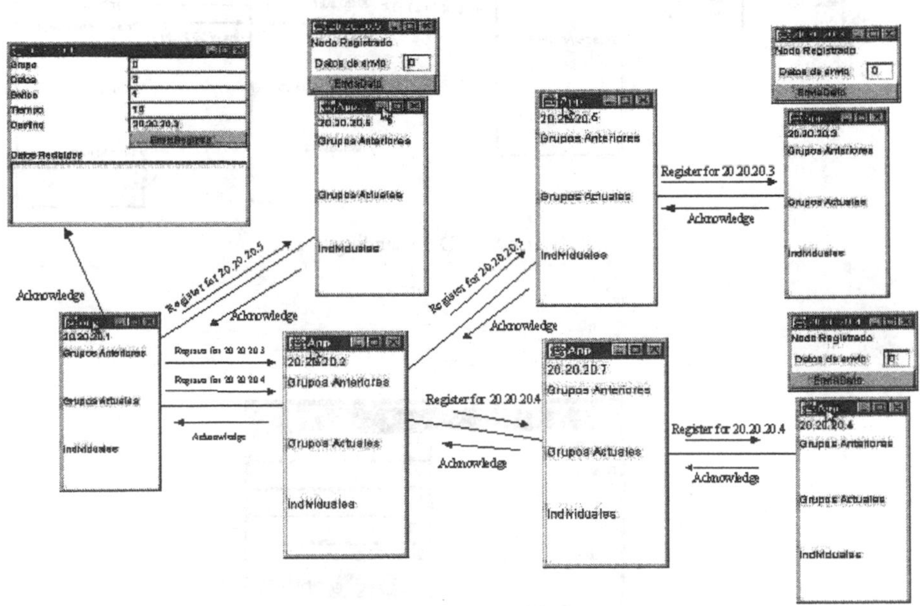

Figure 7. Register Operation

Once all the nodes are registered, data will be grouped by following the schema shown in Figure 8, where one can see that nodes 20.20.20.3 and 20.20.20.4 group their data at 20.20.20.6 and 20.20.20.7 at one jump as it has been indicated at the configuration. The *grouping point* will be the tree intersection (20.20.20.2). Since node 20.20.20.5 is located at distance of one jump from

20.20.20.1, their data will reach the target before grouping, hence data will be received ungrouped (Figure 9).

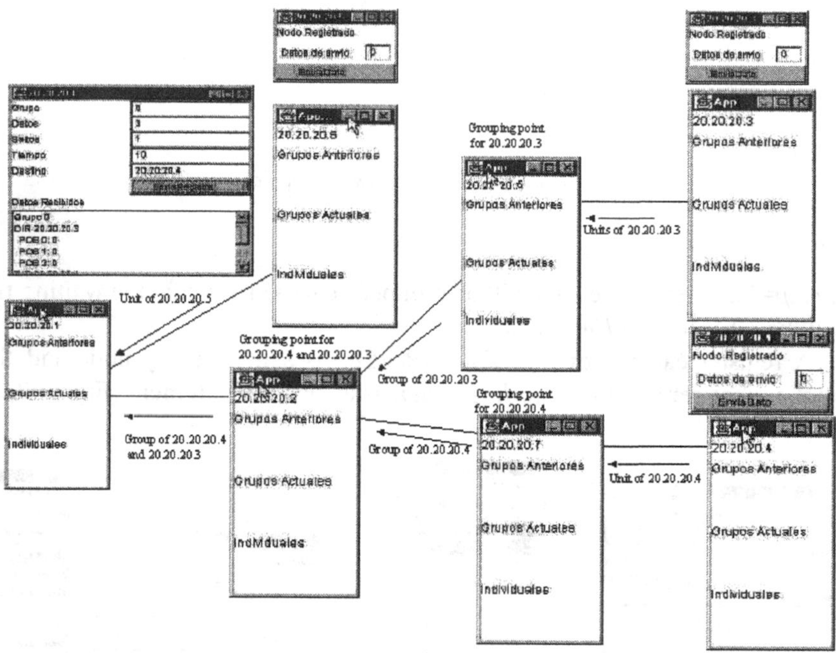

Figure 8. Data sending

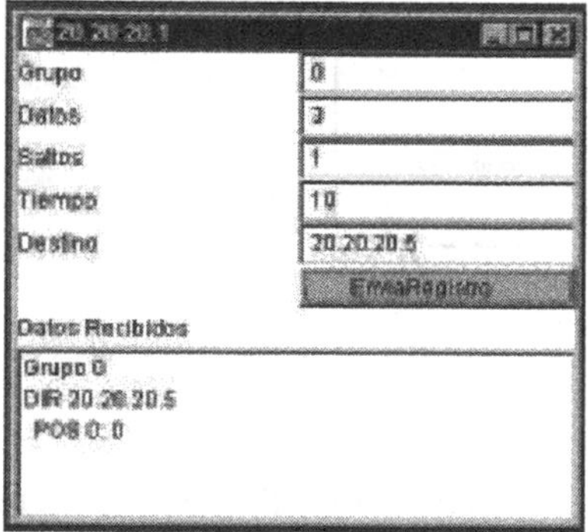

Figure 9. Data from 20.20.20.5

6. CONCLUSIONS

This paper describes an attempt to improve network services by means of active networks. We propose a way to reduce network traffic for possibly low-bandwidth network, thus optimizing their performance and helping to maintain a reasonable response time.

To reach this goal, we have define a protocol that groups data coming from both single and multiple sources. In order to implement the protocol, we have used a tool to construct active networks that is based on the ANTS capsule model. Several applications that use this protocol have been developed, as well as an application used to monitor intermediate nodes. Finally, in order to include timing procedures, it has been necessary to develop an timer extension that can be used in any other research field related to Active Networks.

REFERENCES

R. Gilligan and E. Nordmark. *Transition Mechanisms for IPv6 Hosts and Routers.* Internet Draft, march 1995

D. L. Tennenhouse, J. M. Smith, W. D. Sincoskie, D. J. Wetherall and G. J. Minden. *A Survey of Active Network Research.* IEEE Communications Magazine, Vol. 35, No. 1, january 1997, pp 80-86

D. L. Tennenhouse and D. J. Wetherall. *Towards an Active Network Architecture.* Multimedia Computing and Networking (MMCN 96),january 1996, San Jose, CA: SPIE. A revision of this article appeared in Computer Communication Review, Vol. 26, No. 2 (april 1996). http://www.tns.lcs.mit.edu

D. J. Wetherall, J. Guttag and D. L. Tennenhouse. *ANTS: A Toolkit for Building and Dynamically Deploying Network Protocols.* IEEE OPENARCH'98, San Francisco, CA, april 1998

E. L. Nygren, S. J. Graland and M. F. Kaashoek. *PAN: A High-Performance Active Supporting Multiple Mobile Code Systems.* Proceedings of IEEE OPENARCH'99, march 1999, pp 78-89

P. Tullmann, M. Hibler, and J. Lepreau. *Janos: A Java-oriented OS for Active Networks.* Appears in IEEE Journal on Selected Areas of Communication. Volume 19, Number 3, March 2001.

J. M. Smith, D. J. Farber, C.A. Gunter and S. M. Nettles, D.C. Feldmeier, W. D. Sincoskie. *SwitchWare: Accelerating Network Evolution (White Paper).* 1996

S. Merugu, S. Bhattacharjee, E. W. Zegura and K. L. Calvert. *Bowman: A Node OS for Active Networks.* IFOCOMM,2000

E. Zegura, K.Calvert. *Composable Active Network Elements:Lessons Learned.* ANTETS PI Meeting, may 2000

I. Blasco. *Aplicabilidad de las redes activas a la mejora de los servicios de red.* Technical Report, University of Salamanca, june, 2000

Part Six: Performance Evaluation

AN ALGEBRAIC MODEL OF AN ADAPTIVE EXTENSION OF DIFFSERV FOR MANETs

Osman Salem and Abdelmalek Benzekri
Institut de recherche en informatique de Toulouse, Université Paul Sabatier, 118 Route de Narbonne - 31062 Toulouse Cedex 04 – France; E-mail: {benzekri, osman}@irit.fr

Abstract: In this paper, we propose an extension to DiffServ QoS architecture in order to enhance its performance and its flexibility when used in MANETs and its adaptation to the characteristics of these networks. Then we present a formal model of our proposed extension using stochastic process algebras in order to verify the correctness and the efficiency of the proposed extension.

Key words: QoS; MANET; DiffServ; Admission Control; Stochastic Process Algebras; Markov chain; Performance Evaluation.

1. INTRODUCTION

A Mobile Ad Hoc Networks (MANETs)[1] is an autonomous system of mobile hosts connected by wireless links and forming a temporary network without any pre-existing infrastructure. Each host is directly connected to hosts that are within its range of transmission and reception, and it is free to move randomly in and out of any other host's range. Communication between hosts that are not located in the same covering range can be realized by establishing a multi-hop route through intermediates hosts that act as routers when they forward data for others.

A lot of research has been done in routing area, and today routing protocols are mature enough to face frequently changing network topology. A quick look at intended applications area for MANET shows the need to integrate real time multimedia traffic with data traffic. Many QoS aware routing protocols and models that claim to provide a partial (or complete) solution to QoS routing problems have appeared, e.g. QoS-AODV[2], MP-DSR[3], ASAP[4], CEDAR[5].

In the current days, the Integrated services (IntServ)[6] and the differentiated services (DiffServ)[7] are the two principal architectures proposed to provide QoS in wired network. IntServ suffers from well-known scalability problem caused by massive storage cost at routers when keeping flows' state information. The migration of this architecture to MANETs is judged very heavy because of network's constraint in term of storage capacity, contention of RSVP's out-band signaling packets with data packets and the two ways reservation mechanism of RSVP. The two-way reservation mechanism is inadequate and slow to adapt with the highly dynamic nature of hosts in MANETs, which leads to frequently change in the paths and thus rendering existing reserved resources unusable for some amount of time, in addition to excessive control overhead when path is broken.

DiffServ[7] on the other hand classify flows into several classes whose packets are treated differently in forwarding routers. It was designed to overcome the scalability drawbacks of IntServ. However, the notion of three kinds of nodes (ingress, interior, and egress nodes) and the SLA[7] (service level agreement) do not exist in MANETs. In DiffServ, the edge router is responsible to mark DSCP for each flow according to user profile listed in the SLA that includes the whole or partial traffic conditioning rules used to mark or re-mark traffic streams, discard or shape packets according to the traffic characteristics such as rate, and burst size. To alleviate these problems in MANETs, each host must be able to act as an edge and core router, and each host must be responsible for marking its traffic with the appropriate DSCP according to application's requirements. This means that every host plays the role of ingress router if it is transmitting data, a core router if it is forwarding data and an egress router if it is receiving data.

Several QoS schemes that are either a modification of the conventional IntServ and DiffServ based models have proposed for MANETs, like INSIGNA[8], FQMM[9], and SWAN[10]. SWAN (Service differentiation in stateless Wireless Ad Hoc Networks) differentiates traffic into 2 classes: high priority for real time UDP traffic and low priority for best effort UDP/TCP traffic. SWAN architecture (presented in figure 1) uses traffic differentiation in conjunction with a source based admission control mechanism to provide soft QoS assurances for real time traffic. However, this model differentiates traffic into two classes only; as it serves all real time traffics with equal priority, also it drops real time traffic with equal probability when congestion occurs, regardless their requirement in term of bandwidth and delay.

Many priority levels are required to differentiate important flows from others like in DiffServ, because it supports many classes of traffic. In addition, DiffServ relays on TCP rate control to reduce congestion and it is interesting enough to deserve a closer investigation. We think that the adoption of DiffServ by MANETs is better than IntServ due to characteristics of

MANETs, which can not guarantee any tight bounds on performance measures. It is useless to make a reservation of resources to guarantee a worst case delay for high priority flows in MANETs, because we can not guarantee neither the lifetime of the link or the delay on the link. Consequently, IntServ and reservation based approaches are not a favorite candidate to provide QoS in ad hoc networks. In contrast, DiffServ overcomes these disadvantages; it does not define any absolute guarantee and only proposes differentiations in scheduling when forwarding flows. In addition, extending DiffServ to Ad Hoc networks will provide consistent end-to-end QoS behavior when relaying flows between heterogeneous networks.

However, the differentiated services architecture does not define any scheme for taking corrective action when congestion occurs, and this is why a pure static DiffServ model is not suitable for ad hoc networks. Therefore, it is imperative to use some kind of feedback as a measure of the conditions of the network to dynamically regulate the traffic of the network when using this technology.

Our approach to provide QoS in MANETs is to extend DiffServ by adopting some positive aspects of SWAN and by adding new component to make DiffServ flexible and adaptive with bandwidth variation. The qualitative and quantitative study of our scheme is conducted on a formal description, expressed through stochastic extensions of process algebras that allow us to formally describe both functional and performance aspects.

The rest of this paper is organized as follows. Section 2 gives a brief overview of the SWAN architecture. Our proposed scheme is described in detail in section 3. In section 4 we present a formal specification of our proposed extension to DiffServ using stochastic process algebra and we analyze both qualitative and quantitative aspects of the described model. Finally, in section 5 we conclude the paper with a summary of the results and future direction.

2. SWAN MODEL

The SWAN model presented in Ahn and Campbell,[10] differentiates flows into real time and best effort. This model works as follows: if a real time application at a node wants to communicate with another, it probes the network to obtain the minimal available bandwidth on the path, assuming the QoS aware routing protocol has found a path.

SWAN is composed from a classifier and a shaper located between the IP and the MAC layer as appear in figure 1. Aware that the bandwidth probe is sent at the beginning, so topology changing due to mobility and the variation of channel load conditions, SWAN uses rate control and explicit congestion notification mechanisms to adapt to these variations. It uses rate control for shaping or delaying the UDP/TCP best effort traffics and marks the ECN bits in the packet header to dynamically regulate admitted real-time traffic when temporary overload occurs by asking one or more real time source(s) to find another path. The rate control operation of best effort traffic is performed at every mobile host in a distributed manner, where this rate will vary based on feedback information from MAC layer to maintain delay and bandwidth bounds for real time traffic.

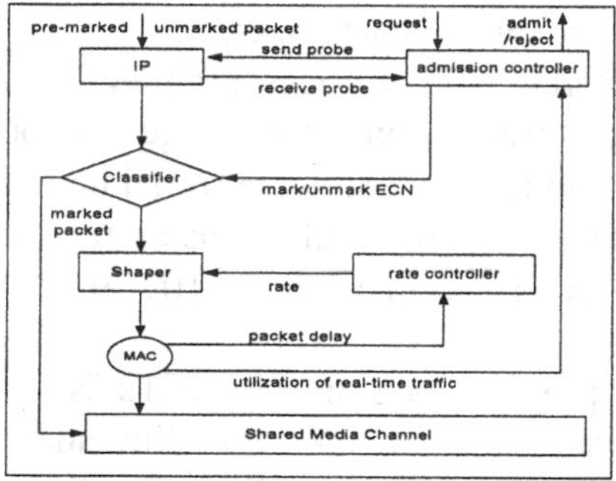

Figure 1. SWAN ARCHITECTURE

3. PROPOSED ARCHITECTURE

Our proposed extension to DiffServ QoS architecture is illustrated in figure 2. This scheme works as follows: applications notify their requirements and send their traffics to DiffServ component, which is responsible for marking and conditioning received packets from high level according to application's requirements. If received packets must be marked with one of the highest priority level, the marker component sends a request to the Call Admission Control (CAC) component. This request contains the amount of bandwidth required by this traffic to work properly. The CAC component verifies if the required QoS can be provided by the network, and this can be done by sending a probe request to routing protocol, which will collect the minimum end-to-end bandwidth (bottleneck bandwidth) along all existing paths from this source toward the specified destination. If the required bandwidth can not be provided by the network, CAC component reject the request and notify the application, which must decide to defer or to send with a medium priority profile.

In MANETs, even though admission control is performed to guarantee enough available bandwidth before accepting high priority flows, the network can still experience congestion due to mobility or connectivity changes. We do not attack the problem of broken link due to mobility, and we rely on the underlying routing protocol to detect and to resolve this problem by finding an alternate path toward the destination. Nevertheless, we try to resolve the problem of invading the channel by others nodes forwarding medium and low priority flows and causing degradation to the QoS of accepted highest priority flows because of the limited bandwidth of this shared media. This is why the congestion control component is extremely important in our model. It monitors the network bandwidth utilization continuously and signal network congestion to the rectification component.

The proposed scheme has six basic components, namely bandwidth estimator, routing protocol, call admission control, DiffServ, congestion control and rectification component.

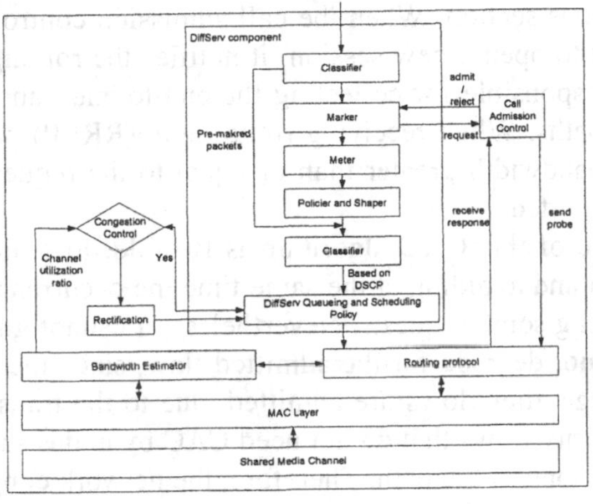

Figure 2. Proposed scheme

3.1 Bandwidth estimator component

The bandwidth estimator will periodically calculate the available bandwidth at each node. This value will be used by QoS aware routing protocol and the call admission control component to determine if flows can be admitted with one of the highest priority classes.

In MANETs, the communication media is shared among neighboring nodes, so to determine available bandwidth capacity in each node, we must take into account the transmissions of all its neighbors. We use the status of the shared channel as our base to calculate the channel utilization rate and the available bandwidth by each node. Using a shared channel allows mobile hosts to listen to packets sent within its radio transmission range and to calculate resource availability. The channel utilization ratio is defined as the fraction of time within which a node is sensing the channel as being utilized.

A node can calculate the utilization ratio of the channel by adding the time when it pumps data in the channel with the time that it finds the channel busy during period T, as R = (channel busy period/T) Then we use the RTT method to estimate the average utilization ratio through the following formula: $Average\_Ratio_t = \alpha \times Average\_Ratio_{t-1} + (1-\alpha) \times measured\_Ratio_t$, where α is a constant belong to [0,1]. After estimating the channel utilization at time t, we are able to calculate the available bandwidth of a node at time t as $BW_t = W \times (1-R_t)$, where W is the raw channel bandwidth (2 Mbps for a standard IEEE 802.11 radio).

3.2 Call admission control component

The bandwidth information is already available at each node like explained in previous section. When the call admission control component receives a request to open a new session, it notifies the routing protocol component that is responsible for collecting the end-to-end bandwidth available along existing paths. When receiving route reply (RREP) packet for a path with available bandwidth greater than or equal to the requested value, then the session is admitted.

The objective of this CAC algorithm is twofold: to grant highest bandwidth utilization and avoiding at the same time the occurrence of congestion events by rejecting some requests. Nevertheless, it cannot guarantee that the bandwidth will not degrade for the admitted flow, since the available bandwidth may change after flows are admitted, due to the transmission of medium or low priority flows that do not need CAC by nodes sharing the media with any node belong to the path. Therefore, the network congestion can still occur and can be detected by the congestion control component.

3.3 Congestion control component

The role of the congestion control component is to detect network congestion, which is very simple to be detected in wired networks by verifying if the queue is overflow or even begin to build up. However, detecting congestion in MANETs is very difficult, because the queue length is no longer a valid indication of congestion. The MAC layer usually retries to transmit a packet for a limited number of times (e.g. default retry time of the IEEE 802.11b DCF is 7) before dropping this packet. Therefore, the queue may not have yet build up at the early stage of congestion. In our scheme, the congestion control component uses the channel utilization ratio provided by the bandwidth estimator component to detect congestion in MANETs. This component contains a predefined threshold value for channel utilization ratio, and compares this value with that provided by the estimator. If the estimated value is larger than this threshold, it assumes that due to congestion and signals congestion occurrence to the rectification component, only if this node contains a high priority packet to forward.

Some neighbors of this congested node may be carrying medium or low priority traffic that reduces the available bandwidth and cause severe performance degradation to the high priority flows forwarded by this node. As they are in the direct reception range, these nodes can be directly informed by the rectification component.

3.4 Rectification component

The rectification component reacts to the detection of bandwidth degradation when there are some accepted flows with high priority crossing this congested node. Generally, when channel utilization ratio exceeds the predefined value, rectification component receives a notification message from the congestion control component and it broadcasts a stop-sending-request with TTL equal to 1. All nodes within direct reception range will receive this message and react accordingly if they forward medium or low priority packet.

Obviously, if all nodes receiving this request stop sending, there will be an under utilization of the channel and this strategy will prohibit these flows from using existing path even if there is enough bandwidth. To fight this problem, when a node receives stop-sending-request message, it does not immediately stop sending medium and low priority packets. Rather, each node chooses an exponential random amount of time from a pre-defined interval and triggers a counter with this value and begins to count down. If the counter reaches zero then this node drops all medium and low priority packets that are in the queue of this node and notify its neighbors that it stops sending. Also, this node (if it is not the source of these flows) mimics a broken link for these traffic by sending route error packet (RERR) to their corresponding sources, and it stops forwarding any new RREQ for a predefined amount of time to make this flows re-routed and dispersed away of the congested area. When neighbor nodes in the same range receive this notification (which can be embed in RERR), they cancel their counter by stops counting down.

3.5 DiffServ component

DiffServ divides traffics into many classes by marking a field in the IP packet header, called the Differentiated Services Code Point (DSCP) field. Its value depends on the traffic profile indicated by the application. The sender of a flow marks all outgoing packet in DS field of their IP headers and intermediate nodes forward these classes with different priorities with respect to the DSCP field content.

DiffServ supports 3 types of services: premium service, assured forwarding and best effort. Premium service or expedited forwarding (EF) class is used for loss and delay sensitive applications such as voice over IP (VoIP). Assured forwarding classes offer a lower priority service from the previous one (EF), and each class of this four is subdivided into three subclasses (3 priority level) [14]. Finally best effort traffic does not require any QoS guarantee. The DiffServ model is composed from a classifier and a traffic conditioner like appears in left part of figure 3. The traffic conditioner presented in right part of figure 3 is composed from a meter, shaper/dropper, maker, separate physical queues for each class of traffic and a scheduler to schedule packets out of queues. DiffServ enqueues flows in separate buffer where packets with high priority will be serviced out of the buffer before than packets marked with low and medium priority. In addition, low priority packets may be selectively dropped prior to dropping packets of medium and high priority when congestion occurs.

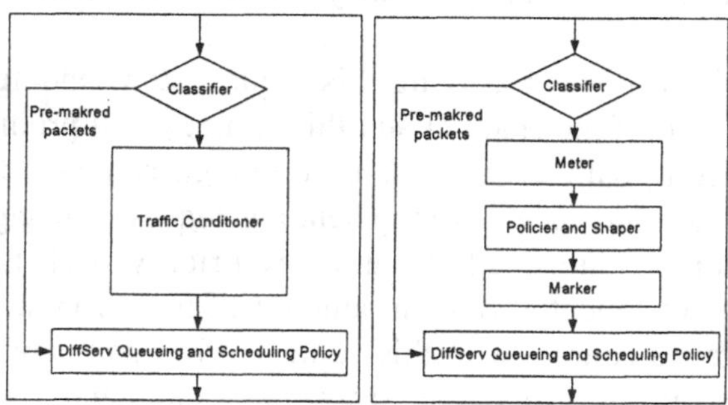

Figure 3. DiffServ router.

3.6 Routing protocol component

The first essential task of routing protocol is to find a suitable path through the network between the source and destination, and to record the minimum available bandwidth for CAC component. Our scheme does not rely on any specific reactive routing protocol and it may use any one able to collect the bottleneck value of the bandwidth along existing path.

4. FORMAL SPECIFICATION

One approach to ensure the correctness of our scheme is to use a formal model that integrates the theory needed for the verification of qualitative requirements together with the expressivity needed to analyze performance aspects. Moreover, a formal integration of these aspects in the initial phase of the system design allows the study of the potential dependencies that may occur between them.

Stochastic process algebras (SPA)[11] are formal specification languages used to describe the behavior of a system in order to derive its functional and performance properties. They are suitable for automatic analysis and verification of the behaviour of systems. Several SPA languages have appeared in the literature; these include PEPA (Performance Evaluation Process Algebra[12]), TIPP (TImed Processes and Performability evaluation[13]) and EMPA (Extended Markovian Process Algebra[14]). They are a high level modeling formalism able to derive the state space structure that consists of all states that a system model can reach. Given the state space of the system, there are many practical algorithms for answering some verification and analysis questions.

These three languages propose the same approach to quantitative analysis, where a random variable is associated with each action to represent its duration. In this paper, we will use EMPA language, which is supported by a tool called TwoTowers. EMPA is inspired from PEPA and TIPP languages and it is considered like an extension to existing languages. The syntax of EMPA can be summarized by the following expression:

$$P = 0 \mid <a, \lambda>.P \mid <a, \infty_{L,W}>.P \mid <a, *>.P \mid P/L \mid P[\varphi] \mid P + Q \mid P \parallel_{s} Q \mid A$$

In the rest of this paper, we suppose that reader is familiar with process algebra. Interested reader must refer to [14] for further details.

We exploit compositionality and abstraction features to specify each component apart by describing tasks that must be accomplished by this component. In addition, for the sake of simplicity and to avoid well-known state space explosion problem, we model only three queues to hold high, medium and low priority flows. Routing protocol component like DSR is modeled in a previous work[15], and a detailed specification of DiffServ router can be found in[16].

The complete model is composed from seven components. Its algebraic specification and the specification of each component are given below:

Adaptive_DiffServ □ (DIFFSERV $\parallel_{COM}$ CAC) $\parallel_{Arr}$ QUEUES $\parallel_{Ser}$ SCHEDULER$\parallel_{SR}$
(Bandwidth_EST $\parallel_{CUR}$ CC $\parallel_{NC}$ RECTIFICATION)

COM = {send_request, send_response}
Serv = {deliver$_H$, deliver$_M$, deliver$_L$}
CUR = {send_CUR_to_cc}
Arr = {arrival$_H$, arrival$_M$, arrival$_L$}
SSR = {stop_sending_request}
NC = {notify_rectification}

- DIFFSERV □ <packet_arrival, λ>.CLASSIFIER
  CLASSIFIER □ <check_header, φ>.<classification, ψ>.CLASSIFIER1
  CLASSIFIER1 □ <not_marked, $\infty_{1,P}$>.MARKER +
  <marked, $\infty_{1,1-P}$ >.SEND_TO_QUEUE
- MARKER □ <τ, $\infty_{1,P}$>.<mark_pkt, $\infty_{1,1}$>.TRAFIC_COND +
  <τ, $\infty_{1,1-P}$>.<send_request, $\infty_{1,1}$>.WAIT_DECISION
  WAIT_DECISION □ <send_response, *>.MARKER1
  MARKER1 □ <mark_pkt, $\infty_{1,p}$>.TRAFIC_COND +
  <notify_application, $\infty_{1,1-p}$>.MARKER
- TRAFIC_COND □ <shaping_and_conditioning, η>. SEND_TO_QUEUE
  SEND_TO_QUEUE □ <arrival$_H$, λ_H>.DIFFSERV + <arrival$_M$, λ_M>.DIFFSERV+
  <arrival$_L$, λ_L>.DIFFSERV
- CAC □ <send_request, *>.<prepare_probe, β>.
  <send_probe, $\infty_{1,1}$>.WAIT_RESP
  WAIT_RESP □ <receive_probe_response, δ>.ADM_MECHANISM
  ADM_MECHANISM □ <comparing_values, ε>.ADM_RESULT
  ADM_RESULT □ <accept, $\infty_{1,P}$>.<send_response, $\infty_{1,1}$>.CAC +
  <reject, $\infty_{1,1-P}$>.<send_response, $\infty_{1,1}$>.CAC
- QUEUES □ Queue_High$_0$ $\|_\varnothing$ Queue_Medium$_0$ $\|_\varnothing$ Queue_Low$_0$
  Queue_High$_0$ □ <arrival$_H$, *>.Queue_High$_1$
  Queue_High$_i$ □ <arrival$_H$, *>.Queue_High$_{i+1}$ +
  <deliver$_H$,*>. Queue_High$_{i-1}$ $1 \leq i \leq N-1$
  Queue_High$_N$ □ <deliver$_H$,*>. Queue_High$_{N-1}$
  Queue_Medium$_0$ □ <arrival$_M$, *>.Queue_Medium$_1$
  Queue_Medium$_j$ □ <arrival$_M$, *>.Queue_Medium$_{j+1}$ +
  <deliver$_M$,*>. Queue_Medium$_{j-1}$ $1 \leq j \leq M-1$
  Queue_Medium$_M$ □ <deliver$_M$,*>. Queue_Medium$_{M-1}$
  Queue_Low$_0$ □ <arrival$_L$, *>.Queue_Low$_1$
  Queue_Low$_k$ □ <arrival$_L$, *>.Queue_Low$_{k+1}$ +
  <deliver$_L$,*>. Queue_Low$_{k-1}$ $1 \leq k \leq P-1$
  Queue_Low$_P$ □ <deliver$_L$,*>. Queue_Low$_{P-1}$
- SCHEDULER □ <deliver$_H$, $\infty_{3,1}$>.<service$_H$, μ>.SCHEDULER +
  <deliver$_M$, $\infty_{2,1}$>.<service$_M$, μ>.SCHEDULER +
  <deliver$_L$, $\infty_{1,1}$>.<service$_L$, μ>.SCHEDULER +
  <stop_sending_request,*>.<service$_{SR}$, $\infty_{1,1}$>.SCHEDULER
- Bandwidth_EST □ <listen_to_channel, ω>.<cal_ch_util_ratio,θ>.
  <send_CUR_to_cc, $\infty_{1,1}$>.Bandwidth_EST
- CC □ < send_CUR_to_cc, *>.<compare_with_thrsh, $\infty_{1,1}$>.
  (<lower, $\infty_{1,P}$ >.CC + <higher, $\infty_{1,1-P}$ >.Congested_Channel)
  Congested_Channel □ <check_high_priority_queue, $\infty_{1,1}$>.(<empty, $\infty_{1,P}$ >.CC +
  <not_empty, $\infty_{1,1-P}$ >.<notify_rectification, $\infty_{1,1}$>.CC)

- RECTIFICATION □ <notify_rectification, *>.
  <stop_sending_request, $\infty_{4,1}$>.RECTIFICATION

4.1 Function analysis

The functional analysis aims at verifying the correctness of the designed model and at detecting conceptual error in its behaviour. An important advantage of using TwoTowers is the possibility of exploiting well knows formal verification techniques to investigate functional properties of the model. Indeed, with respect to conventional simulation environments (e.g. Network Simulator NS-2), TwoTowers provides verification of the satisfaction of correctness properties. Prominent examples of functional verification are checking that specified model does not lead neither to deadlock nor to livelock, controlling that certain activities are carried out according to a given order, or ensuring that certain resources are used in a mutually exclusive way.

The technique adopted in this tool to provide functional verification support is based on temporal model checking logic like μ-calculus/computational tree logic (CTL). With model checking, we refer to the possibility to express functional requirements by means of a set of formulas and verifying the satisfaction of desired properties by the algebraic specification. We start our analysis by verifying some behavioural requirements that we have formalized through the following formulas:

- $\text{No\_deadlock} = (\min X = [-]ff \vee \langle - \rangle X)$
- $AG([arrival_H]ff \vee [arrival_H]A(([deliver_M]ff \wedge [deliver_L]ff)\ W < deliver_H > tt))$
- $AG([send\_stop\_sending\_request]$
  $A([send\_stop\_sending\_request]ff\ \ W < notify\_rectification > tt)$
  $\wedge\ [notify\_rectification]A([notify\_rectification]ff\ \ W < higher > tt))$

The first equation verifies the freedom of our model from deadlock. Equation 2 ensures if there is a high priority packet in the queue, it will be served before low and medium priority packets. Finally, equation 3 ensures that stop_sending_request packet can not be transmitted by the rectification component before the notification of the congestion control component, and this notification can not be transmitted before the occurrence of congestion , which can be detected when the available bandwidth is great than a predefined threshold.

4.2 Performance analysis

QoS characterizes the non-functional properties of a system; it is expressed in terms of a number of quantifiable parameters that can be easily evaluated after the derivation of CTMC from algebraic specification of the model, via a structured operational semantics rules. This chain can be used for constructing the infinitesimal generator matrix which is used to calculate the steady-state and transient probability distribution for the system. These probabilities will be used to derive the desired performance parameters by assigning a number describing a weight (usually called a reward) is attached to every state of the CTMC, and the performance parameter is defined as the weighted sum of the steady state probabilities of the Markovian model, like shows the following formula $\sum r_i \times \pi_i$

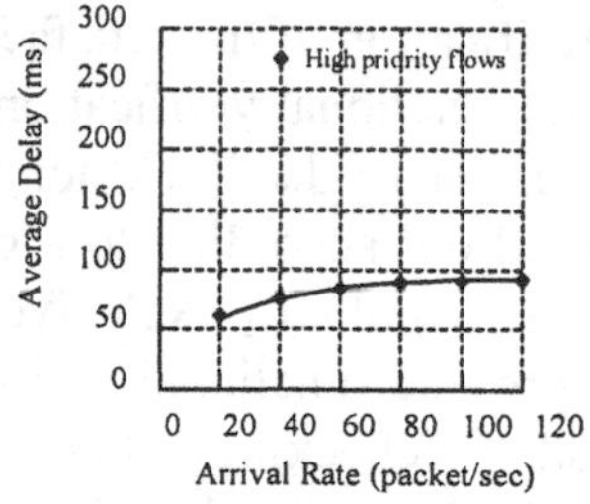

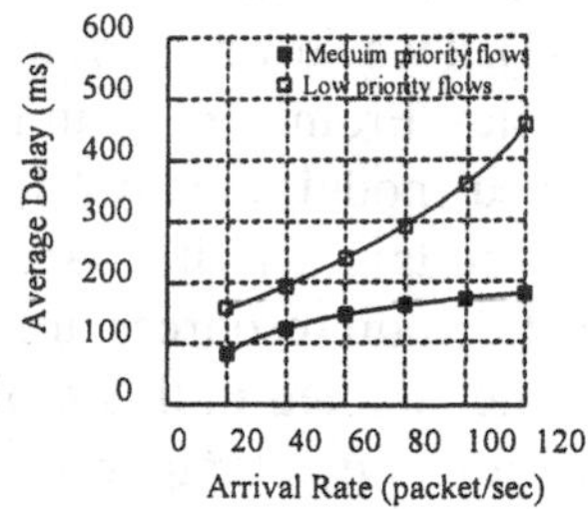

Figure 4. Average delay vs arrival rate

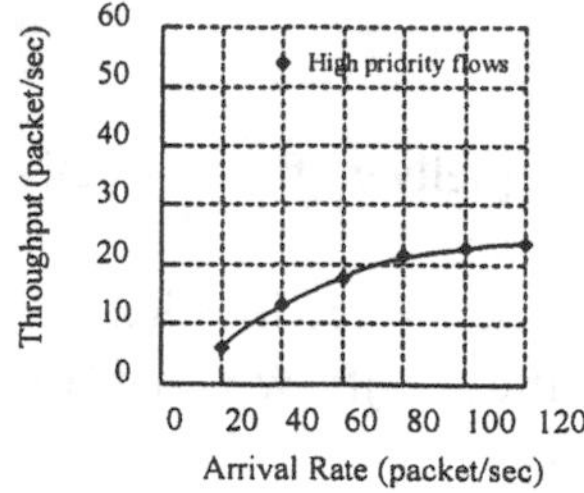

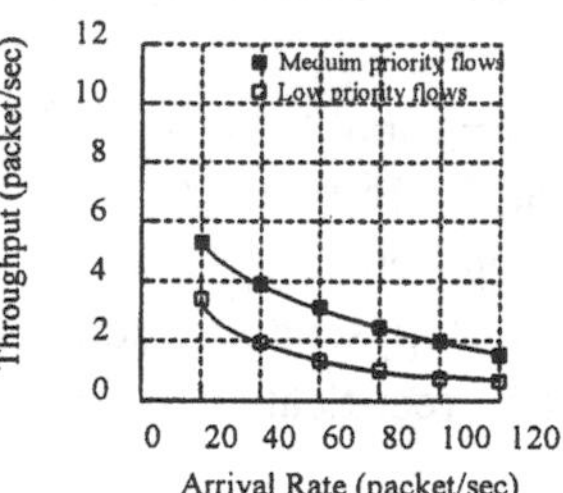

Figure 5. Throughput vs arrival rate

Where N is the size of CTMC, π_i is the steady state probability of state i, and r_i is the reward attached to state i. EMPA provides an automatic way to calculate performance parameters using the preceding formulas, in addition to the implemented simulation routine. Using these techniques, we evaluate the performance of our algebraic model by focusing at the variation of average delay experienced by a packet and the throughput of the three modeled classes. We have increased the arrival rate of flows that pass through the model, and we divide this rate equally between all classes.

The variation curves that are presented in figures 4 and 5 show that our model can provide an acceptable delay and throughput for packets within highest priority range. The accuracy of these results is going to be depending on the detail that we have invested in the algebraic model, on the service rate of the scheduler, on the restriction of activities duration in SPA to exponential delay, and finally to the simple priority scheduler mechanism for queues instead of RED (Random Early Detection) usually used in DiffServ. These approximations are used to deal with the trade-off between accuracy of the model and the ability of the tool to verify functional behaviour due to limited memory with respect to the number of states in functional transition diagram (FTD), and to avoid state space explosion problem when evaluating performance parameters by using CTMC. The used scheduling mechanism is the responsible of the fact that medium and low priority traffics experience a long delay and their throughputs degrade exponentially to near zero.

5. CONCLUSION AND FURTHER RESEARCH

In this paper, we have described a new scheme to create an adaptive extension to DiffServ for MANETs after investigating the suitability of the existing QoS wired technologies (IntServ and DiffServ) to the characteristics of these networks. Our scheme is based on dynamic estimation of the available bandwidth by each node in its shared media in a completely distributed manner, and adopts the call admission control mechanism used in SWAN to gather minimum available bandwidth along existing paths towards a specified destination in route discovery phase. These additional components make DiffServ treats highest priority classes in an elegant and dynamic way while remaining stateless and lightweight.

Stochastic process algebras allow only the description of the task that must be accomplished by each component with an exponentially distributed duration in order to derive the CTMC from the described model. However, this is seen unrealistic for components that need deterministic duration and a description of this scheme with generally distributed activities will provide more accuracy while evaluating its quantitative parameters. This is why we focus in our current and future work at specifying our proposed scheme with generally distributed activities. In addition, some tasks need to be refined like the mechanism executed by nodes when receiving stop-sending-request, because the naïve suppression of any routing information by the node that decides react face to network overload is not a good strategy and may prevent the use of the only possible path between two hosts. This why there is a needs to enhance this mechanism in our future work.

REFERENCES

1. S. Corson and J. Macker, "Mobile Ad hoc Networking (MANET): Routing Protocol Performance Issues and Evaluation Considerations", RFC 2501, January 1999.
2. C. E. Perkins and E. M. Belding-Royer, "Quality of Service for Ad Hoc On-Demand Distance Vector", Internet Draft, draft-perkins-manet-aodvqos-02.txt, 14 October 2003.
3. R. Leung, J. Liu, E. Poon, A. C. Chan, B. Li, "MP-DSR: A QoS Aware Multi Path Dynamic Source Routing Protocol For Wireless Ad Hoc Networks", in the Proceedings of the 26th Annual IEEE Conference on Local Computer Networks (LCN'01), November 14 - 16, 2001, Tampa, Florida.
4. J. Xue, P. Stuedi and G. Alonso, "ASAP: An Adaptive QoS Protocol for Mobile Ad Hoc Networks", in the Proceeding of 14th IEEE International Symposium on Personal, Indoor and Mobile Radio Communications (IEEE PIMRC 2003), 7-10 September 2003, Beijing, china.
5. R. Sivakumar, P.S Inha and V. Bharghavan, "CEDAR: A Core-Extraction Distributed Ad Hoc Routing algorithm", IEEE Journal on Selected Areas in Communications, vol. 17, no. 8, pp. 1454-1465, August 1999.
6. R. Braden, D. Clark, and S. Shenker, "Integrated Services in the internet architecture: an overview", RFC 1633, USC/Information Sciences Institute, MIT, Xerox PARC, June 1994.
7. S. Blake, D. Black, M. Carlson, E. Davies, Z. Wang, and W. Weiss, "An Architecture for Differentiated Services", RFC 2475, December 1998.
8. S-B. Lee and A.T. Campbell, "INSIGNIA: In-band Signaling Support for QoS in Mobile Ad Hoc Networks", in the Proceedings of 5th International Workshop on Mobile Multimedia Communications (MoMuC, 98), Berlin, Germany, October 1998.
9. H. Xiao, W.K.G. Seah, A. Lo and K.C. Chua, "A Flexible Quality of Service Model for Mobile Ad Hoc Networks", in the Proceedings of IEEE Vehicular Technology Conference (IEEE VTC2000-spring), 15-18 May 2000, Tokyo, Japan, pp 445-449.
10. G–S. Ahn, A. T. Campbell, A. Veres, and Li-Hsiang Sun, "SWAN: Service Differentiation in Stateless Wireless Ad Hoc Networks", in the Proceedings of IEEE INFOCOM 2002, June 2002.
11. E. Brinksma and H. Hermanns 2001, "Process Algebra and Markov Chains", Lecture on Formal Methods and Performance Analysis, Nijmegen, pp183–231.
12. J. Hillston, "PEPA Performance Evaluation Process Algebra", Technical Report of Computer Science, Edinburgh University, March 1993.
13. U. Herzog 1993, "TIPP: A Language for Timed Processes and Performance Evaluation", Proceedings of the First International Workshop on Process Algebra and Performance Modelling, University of Edinburgh, UK.
14. M. Bernardo and R. Gorrieri 1998b, "A Tutorial on EMPA: A Theory of Concurrent Processes with Nondeterminism, Priorities, Probabilities and Time", Theoretical Computer Science, pp1-54.
15. A. Benzekri and O. Salem, "Modelling and Analyzing Dynamic Source Routing Protocol with General Distributions", In the Proceedings of the 11th International Conference on Analytical and Stochastic Modelling Techniques and Applications (ASMTA04), 13-16 June 2004.
16. A. Benzekri and O. Salem, "Functional Modelling and Performance Evaluation for Two Class DiffServ Router using Stochastic Process Algebra", in the Proceeding of the 17th European Simulation Multiconference, Nottingham - UK, SCS-European Publishing House , PP. 257-262, 10 June 2003.

CROSS-LAYER PERFORMANCE EVALUATION OF IP-BASED APPLICATIONS RUNNING OVER THE AIR INTERFACE

Dmitri Moltchanov, Yevgeni Koucheryavy, and Jarmo Harju
Tampere University of Technology, Institute of Communications Engineering, P.O.Box 553, FIN-33101, Tampere, Finland. {*moltchan,yk,harju*}*@cs.tut.fi*

Abstract: We propose a novel cross-layer analytic approach to performance evaluation of delay/loss sensitive IP-based applications running over the wireless channels. We firstly extend the small-scale propagation model representing the received signal strength to IP layer using the cross-layer mapping. Then, we replace the resulting IP layer wireless channel model by an artificial equivalent arrival process using the error/arrival mapping. To get performance parameters of interest we use this process together with arrival process modeling the traffic source as an input to the queuing system with deterministic service time, limited number of waiting position and non-preemptive priority discipline representing the IP packet service process of the wireless channel.

Keywords: IP layer wireless channels model, cross-layer mapping, performance evaluation.

1. INTRODUCTION

To predict performance of wireless channels, propagation models representing the received signal strength are often used. However, these models cannot be directly used in performance evaluation studies and must be previously extended to the IP layer at which QoS is defined. For such extension to be accurate, we have to take into account specific peculiarities of underlying layers including modulation schemes at the physical layer, data-link error concealment techniques and segmentation procedures between different layers. Wireless channel model for performance evaluation studies must be cross-layer complex function of propagation characteristics at the layer of interest.

Up to date only a few studies devoted to IP layer performance of applications running over the air interface have been published. Most studies were devoted to performance of the data-link layer protocols (Krunz and Kim, 2001; Zorzi et al., 1997) and rely on approaches specifically developed for wireless transmission medium. Hence, they often require new approximations, algorithms, stable recursions etc. Additionally, most of them adopt restrictive assumptions regarding the performance of wireless channels at layers above physical which may lead to incorrect estimation of IP layer performance parameters.

In this paper we propose a cross-layer analytic approach to performance evaluation of delay/loss sensitive IP-based applications running over the wireless channels. To use results available in queuing theory we show how to represent the problem of unreliable transmission over the wireless channel using a simple queuing model with two inputs. To achieve this goal we firstly extend the small-scale propagation model representing the received signal strength to the IP layer using the cross-layer mapping. Then, we replace the IP layer wireless channel model by an artificial equivalent arrival process using the error/arrival mapping. We use this process together with the arrival process modeling the traffic source as an input to the queuing system with deterministic service time, limited number of waiting position and non-preemptive priority discipline modeling the service process of the wireless channel.

Our paper is organized as follows. In Section 2 we review propagation characteristics of wireless channels and models used to capture them. In section 3 we propose an extension to the IP layer. Performance evaluation is carried out in Section 4. Then, in Section 5 we provide numerical examples. Conclusions are drawn in the last section.

2. WIRELESS PROPAGATION CHARACTERISTICS

To represent performance of wireless channels propagation models are used. Usually, we distinguish between large-scale and small-scale propagation models (Rappaport, 2002). The latter ones capture characteristics of wireless channel on a finer granularity than large-scale ones and implicitly take into account movements of users over short travel distances (Saleh and Valenzuela, 1987; Durgin and Rappaport, 2000). To capture small-scale propagation characteristics we have to take into account the presence of LOS between the transmitter and a receiver. In the presence of dominant non-fading component the small-scale propagation distribution is Rician. As the dominant component fades away Rician distribution degenerates to Rayleigh one.

Model of small-scale propagation characteristics

Assume a discrete-time environment, i.e. time axis is slotted, the slot duration is constant and given by $\Delta t = (t_{i+1} - t_i)$, $i = 0, 1, \ldots$. We choose Δt such that it equals to the time to transmit a single bit at the wireless channel. Hence, the choice of Δt explicitly depends on properties of the physical layer.

Small-scale propagation characteristics are often represented by the stochastic process $\{L(n), n = 0, 1, \dots\}$ modulated by the discrete-time Markov chain $\{S_L(n), n = 0, 1, \dots\}$, $S_L(n) \in \{1, 2, \dots, M\}$ each state of which is associated with conditional probability distribution function of the signal strength $F_L(k\Delta f|i)(\Delta f) = Pr\{L(n) = k\Delta f|S_L(n) = i\}$, $k = 1, 2, \dots, N$, $i = 1, 2, \dots, M$, where N is the number of bins to which the signal strength is partitioned and Δf is the discretization interval (Zhang and Kassam, 1999; Swarts and Ferreira, 1999). Underlying modulation allows to capture autocorrelation properties of the signal strength process. Since it is allowed for the Markov process $\{S_L(n), n = 0, 1, \dots\}$ to change the state in every time slot, every bit may experience different signal strengths.

Let D_L and $\vec{\pi}_L = (\pi_1, \pi_2, \dots, \pi_M)$ be the one-step transition probability matrix and stationary probability vector of $\{S_L(n), n = 0, 1, \dots\}$ respectively. Parameters M, D_L, $F_L(k\Delta f|i)(\Delta f)$, must be estimated from statistical data (Zhang and Kassam, 1999; Swarts and Ferreira, 1999). For the ease of notation we will use $F_L(k|i)$ instead of $F_L(k\Delta f|i)(\Delta f)$.

3. WIRELESS CHANNEL MODEL AT THE IP LAYER

The small-scale propagation model of the received signal strength defined in the previous section cannot be directly used in performance evaluation studies and must be properly extended to the IP layer at which QoS is defined. To do so we have to take into account specific peculiarities of layers below IP including modulation schemes at the physical layer, data-link error concealment techniques and segmentation procedures between different layers.

In the following we define models of incorrect reception of the protocol data units (PDU) at different layers. For this purpose we implicitly assume that the appropriate PDUs are consecutively transmitted at corresponding layers.

Bit error process

Consider a certain state i of the Markov chain $\{S_L(n), n = 0, 1, \dots\}$. Since the probability of a single bit error is the deterministic function of the received signal strength (Rappaport, 2002), all values of $F_L(k|i)$ that are less or equal to a computed value of so-called bit error threshold B_T cause bit error. Those values which are greater than B_T do not cause bit error. Thus, each state of the Markov process $\{S_L(n), n = 0, 1, \dots\}$ can be associated with the following bit error probability:

$$p_{E,i} = Pr\{E(n) = 1|S_E(n) = i\} = \sum_{k=1}^{B_T} Pr\{L(n) = k|S_L(n) = i\}, \quad (1)$$

where $\{E(n), n = 0, 1, \dots\}$, $E(n) \in \{0, 1\}$ is the bit error process for which 1 denotes incorrectly received bit, 0 denotes correctly received bit, $\{S_E(n), n = 0, 1, \dots\}$ is the underlying Markov chain of $\{E(n), n = 0, 1, \dots\}$. Note that $\{S_L(n), n = 0, 1, \dots\}$ and $\{S_E(n), n = 0, 1, \dots\}$ are the same and $\vec{\pi}_E = \vec{\pi}_L$, $D_E = D_L$, where D_E and $\vec{\pi}_E$ are one-step transition probability matrix and stationary distribution vector of $\{S_E(n), n = 0, 1, \dots\}$ respectively. B_T must be estimated based on a modulation scheme and other specific features of the physical layer utilized at a given wireless channel (Rappaport, 2002).

Denote by $d_{E,ij}(k) = Pr\{E(n) = k, S_E(n) = j|S_E(n-1) = i\}, k = 0, 1$, the transition probability from state i to state j with correct and incorrect bit reception respectively. These probabilities are represented in a compact form using matrices $D_E(1)$ and $D_E(0)$, $D_E(1) + D_E(0) = D_E$. The state from which the transition occurs completely determines the bit error probability. The state to which transition occurs is used for convenience of matrix notation.

Frame error process without FEC

Assume that the length of the frame is constant and equal to m bits. Sequence of consecutively transmitted bits, denoted by gray rectangles, is shown in Fig. 1, where $(l-1)$, l, $(l+1)$ denote time intervals whose length equal to the time to transmit one frame; k, i, j, denote the state of $\{S_E(n), n = 0, 1, \dots\}$ in the beginning of these intervals.

Consider the stochastic process $\{N(l), l = 0, 1, \dots\}$, $N(l) \in \{0, 1, \dots, m\}$, describing the number of incorrectly received bits in consecutive bit patterns of the length m. This process is doubly-stochastic with underlying Markov chain $\{S_N(l), l = 0, 1, \dots\}$ and can be defined via the bit error process.

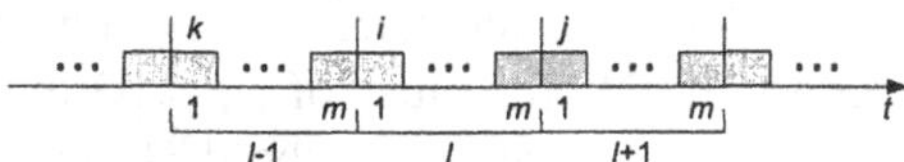

Figure 1. Sequence of consecutively transmitted bits at the wireless channel.

Denote the probability of going from state i to state j for the Markov chain $\{S_N(l), l = 0, 1, \dots\}$ with exactly k, $k = 0, 1, \dots, m$ incorrectly received bits in a bit pattern of length m by $d_{N,ij}(k) = Pr\{N(l) = k, S_N(l) = j|S_N(l -$

$1) = i\}$. These probabilities can be found using $D_E(k)$, $k = 0, 1$, $\vec{\pi}_E$:

$$d_{N,ij}(0) = \vec{\pi}_E D_E^m(0)\vec{e},$$

$$d_{N,ij}(1) = \vec{\pi}_E \sum_{k=m-1}^{0} D_E^{m-k-1}(0) D_E(1) D_E^k(0)\vec{e},$$

$$\dots$$

$$d_{N,ij}(m) = \vec{\pi}_E D_E^m(1)\vec{e}, \tag{2}$$

where $\vec{e}$ is the vector of ones of appropriate size.

Let us introduce the frame error process $\{F(l), l = 0, 1, \dots\}$, $F(l) \in \{0, 1\}$, where 0 indicates the correct frame reception, 1 denotes incorrect frame reception. Process $\{F(l), l = 0, 1, \dots\}$ is modulated by underlying Markov chain $\{S_F(l), n = 0, 1, \dots\}$. Note that $\{S_F(l), l = 0, 1, \dots\}$ and $\{S_N(l), l = 0, 1, \dots\}$ are the same. Let us denote the probability of going from state i to state j for the Markov chain $\{S_F(l), l = 0, 1, \dots\}$ with exactly k, $k = 0, 1$ incorrectly received frames by $d_{F,ij}(k)$. Process $\{N(l), l = 0, 1, \dots\}$ describing the number of bit errors in consecutive frames can be related to the frame error process $\{F(l), l = 0, 1, \dots\}$ using the so-called frame error threshold F_T:

$$d_{F,ij}(0) = \sum_{k=0}^{F_T-1} d_{N,ij}(k), \qquad d_{F,ij}(1) = \sum_{k=F_T}^{m} d_{N,ij}(k). \tag{3}$$

Expressions (3) are interpreted as follows: if the number of incorrectly received bits in a frame is greater or equal to a computed value of the frame error threshold ($k \geq F_T$) the frame is incorrectly received and $F(l) = 1$, otherwise ($k < F_T$) the frame is correctly received and $F(l) = 0$.

Assume that FEC is not used at the data-link layer. It means that every time a frame contains at least one bit error, it is received incorrectly ($F_T = 1$). Thus, transition probabilities (3) of the frame error process take the following form:

$$d_{F,ij}(0) = d_{N,ij}(0), \qquad d_{F,ij}(1) = \sum_{k=1}^{m} d_{N,ij}(k) = 1 - d_{F,ij}(0). \tag{4}$$

The slot durations of $\{N(l), l = 0, 1, \dots\}$ and $\{F(l), l = 0, 1, \dots\}$ are the same $\Delta t'$ and related to the slot duration of received signal strength process $\{L(n), n = 0, 1, \dots\}$ as $\Delta t' = nl\Delta t$, $n = 0, 1, \dots$.

Frame error process with FEC

F_T depends on FEC correction capabilities. Assume that the number of bit errors that can be corrected by a FEC code is l. Then, $F_T = (l+1)$ and a frame is incorrectly received when $k \geq (l+1)$. Otherwise, it is correctly received. Thus, transition probabilities (3) take the following form:

$$d_{F,ij}(0) = \sum_{k=0}^{l} d_{N,ij}(k), \qquad d_{F,ij}(1) = \sum_{k=l+1}^{m} d_{N,ij}(k). \tag{5}$$

IP packet error process

Assume that every IP packet is segmented into z frames of equal size at the data-link layer. Sequence of consecutively transmitted frames, denoted by gray rectangles, is shown in Fig. 2, where $(h-1)$, h, $(h+1)$ denote time intervals whose length equal to the time to transmit one packet; k, i, j, denote the state of the Markov chain $\{S_F(n), n = 0, 1, \dots\}$ in the beginning of these intervals. Assumption of the constant frame size does not restrict the generality of results as long as only one traffic source is allowed to be active at any instant of time.

Let the stochastic process be $\{M(h), h = 1, 2, \dots\}$, $M(h) \in \{0, 1, \dots, z\}$, describing the number of incorrectly received frames in a consecutive frame patterns of the length z. The process, modulated by Markov chain $\{S_M(h), h = 0, 1, \dots\}$, can be completely defined via the frame error process.

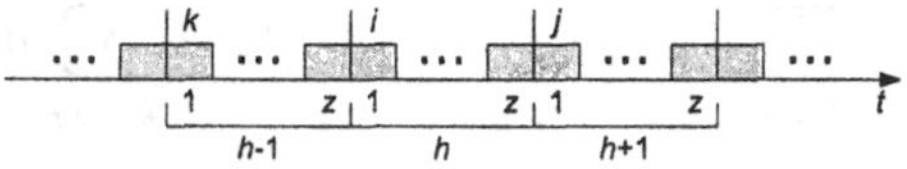

Figure 2. Sequence of consecutively transmitted frames at the wireless channel.

Denote the probability of going from state i to state j for the Markov chain $\{S_M(h), h = 0, 1, \dots\}$ with exactly k, $k = 0, 1, \dots, z$ incorrectly received frames in a frame pattern of length z by $d_{M,ij}(k) = Pr\{M(h) = k, S_M(h) = j | S_M(h-1) = i\}$. These transition probabilities can be found using $D_F(k)$, $k = 0, 1$ and $\vec{\pi}_F$ of $\{F(l), l = 0, 1, \dots\}$ as follows:

$$d_{M,ij}(0) = \vec{\pi}_F D_F^z(0)\vec{e},$$

$$d_{M,ij}(1) = \vec{\pi}_F \sum_{k=z-1}^{0} D_F^{z-k-1}(0) D_F(1) D_F^k(0)\vec{e},$$

$$\dots$$

$$d_{M,ij}(z) = \vec{\pi}_F D_F^z(1)\vec{e}, \tag{6}$$

where $\vec{\pi}_F$ is the stationary distribution vector of $\{S_F(l), l = 0, 1, \dots\}$.

Let us introduce the packet error process $\{P(h), h = 0, 1, \dots\}$, $P(h) \in \{0, 1\}$, where 0 indicates correct packet reception, 1 denotes incorrect packet reception. Process $\{P(h), h = 0, 1, \dots\}$ is modulated by underlying Markov chain $\{S_P(h), h = 0, 1, \dots\}$. $\{S_P(h), h = 0, 1, \dots\}$ and $\{S_M(h), h = 0, 1, \dots\}$ are the same. Denote the probability of going from state i to state j for the Markov chain $\{S_P(h), h = 0, 1, \dots\}$ with exactly k, $k = 0, 1$ incorrectly received packets by $d_{P,ij}(k)$. Process $\{M(h), h = 0, 1, \dots\}$ describing the number of incorrectly received frames in consecutively transmitted packets can be related to the packet error process $\{P(h), h = 0, 1, \dots\}$ using the so-called packet error threshold P_T:

$$d_{P,ij}(0) = \sum_{k=0}^{P_T-1} d_{M,ij}(k), \qquad d_{P,ij}(1) = \sum_{k=P_T}^{z} d_{M,ij}(k) = 1 - d_{M,ij}. \quad (7)$$

Expressions (7) are interpreted as follows: if the number of incorrectly received frames in the packet are greater or equal to a computed value of the packet error threshold ($k \geq P_T$) the packet is incorrectly received ($P(h) = 1$). Otherwise, it is correctly received ($P(h) = 0$). Since no error correction procedures are defined for IP layer, $P_T = 1$ and only $d_{M,ij}(0)$ must be computed in (6). That is every time a packet contains at least one incorrectly received frame, the whole packet is received incorrectly.

Slot durations of $\{P(h), h = 0, 1, \dots\}$ and $\{M(h), h = 0, 1, \dots\}$ are the same $\Delta t''$ and related to the slot duration of the received signal strength process as $\Delta t'' = nlh\Delta t$, $n = 0, 1, \dots$.

Illustration of the proposed extension

An illustration of the proposed cross-layer mapping is shown in the Fig. 3 where time diagrams of $\{L(n), n = 0, 1, \dots\}$, $\{E(n), n = 0, 1, \dots\}$, $\{N(l), l = 0, 1, \dots\}$, $\{F(l), l = 0, 1, \dots\}$, $\{M(h), h = 0, 1, \dots\}$, $\{P(h), h = 0, 1, \dots\}$ are shown. To define the model of incorrect reception of PDUs at different layers we implicitly assumed that appropriate PDUs are consecutively transmitted at corresponding layers. Hence, the IP packet error process is conditioned on the event of consecutive transmission of packets.

4. PERFORMANCE EVALUATION

In this subsection we provide an analytical technique for derivation of various performance parameters of the packet service process at the wireless channel. The proposed method is entirely based on classic queuing theory and applicable for delay/loss sensitive applications for which only FEC is implemented at the data-link layer.

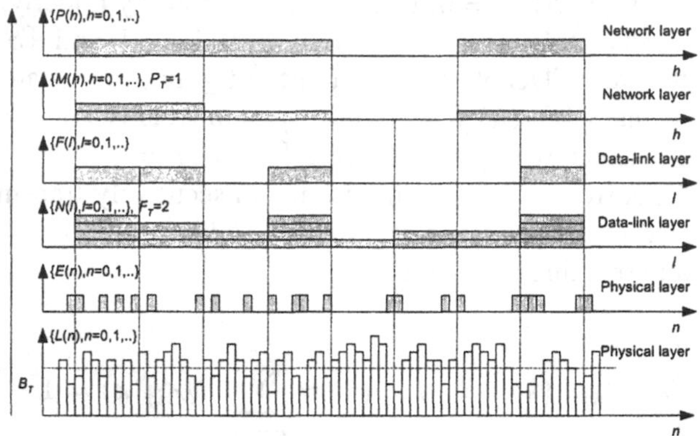

Figure 3. An illustration of the proposed cross-layer mapping.

Artificial arrival process. Given a Markov chain each state of which is associated with a certain packet error probability we propose to approximate the complexity of the packet error process by the so-called 'artificial' Markov modulated arrival process using error/arrival mappings. The mapping is straightforward: every time when the packet is incorrectly received with a certain probability, an arrival occurs with the same probability. Since it is not required to change the underlying Markovian structure and parameters of the IP layer wireless channel model, new parametrization procedure is not needed. The resulting arrival process is classified as a discrete-time Markovian arrival process (D-MAP). We denote it by MAP$_E$.

Tagged arrival process. Assume that the arrival process from user's traffic source is represented by D-MAP $\{W_T(n), n = 0, 1, \dots\}$ with underlying Markov chain $\{S_T(n), n = 0, 1, \dots\}$, $S_T(n) \in \{1, 2, \dots, M_T\}$. According to it, each pair of states of the set $\{1, 2, \dots, M_T\}$ is associated with probability of one arrival, i.e. given a pair (i, j), $i, j = 1, 2, \dots, M_T$ one arrival occurs in a slot with probability $p_{T,ij} = Pr\{W_T(n) = 1, S_T(n) = i | S_T(n-1) = i\}$ and no arrival occurs with complementary probability $1 - p_{T,ij}$. We denote this process by MAP$_T$

Description of the queuing system. We represent the service process of the wireless channel by the queuing system MAP$_T$+MAP$_E$/D/1/K where arrivals coming from MAP$_E$ are of higher priority than those ones coming from MAP$_T$ and every arrival requires a service time of exactly one slot $\Delta t''$. The system operates as follows. At each slot boundary the server tries to serve an arrival from MAP$_E$, if any. If not, it serves an arrival from MAP$_T$. It is sufficient to assume a non-preemptive priority discipline for our system. Indeed, any arrival coming from MAP_E does not wait in the buffer and scheduled for service at the beginning of the nearest slot. Since at most one arrival may occur per a slot from MAP_E and these arrivals do not wait for service, waiting positions are only occupied by low priority arrivals. The service discipline is FCFS for low priority arrivals.

Queuing analysis. The loss probability is an important parameter for delay/loss sensitive applications. Given previously defined queuing model we have to determine packet losses experienced by MAP_T due to buffer overflow.

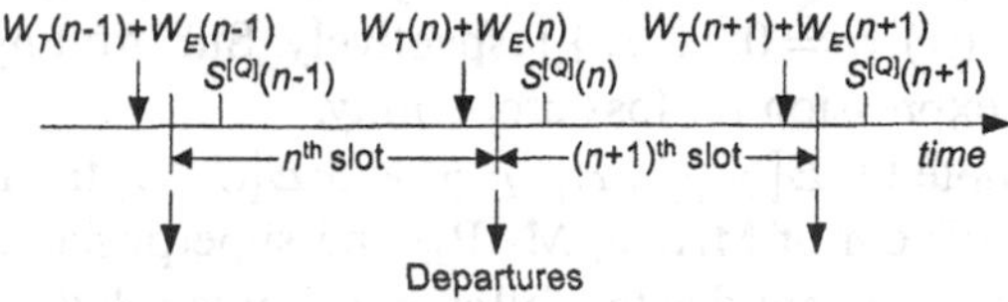

Figure 4. Time diagram of the queuing system.

Time diagram of the queuing system is given in the Fig. 4. Complete description of the queuing system requires a two-dimensional Markov chain $\{S_A(n), S_Q(n), n = 0, 1, \dots\}$ embedded at the moments of departures, where $S_A(n) = S_E(n) \times S_T(n)$ is the state of superposition of MAP_T and MAP_E, $S_Q(n) \in \{0, 1, \dots, K\}$ is the number of customers in the system at the moments of departures. Let T be the transition matrix of such Markov chain and $\vec{x} = (x_{1,0}, .., x_{M_A,K})$, $M_A = M_E M_T$ be the row array containing stationary probabilities. One can solve for $\vec{x}$ using $\vec{x}T = \vec{x}$, $\vec{x}\vec{e} = 1$.

Let us determine the loss probability L_T for MAP_T as a limiting probability of ratio between the number of lost arrivals of MAP_T in a slot and the number of arrivals from this process in a slot:

$$L_T = \lim_{n \to \infty} \frac{L_T(n)}{W_T(n)}. \tag{8}$$

Since at most one arrival is allowed in a slot from each of arrival processes the loss of one customer from MAP_T always occurs when the state of the queuing system is $(S_A = i, S_Q = K)$, $i = 1, 2, \dots, M_E M_T$ and may occur when the state of the system is $(S_A = i, S_Q = K - 1)$. We have:

$$\begin{aligned} L_T(n) = & \sum_{i=1}^{M_E} \sum_{j=1}^{M_T} x_{(i,j),K} \sum_{l=1}^{M_T} p_{T,jl} + \\ & + \sum_{j=1}^{M_E} \sum_{i=1}^{M_T} x_{(i,j),K-1} \sum_{m=1}^{M_E} p_{E,im} \sum_{l=1}^{M_T} p_{T,jl}. \end{aligned} \tag{9}$$

where $x_{(i,j),K}$ and $x_{(i,j),K-1}$ are $(i+j, K)$ and $(i+j, K-1)$ elements of stationary distribution of $\{S_A(n), S_Q(n), n = 0, 1, \dots\}$, $p_{T,ij}$, $i, j = 1, 2, \dots, M_T$ are probabilities of arrival from MAP$_T$ going from state i to state j, $p_{E,ij}$, $i, j = 1, 2, \dots, M_E$ are probabilities of arrival from MAP$_E$ going from state i to state j. The mean arrival rate of MAP$_T$ is given by:

$$E[W_T] = \sum_{i=1}^{M_T} \pi_{T,i} \sum_{j=1}^{M_T} p_{T,ij}, \tag{10}$$

where $\pi_{T,i}$ and $\pi_{E,i}$ are elements of stationary probability vectors of $\{S_T(n), n = 0, 1, \dots\}$ and $\{S_T(n), n = 0, 1, \dots\}$ respectively. Substituting (10) and (9) in (8) we get a final expression for loss probability.

Let us now denote by $E[S_{T,Q}]$, $E[S_{E,Q}]$ and $E[S_{A,Q}]$ the mean number of customers in the system of MAP$_T$, MAP$_E$ and superposed arrival processes respectively. For these quantities the following simple relation holds:

$$E[S_{T,Q}] = E[S_{A,Q}] - E[S_{E,Q}]. \tag{11}$$

Since MAP$_E$ has an absolute priority over MAP$_T$, $E[S_{T,Q}]$ can be estimated from queuing system MAP$_E$/D/1/K neglecting those arrivals coming from MAP$_T$. For arrivals from MAP$_E$ in MAP$_E$/D/1/K queuing system the Little's result gives:

$$E[S_{E,Q}] = E[W_E]E[D_E], \tag{12}$$

where $E[D_E]$ is the mean sojourn time in the system of arrivals from MAP$_E$. One may note that this time is always equal to the service time (one slot). Indeed, at most one arrival from MAP$_E$ is allowed and this arrival is always served without waiting for service. Therefore, we have:

$$E[S_{E,Q}] = \sum_{i=1}^{M_E} \pi_{E,i} \sum_{j=1}^{M_E} p_{E,ij}, \tag{13}$$

where $\pi_{E,i}$ is the ith element of steady-state distribution of $\{S_E(n), n = 0, 1, \dots\}$. $E[S_{A,Q}]$ can be estimated from the stationary distribution $\vec{x}$ as:

$$E[S_{A,Q}] = \sum_{i=1}^{M_E M_T} \sum_{k=0}^{K} x_{i,k} k. \tag{14}$$

Finally, substituting (14) and (13) to (11) we get:

$$E[S_{T,Q}] = \sum_{i=1}^{M_E M_T} \sum_{k=0}^{K} x_{i,k} k - \sum_{i=1}^{M_E} \pi_{E,i} \sum_{j=1}^{M_E} p_{E,ij}. \tag{15}$$

5. NUMERICAL EXAMPLES

To provide an example of IP layer performance evaluation let us assume that the wireless channel at the physical layer is represented by the Markov chain with $M = 4$, $p_{E,i} = 0$, $i = 1, 2, 3$, $p_{E,4} = 0.01$ and the following transition probability matrix:

$$P_E = \begin{pmatrix} 0.42 & 0.18 & 0.24 & 0.16 \\ 0.18 & 0.42 & 0.04 & 0.36 \\ 0.07 & 0.03 & 0.54 & 0.36 \\ 0.03 & 0.07 & 0.09 & 0.81 \end{pmatrix}. \quad (16)$$

Consider how the segmentation procedures between the data-link and IP layers affect the mean number of incorrectly received packets at the IP layer. Estimated values of the mean number of incorrectly received packets, $E[P]$, for different values of the number of frames to which the IP packet is segmented, z, number of bits in a single frame, m, number of incorrectly received bits that can be corrected, l, and Bernoulli arrival process with probability of one arrival set to 0.7 are shown in Fig. 5. Observing the obtained results we conclude that $z = 1$ gives the best possible conditions at the IP layer. Note that setting $z = 1$, $E[P]$ provides the number of incorrectly received frames at the data-link layer.

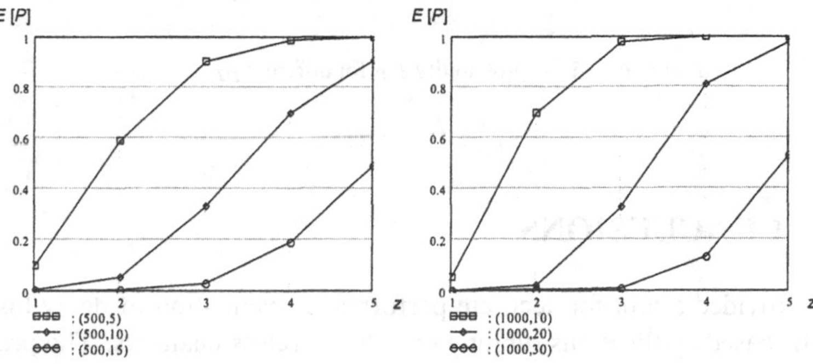

Figure 5. Mean number of lost IP packets for different values of z, m and l.

Assume now that the wireless channel at the physical layer is represented by the Markov chain with $M = 3$, $p_{E,1} = 0$, $p_{E,2} = p_{E,3} = 1$, and the following transition probability matrix:

$$P_E = \begin{pmatrix} 0.6 & 0.4 & 0 \\ 0 & 0.2 & 0.8 \\ 0.8 & 0 & 0.2 \end{pmatrix}. \quad (17)$$

We set $z = 1$ and choose the arrival process from the traffic source to be Bernoulli with p_T as the probability of one arrival. In the left part of the Fig.

6 the packet loss probabilities are shown for different values of p_T and different capacity of the system K. One may notice that given a wireless channel characteristics and $p_T > 0.3$ packet losses become significant.

Let us now assume that it is possible to adjust parameters of environment such that $p_{E,2} = 0$. For example, it can done by proper displacement of the transmitter antenna. Loss probabilities for $p_{E,2} = 0$ and $p_{E,2} = 1$ are shown in the right part of Fig. 6. One may notice that the probability of loss decreases significantly when $p_{E,2} = 0$. For example, for $p_T = 0.3$ and $K = 9$, L_T reaches 10^{-6}.

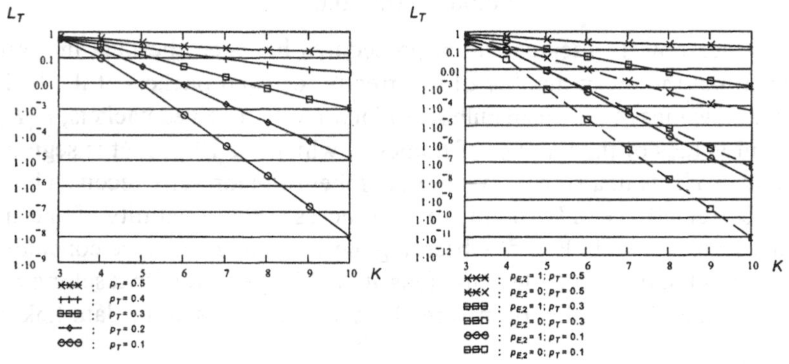

Figure 6. Loss probability L_T for different p_T and $p_{E,2}$.

6. CONCLUSIONS

We provided a tool for accurate performance evaluation of delay/loss sensitive IP-based applications running over the wireless channels. The proposed wireless channel model can be used for performance evaluation at the IP layer using the classic queuing theory.

Despite of abovementioned advantages, the proposed model entirely relies on classic small-scale propagation model and does not take into account signal strength attenuation caused by movements over larger distances. The aim of our further work is to relax this assumption.

REFERENCES

Durgin, D. and Rappaport, T. (2000). Theory of multipath shape factors for small-scale fading wireless channels. *IEEE Trans. on Ant. and Propag.*, 48:682–693.

Krunz, M. and Kim, J.-G. (2001). Fluid analysis of delay and packet discard performance for QoS support in wireless networks. *IEEE JSAC*, 19(2):384–395.

Rappaport, T. (2002). *Wireless communications: principles and practice*. Communications engineering and emerging technologies. Prentice Hall, 2nd edition.

Saleh, A. and Valenzuela, R. (1987). A statistical model for indoor multipath propagation. *IEEE JSAC*, 5(2):128–137.

Swarts, J. and Ferreira, H. (1999). On the evaluation and application of markov channel models in wireless communications. In *Proc. VTC'99*, pages 117–121.

Zhang, Q. and Kassam, S. (1999). Finite-state markov model for Rayleigh fading channels. *IEEE Trans. on Comm.*, 47(11):1688–1692.

Zorzi, M., Rao, R., and Milstein, L. (1997). ARQ error control for fading mobile radio channels. *IEEE Trans. on Veh. Tech.*, 46(2):445–455.

COLLISION AVOIDANCE IN METROPOLITAN OPTICAL ACCESS NETWORKS

Nizar Bouabdallah [1,3], Andre-Luc Beylot [2] and Guy Pujolle [1]

[1] LIP6, University of Paris 6, 8 rue du Capitaine Scott, F-75015 Paris, France; [2] ENSEEIHT - IRIT/TeSA Lab., 2, rue C. Camichel, BP7122, F-31071 Toulouse; [3] Alcatel Research & Innovation, Route de Nozay, F-91460 Marcoussis, France

Abstract: Packet-based optical ring becomes the standard access medium in metropolitan networks. Its performance depends mainly on how optical resource sharing, among different competing access nodes, takes place. This network architecture has mostly been explored in regard to synchronous transmission. However, in the present paper, we focus on the performance of asynchronous transmission-based metropolitan networks with variable packet sizes. Analytical models are presented in an attempt to provide explicit formulas that express the mean access delay of each node of the bus-based optical access network. In addition, we prove that in such a network, fairness problems are likely to arise between upstream and downstream nodes sharing a common data channel. Furthermore, we show that sharing the available bandwidth fairly and arbitrarily between access nodes, as in slotted WDM rings, does not resolve the fairness problem in asynchronous system. A basic rule, in order to achieve fairness, consists in avoiding random division of the available bandwidth caused by the arbitrary transmission of the upstream nodes.

Key words: Bus-based optical access network, Medium Access Control (MAC) Protocol, Fairness control, Access delay evaluation.

1. INTRODUCTION

In next-generation metropolitan networks, internet traffic is deemed to be stamped by three important characteristics. In fact, packet-based data traffic of bursty nature will become prevalent 1. Moreover, it is believed that traffic will fluctuate heavily and on a random basis. Finally, internet traffic will keep on growing in the next few years up to, and eventually beyond, 1 Tbit/s. The architecture of next-generation metro networks must consequently evolve enabling to tackle the new challenges, which are set by the aforementioned characteristics. In this regard, three major enabling factors can be identified as crucial for the evolution process: optics, packet switching and protocol convergence.

In the metropolitan segment, infrastructures are generally organized over a ring topology. We have proposed a new architecture named DBORN (Dual Bus Optical Ring Network), which satisfies all the requirements of next-generation metro networks. The DBORN architecture will be described in this paper. For more detailed information about this architecture the reader is invited to refer to 2. Nonetheless, the work presented in this study, is more pertaining to the design of the media-access-control (MAC) protocol planned for the bus-based optical access networks such as DBORN. This protocol is designed for efficient transport of variable-sized IP packets, whereas it does not address the inherent fairness control issue, characteristic of shared medium networks.

Generally, in order to avoid collisions on the individual WDM channels of such networks and arbitrate the bandwidth access, MAC protocols are needed. In the mean time, several access protocols for all-optical slotted WDM rings have been proposed in the literature 3, 4, 5. Most of them consider as many wavelength channels as nodes in the network, resulting in serious scalability issues, especially for MANs (Metropolitan Area Networks). Moreover, some proposals require transmitter/receiver arrays at each node leading to high equipment costs and control complexity 5. In order to deal with the aforementioned limitations, we proposed a novel access protocol for a packet-based optical metropolitan network supporting much more ring nodes than the available wavelengths in the network. The proposed MAC protocol addresses the case of non slotted WDM rings.

Since several source nodes share a common channel, one upstream node can grab all the available bandwidth, and possibly starve downstream nodes competing to access the same channel. Protocols at various levels (such as MAC or CAC – Call Admission Control) must be introduced to ensure good utilization of transmission resources and alleviate fairness problems. In general, fairness control mechanisms limit the transmission of upstream nodes in an attempt to leave enough bandwidth for downstream stations 6,7. These schemes may be efficient in the case of slotted WDM rings (i.e. synchronous transmission). However, they do not perform well in the case of asynchronous transmission based architectures like DBORN. We present here analytical models that aim to illustrate this issue. Despite its importance and up to now, the analytical study of asynchronous transmission in bus-based optical access networks has not been tackled.

The key behavior metric in such networks is the access delay at each node competing to access the shared data medium. By presenting a specific two-nodes bus as a first case study, we examine the average access delay of each node thanks to an exact analytical model. Afterwards, approximate models handling the general case, with much more nodes, are developed. The fairness issues are also dealt with in the proposed models. Simulation results show that the analytical models remain highly accurate under various traffic loads.

The remaining parts of this article are organized as follows. Section II focuses on the MAC context including a description of the network and node architectures along with the main features. Analytical models for evaluating the access delay performance of each ring node are developed in Section III. Then, section IV validates the accuracy of the models by comparing the analytical results with that obtained by means of simulations, and it discusses the effects of unfair access to the data channel. Finally, some conclusions are drawn in section V.

2. NETWORK ARCHITECTURE AND MAC DESIGN

This section describes the DBORN architecture and the proposed MAC protocol. DBORN can be described as a unidirectional fiber split into downstream and upstream channels spectrally disjoint (i.e. on different wavelengths) as shown in figure 1. The downstream bus, initiated at the hub node, is a medium shared in reading, while the upstream bus, initiated in the ring nodes, is a multiple access-writing medium.

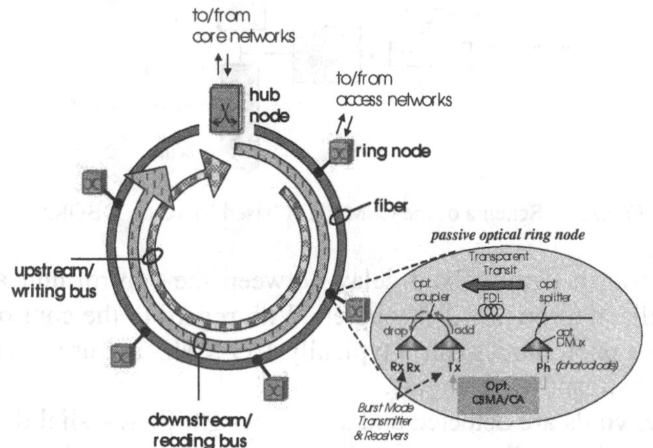

Figure 1. Overview of DBORN network and node architecture

In terms of logical performance, the main issue is related to the collision-free packet insertion on a shared writing bus. Since the transit path remains transparent and passive, no packet is dropped once transmitted on the ring (optical memory is still in the research stage). Hence, traffic control mechanisms are required at the electronic edge of the ring nodes to regulate data emission. In this regard, each DBORN ring node is equipped with void/null-detection mechanism in its upstream operating plane. This mechanism tends to retain the upstream traffic flow within the optical layer while monitoring the medium activity.

In a fixed-slotted ring system with fixed-packet size, void (i.e. slot) filling can be carried out immediately upon its detection, since the void duration is either one or multiple series of fixed-packet size duration. The detected void is therefore guaranteed to have a minimum duration of one fixed-packet length. However in non slotted ring systems with variable packet length and arbitrary void duration, it is very likely for a collision to occur if a packet is immediately transmitted upon detecting the edge of a void.

To avoid the abovementioned problem, a very simple collision avoidance system is implemented through photodiode power detection on each locally accessible upstream wavelength (figure 2). So, ring nodes first use an optical coupler to separate an incoming signal into two identical signals: the main transit signal and its copy used for control. A Fiber Delay Line (FDL) cre-

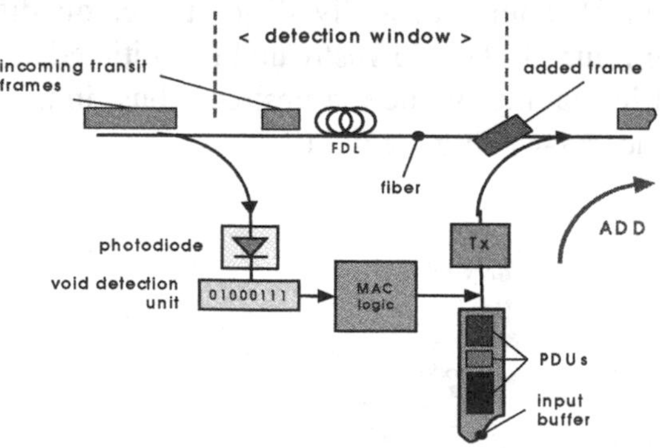

Figure 2. Schema of the CSMA/CA based MAC of DBORN

ates on the transit path a fixed delay between the control unit and the add function realized through a 2:1 coupler. With regard to the control part, as in 8, low bit rate photodiodes (ph) –typically 155 MHz- are used to monitor the activity on upstream wavelengths.

This way, voids are detected and a fixed length FDL – slightly larger than the MTU (Maximum Transmission Unit) size allowed on the network – ensures collision free packet insertion on the upstream bus from the add port. The introduction of a FDL delays the upstream flow by one maximum frame duration plus the information processing time, so that the MAC unit will have sufficient time to listen and measure the medium occupancy. The ring node will begin injecting a packet to fill the void only if the null period is large enough (i.e. at least equal to the size of the packet to be inserted). Undelivered data will remain buffered in the electronic memory of the ring node until a sufficient void space is detected.

However, considering only this basic mechanism, HOL (Head Of the Line) blocking and fairness issues arise. A direct resulting effect is performance degradation for ring nodes that are close to the hub node on the upstream bus. Additional flow control mechanisms have thus to be considered, both at the MAC layer and in upper layers at edge nodes.

3. ANALYTICAL MODELS

3.1 Framework

In this section, we will analyze the performance of the network in term of access delay. The proposed MAC protocol, which is based on CSMA/CA principle, avoids collision between local and transient packets competing to access the shared medium. As described earlier, the MAC protocol detects a gap between two packets on the optical channel, then it tries to insert a local packet into the perceived gap. However, in such an environment, fairness issues could arise.

In this study, the network is composed of N ring nodes sharing a common medium (e.g. one wavelength) used to contact the hub. Packets arrive to each node according to a Poisson process with an arrival rate λ (the analysis presented in this paper can easily be extended to unbalanced traffic conditions). We assume that the transmission time of the packets S forms a sequence of iid random variables, distributed according to some common distribution function f_S with a mean $E[S]$, a second moment $E[S^2]$ and a Laplace transform B^*. Moreover, we assume that the length of the packets emitted by the different nodes has the same distribution. The input load ρ_i of a node i ($i = 1,\dots, N$) is consequently equal to:

$$\rho_i = \rho = \lambda E[S] \tag{1}$$

The aim of this study is to determine the mean waiting time (or the access delay) of the different nodes $E[W_i]$, defined as the time spent by a packet in the queue i until successfully starting its transmission. Once a packet is emitted, it will not be blocked anymore and will only experience constant delays up to the hub.

We will first study the performance of the first two nodes. An exact model is presented. Approximate analytical methods are then proposed to extend the results to the following nodes, giving upper and lower bounds of the waiting time.

3.2 Analysis of the first two nodes

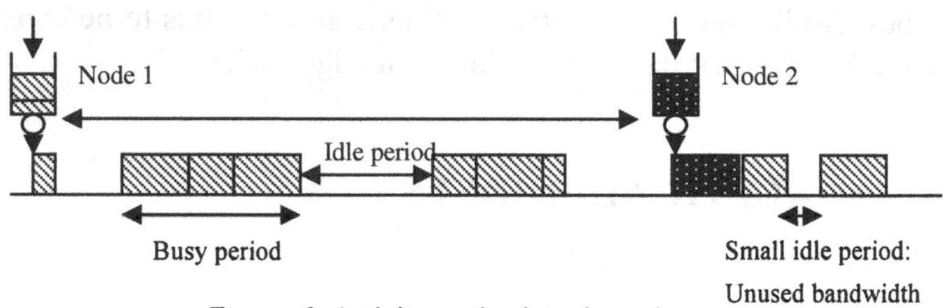

Figure 3. Activity on the data channel

In order to simplify the analysis, let us primarily consider the first two nodes. The traffic of the first node has a higher priority to access the medium. The head-of-line packet of the second queue can only access the channel if the medium is free for a sufficient time period, larger than its transmission time (figure 3). So, the emission process of the second node depends on the activity of the first one. The first queue can be simply modeled by an M/G/1 queue. Hence, the waiting time of the first node is given by:

$$E[W_1] = \frac{\lambda E[S^2]}{2(1-\rho)} \tag{2}$$

So in the remainder study, we will focus on the second queue analysis. This method will be iterated to determine the performance of the other nodes. In this paper, the "link state" refers to the state of the link when a packet, from a ring node, attempts to access the data channel. The wavelength channel can be in one of two states: free (idle) or occupied (busy). It is obvious that for packets from the upstream node, i.e. node 1, the channel is always idle. However, when packets from the downstream node, i.e. node 2, try to access the channel, the latter can be either free or occupied by upstream traffic. It is important to note in this regard that the state of the medium, as seen by node 2, alternates continuously between an idle and a busy period.

Let $\{A_i(t), t \in \mathbf{R}\}$ denote the arrival process of packets to the queue i. As the delays due to the propagation from node 1 to node 2 (Δ_1) and to the FDL (Δ_2) are constant, the whole system can be analyzed as a priority queue with Preemptive Repeat Identical (PRI) discipline 9 (i.e. if a packet can not be sent because the idle period is not long enough, the packet size will not change). The arrival process of packets is defined as, $A(t) = A_1(t-\Delta) + A_2(t)$, where traffic 1 has the higher priority and $\Delta = \Delta_1 + \Delta_2$. The workload for the queue consists of two classes of jobs. The objective is to determine the average waiting time for jobs of each class in the queue. Note that the waiting time of the higher priority class, $E[W_1]$, is simply the waiting time in an ordinary M/G/1 queue as described in (2). Below, we will focus on the waiting time of the class i customers where $i \geq 2$.

Under a preemptive repeat policy, service is interrupted whenever an arriving customer has higher priority than the one in service. The new arrived customer begins service at once. A preempted job will restart service from the beginning as soon as there are no higher priority jobs remaining in the queue. In other words, the preemptive repeat strategy stipulates that the work already done on an interrupted job is lost. In this case, the transmission time of the interrupted packet may be re-sampled according to the service time distribution after every preemption (preemptive repeat different discipline) or it may be the same as in the first service attempt (preemptive repeat identical discipline). In this study, we adopt the PRI discipline since it coincides with the real behavior of the network.

We can consequently apply the results presented in 9 based on 10. Let C_i denote the completion time of a class i customer (i.e. the time between starting and finishing service, including the preemption time). Let S_2 be the transmission time of the packet of class 2 that is chosen first. Suppose that $\tilde{n}$ preemptions occur because of the arrival of packets of class 1. Let $I(n)$ be the service time futilely expended due to the nth preemption, and $B(n)$ be the duration of the nth preemption. Note that $I(n)$ is the nth unusable idle period encountered by the packet while trying to access the data channel and $B(n)$ is the nth busy period of packets of class 1. The completion time C_2 for a packet of class 2 can be written as:

$$C_2 = S_2 + \sum_{n=1}^{\tilde{n}} I(n) + \sum_{n=1}^{\tilde{n}} B(n) \tag{3}$$

Which leads to:

$$E[C_2] = \frac{\overline{X}_{1,1}}{1-\rho}, \text{ with } \overline{X}_{1,1} = \frac{B^*(-\lambda)-1}{\lambda} \tag{4}$$

The mean waiting time may be derived as follows. In the book referenced above, the mean waiting time $E[Z_2]$ is the time spent by a class 2 packet from its arrival until service begins. It does not include the completion time. The mean response time $E[R_2]$ is consequently equal to:

$$E[R_2] = E[Z_2] + E[C_2] \tag{5}$$

As explained before, we refer in this paper to the waiting time, as the time spent by a packet in the queue until its transmission successfully begins. The mean waiting time $E[W_2]$ can be written as:

$$E[W_2] = E[R_2] - E[S] \tag{6}$$

In the case of p traffic classes, we have:

$$E[Z_p] = \frac{\sum_{k=1}^{p} \lambda \left(\overline{X}_{k,2}\left(1-\rho_{k-1}^{+}\right) + \frac{2\left(B^*(-2(k-1)\lambda) - 2B^*(-(k-1)\lambda) + 1\right)\rho_{k-1}^{+}}{((k-1)\lambda)^2} \right)}{2\left(1-\rho_{p-1}^{+}\right)\left(1-\rho_p^{+}\right)} \tag{7}$$

with

$$\rho_k^{+} = \sum_{i=1}^{k} \lambda \overline{X}_{k,1} \tag{8}$$

$$\overline{X}_{k,1} = \frac{B^*(-(k-1)\lambda) - 1}{(k-1)\lambda}, \quad \overline{X}_{1,1} = E[S] \tag{9}$$

$$\overline{X}_{k,2} = \frac{2\left\{B^*(-2(k-1)\lambda) - B^*(-(k-1)\lambda) + (k-1)\lambda B'^*(-(k-1)\lambda)\right\}}{(k-1)\lambda} \tag{10}$$

$$\overline{X}_{1,2} = E\left[S^2\right] \tag{11}$$

Solving (7) for p=2, we can determine the mean waiting time of the second node queue, which is given by:

$$E[W_2] = E[Z_2] + E[C_2] - E[S] \tag{12}$$

3.3 Extension to N nodes

3.3.1 An upper bound for the mean waiting time

Unfortunately, the previous method can not be applied to the following nodes. Indeed, in the single priority queue with PRI discipline, the emission time already elapsed on an interrupted job is lost and can not be used anymore by lower priority jobs ($i+1,\ldots,N$). However, in reality, if the idle period is not long enough to support the queue i head-of-line packet, the medium remains free and this idle period can be used by downstream nodes.

Using this method, it can be shown that the analysis of the system with a single priority queue will lead to an upper bound of the mean waiting time for the downstream node k where $k > 2$:

$$E[W_k] \le E[W_k^+] = E[Z_k] + E[C_k] - E[S] \tag{13}$$

Where $E[Z_k]$ is derived using (7) and

$$E[C_k] = \frac{\overline{X}_{k-1,1}}{1-\rho_{k-1}} \tag{14}$$

3.3.2 A lower bound for the mean waiting time

Conversely, the following method leads to a lower bound for the waiting time. In each node, the upstream traffic has a higher priority than the local traffic. So, the emission process of the local queue depends only on the activity of the upstream nodes and the profile of busy and idle periods generated by upstream flows. The method consists on aggregating all the upstream traffics in a single flow. The packets of the aggregated flow arrive according to a Poisson process. Then, we analyze each node as a single queue with two traffic classes under PRI priority discipline where the local traffic has the lower priority.

This approximate analysis leads to an underestimation of the mean response time because it may cause longer busy period duration and consequently longer idle period duration as well. This is true as long as the distribution of the packet length is the same in the different nodes (on average, if a packet of a previous node can not be sent in the medium because it is too large, a packet of the local node will not pass either). In reality, the free bandwidth seen by a downstream node is much more fragmented than the one generated by the aggregated upstream flow. One then obtains the following results by applying the method of paragraph 3.2. to each node k with two flow classes (i.e. upstream and local traffic) with respective arrival rates:

$$\lambda_{k-1}^- = (k-1)\lambda, \quad \lambda_k = \lambda \tag{15}$$

It corresponds to "equivalent loads":

$$\rho_{k-1}^- = \lambda_{k-1}^- E[S], \quad \rho_k^- = \lambda_{k-1}^- E[S] + \frac{B^*\left(-\lambda_{k-1}^-\right) - 1}{(k-1)} \tag{16}$$

The lower bound of the waiting time is given by:

$$E[W_k^-] = E[Z_k^-] + E[C_k^-] - E[S] \tag{17}$$

Where $E[Z_k^-]$ is derived using (7) and

$$E\left[C_k^-\right]=\frac{B^*\left(-\lambda_{k-1}^-\right)-1}{\lambda_{k-1}^-\left(1-\rho_{k-1}^-\right)} \tag{18}$$

3.4 Example

Different packet length distributions can be considered. In the present paper, we consider packets of variable length (50, 500 and 1500 bytes) more or less representative of the peaks in packet size distribution in Ethernet.

Let p_i be the probability of the different packet sizes and d_i the corresponding emission time.

The mean waiting time of the first queue (2), of the second queue (12) and the bounds on the waiting times for the following nodes (13) (17) can be derived using the following parameters:

$$E[S^k]=\sum_i p_i d_i^k,\ B^*(s)=\sum_i p_i e^{-sd_i} \tag{19}$$

4. NUMERICAL RESULTS

To evaluate the accuracy of the proposed analytical models, we compare their results with those obtained from a simulation conducted on, network simulator 2. In the following, only a subset and a synthesis of the results are presented. In all our simulations, unless otherwise specified, we assume that (1) all the ring nodes share a common upstream wavelength modulated at 1 Gbit/s ; (2) the packets arrive according to a Poisson process; (3) the arrival rate of the packets to each node is the same in order to highlight the fairness issues; and (4) all the ring nodes transmit only to the hub. In all the figures depicting the simulation results, the traffic load on x-axis denotes the average traffic load ρ sourced from every node to the hub.

The analysis results of access delay for the first two nodes are presented in figure 4, revealing a perfect match with the simulation results: analytical results practically coincide with the simulation results. We consider packets of variable length (50, 500 and 1500 bytes) more or less representative of the peaks in packet size distribution in Ethernet. The total traffic volume comprises 50% of 1500 Bytes, 40% of 500 Bytes and 10% of 50 Bytes packets size. We observe that:

- Under light traffic load, the access delay of the downstream node is more important than upstream node one. As a result, the Fairness issue is pronounced even under light traffic load.

• Under high traffic load, the difference between the performance of upstream and downstream nodes sharing the optical channel increases. The

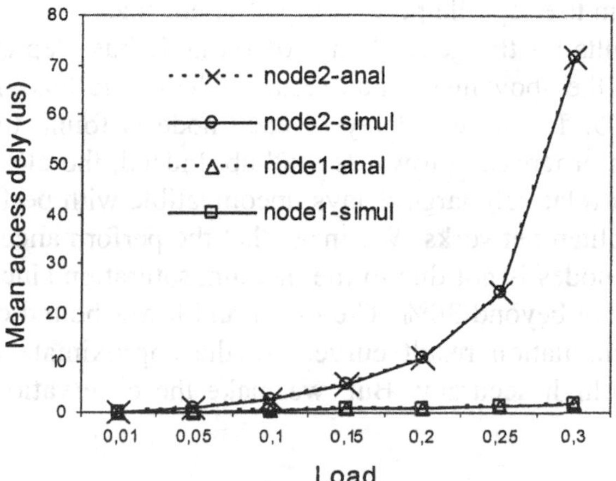

Figure 4. Mean access delay of the first two nodes with variable-packet size traffic

main reason is that upstream node grabs more bandwidth thus leaving less capacity to the downstream node.

The analysis results in this special scenario are so significant. We observe that even when the upstream node uses a small part of the available bandwidth, the downstream nodes performance is strongly affected. The fairness issue is always present in shared medium networks. This is mainly due to the lack of organization of the emission process in the network and the absence of control mechanisms. In fact, the mismatch, between the idle period distribution resulting from upstream node utilisation and the packets' size distribution of the downstream node, often leads to bandwidth waste as well as fairness problems with regard to resource access.

Once a packet of maximum size is at the head of the insertion buffer, it blocks the emission process until finding an adequate void: this is the well-known HOL blocking problem. Thus, sharing the bandwidth fairly and arbitrarily between nodes is not sufficient to ensure satisfying results. The sharing process must thus be done smartly in order to preserve a maximum of useful bandwidth for downstream nodes. In general, fairness control mechanisms limit the transmission of upstream nodes to keep enough bandwidth for downstream stations. These schemes may be efficient in the case of slotted WDM rings. However, they don't perform well in the case of asynchronous transmission based architectures like DBORN.

Hence, we suggest preserving bandwidth (represented by idle periods) by upstream nodes in order to satisfy downstream nodes requirements in an organized way. A basic rule consists in avoiding random division of the resource, which would lead to inadequacy between idle periods length and the layer 2 PDUs (Protocol Data Units) size. Therefore the control mechanism has to prevent greedy upstream stations from taking more than their fair share by forcing them to keep idle periods of sufficient size.

The analysis results for the general case of six-node bus, depicted in the figure 5, emphasize the abovementioned results. The traffic load ρ sourced by each node is 0,05. The access delay of each node is found to increase monotonically when progressing towards the hub. Indeed, the closest nodes to the hub encounter relatively large delays, incompatible with performances expected in metropolitan networks. We insist that the performance degradation of downstream nodes is not due to the medium saturation since the medium occupation is not beyond 30%. The upper and lower bound curves are very close to the simulation result curve. So, the approximate analytical models can achieve high accuracy. But, we make the observation that the

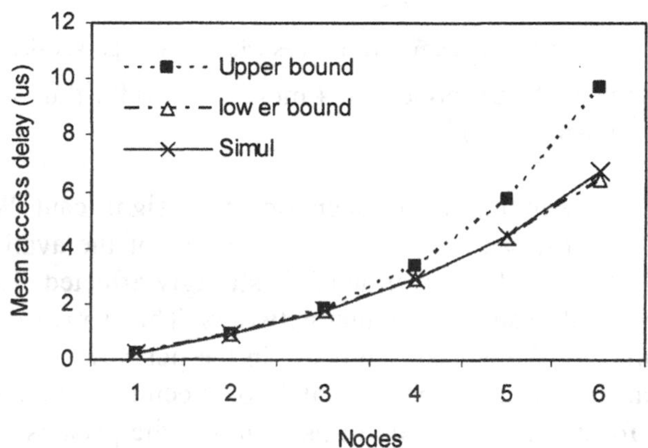

Figure 5. Mean access delay of the six-node bus with variable-packet size traffic

bounds become less accurate for the closest nodes to the hub, especially the upper one.

5. CONCLUSION

This paper, to the author's knowledge, provides the first analysis of shared bus network behavior with asynchronous transmission. We analyzed the system performance in terms of access delay required by each node to inject a packet on the shared medium. The analysis results showed that fairness issues are likely to arise between upstream and downstream nodes even under light loads. We observed that sharing the available bandwidth fairly and arbitrarily between nodes does not resolve the fairness problem. Consequently, additional flow control mechanism has to be considered, not only to limit the transmission of the upstream nodes but also to organize their emission process. Simulations results showed that the proposed analytical models are extremely accurate under various traffic loads.

REFERENCES

1. M. J. O'Mahony, D. Simeonidou, D. K. Hunter, A. Tzanakak, "The Application of Optical Packet Switching in Future Communication networks", IEEE Commun. Mag., pp. 128-135, March 2001.
2. N. Le Sauze et al., "A novel, low cost optical packet metropolitan ring architecture", Proc. Of ECOC '01, Amsterdam, Netherlands, Vol. 4, pp. 66-67, October 2001.
3. M. A. Marsan, A. Bianco, E. Leonardi, M. Meo, and F. Neri, "MAC protocols and fairness control in WDM multirings with tunable transmitters and fixed receivers", IEEE/OSA J. Ligh. Tech., vol. 14, pp. 1230-1244, June 1996.
4. J. Fransson, M. Johansson, M. Roughan, L. Andrew, and M. A. Summerfield, "Design of a medium access control protocol for a WDMA/TDMA photonic ring network", Proc. of GLOBECOM '98, Sydney, Australia, Vol. 1, pp. 307-312, November 1998.
5. A. Fumagalli, M. Johansson, and M. Roughan, "A token-based protocol for integrated packet and circuit switching in WDM rings", Proc. of GLOBECOM '98, Sydney, Australia, pp. 2339-2344, November 1998.
6. M. A. Marsan et al., "Metaring Fairness Control Schemes in All-Optical WDM Rings", Proc. of INFOCOM '97, Kobe, Japan, vol. 2, pp. 752-760, April 1997.
7. J. S. Yih, C. S. Li, D. D. Kundlur, and M. S. Yang, "Network access fairness control for concurrent traffic in gigabit LANs", Proc. of INFOCOM '93, San Francisco, California, vol. 2, pp. 497-504, March 1993.
8. R. Gaudino et al., "RINGO: a WDM Ring Optical Packet Network Demonstrator", Proc. of ECOC '01, Amsterdam, Netherlands, Vol. 4, pp. 620-621, September 2001.
9. H. Takagi, Queueing Analysis Vol I: Vacation and Priority Systems Part I, North Holland, 1991.
10. N.K. Jaiswal, Priority Queues, Academic Press, 1968.

Part Seven: Posters

TOWARD AN INTELLIGENT BANDWIDTH BROKER MODEL FOR RESOURCES MANAGEMENT IN DIFFSERV NETWORKS

R. Nassrallah, M. Lemercier, D. Gaïti
Institut of Computer Science and Engineering of Troyes (ISTIT), University of Technology of Troyes (UTT), 12 rue Marie Curie, BP 2060, Troyes cedex, France
(nassrallah,lemercier,gaiti)@utt.fr

Abstract: DiffServ model is known to be the most used model to handle QoS over IP networks. Moreover this model has the advantage to be appropriate for use on large network contrary to the IntServ model that suffers from scalability problem. However, the DiffServ model has two major difficulties: routers' configuration and resources' allocation problems. In this paper, we introduce a new approach based on customers' Service Level Agreements (SLA) declaration. The resource allocation is done by a federal entity called Bandwidth Broker implemented using Web-Services. Our proposal avoids the use of signaling protocol between the Bandwidth Broker and the core routers when establishing a new flow. Thus, core routers do not have the responsibility to store the customers' traffics information and therefore, we respect the DiffServ model philosophy. Our tool provides the admission control and resource allocation management using overbooking techniques which guarantees the performances of priority traffics.

Keywords: QoS, DiffServ, SLA, Bandwidth Broker, Intelligent agent, Web-Services

1. INTRODUCTION

The IETF proposed two models to handle the QoS over IP networks. The Intserv model adopts a per-flow approach, which means that each traffic flow is handled separately at each router. thus, resources can be allocated individually to each flow using RSVP (Reservation Protocol). It has been recognized that such a per-flow approach is affected by scalability problems which prevents from its applicability to large networks.

DiffServ is the second model[1] and it responds better to the QoS problem over IP networks. It aims at providing QoS on a per-aggregate basis. It offers services differentiation mechanisms, which allow packets classification. The DSCP (DiffServ Code Point) is a six-bit field in the IP packets header. It allows the classification of 64 different service classes[2]. The DiffServ model defines some standard service classes. The Premium service is suitable for real time applications (Voice over IP, Videoconference) that need lower transfer delay and jitter. The Assured Service suites for non real-time applications. It is characterized by its reduced packet loss rate and its reasonable transfer delay. The best effort service is a none-guaranteed service. Best effort packets are always accepted in the network and they do not affect higher priority packets. They are the first packets to be lost in case of congestion. Finally, eight values of DSCP were reserved to assure the compatibility with the previous TOS filed in IPv4. Those values constitute the CS (Class Selector) service.

To assure theses services, the DiffServ routers must support a set of predefined behaviors called PHB (Per Hop Behavior). The internal routers handle packets according to the PHB identifier, and do not distinguish the individual flows. Then a Premium service is assured by the EF (Expedite Forwarding) PHB[3]. Similarly, the Assured Services are handled by the AF (Assured Forwarding) PHB[4].

DiffServ defines the network architecture inside a DS domain. Each domain is a set of interior routers (core routers) enclosed by another set of boundary routers (edge router). The edge routers handle the packet classification and the traffic monitoring functions. They control the incoming traffic to see if the access contracts are respected (flow, peak rate, packets size, etc.). However, the core routers assure basic functions such as queuing and scheduling according to packets priorities without having to know the contracts characteristics. Consequently, the DiffServ model pushes back the network management complexity to the edge, leaving relatively simple tasks to core routers.

The DiffServ operation can be guaranteed only if the incoming traffics respect a set of predefined constraints. Thus, each flow has to state its characteristics to enable an optimal configuration of the network devices. It is therefore of primary importance to assure the DiffServ policing at the level of the edge routers. The policing control takes into consideration certain pre-stated parameters. Many algorithms were proposed, however this debate seems to be closed because the IETF selected and published algorithms for this function. The Token Bucket (TB) is a standardized mechanism that allows identifying non-conforming packets[5]. It has two parameters (token depth b, token rate r), a queue for packets and a bucket containing b tokens. Each token represents the right to emit a byte. Thus packets arriving at the TB are conform if their size is equal or lower than the number of available tokens. The rate of the outgoing flow is fixed by r. In addition, if the number of token in the bucket is sufficient a small amount of packets could be emitted at peak rate flow. A packet is declared to be non-conform if there are no sufficient tokens. In this case a second mechanism takes place and chooses the intervention method to be applied. Three methods exist: maker, shaper and dropper.

The RFC 2698[6] proposes an algorithm "A Two Rate Three Color Marker" that handle three levels of policing control. The Two Rate Three Color Marker meters an IP packet stream and marks its packets green, yellow or red. A packet is marked red if it exceeds the peak rate. Otherwise it is marked either yellow or green depending on whether it exceeds or doesn't exceed the TB rate (r). The TRTCM is useful, for example, for ingress policing of a service, where a peak rate needs to be enforced separately from a committed rate.

In the core routers, packets can be buffered into different queues according to the DSCP. The queues being of limited size, a packet is rejected when it overflows. A mechanism of congestion control can be applied for each one of these queues. Several mechanisms were proposed to anticipate congestion problem: RED (Random Early Detection), WRED (Weighted RED), etc.

Each router of the network uses a scheduling policy to determine in which order the packets will be transmitted. There are several algorithms which aim at solving this problem. PQ (Priority Queuing), CBWFQ (Class-Based Weighted Fair Queuing), WRR (Weighted Round Robin), GDR (Deficit Round Robin), WFQ (Weighted Fair Queuing), WF2Q (Worst-Case Fair Weighted Fair Queuing), etc.

Some algorithms provide a static scheduling function, which implies that resources can not be freed if they are not totally used, to be then reallocated to other classes.

Actually, algorithms allowing a dynamic allocation of the resources are privileged in DiffServ networks.

We presented the main functions of a DiffServ router. The IETF recommendations leave many questions about monitoring and scheduling functions without any response. How to configure in an optimal way the policing control algorithm parameters for a given customer using several types of traffic? How to choose the intervention method in case of traffic in excess? As for core routers which will be the configuration allowing the management of EF, AF, CS and BE PHBs? How to define the solution if it is based on a combination of different scheduling mechanisms (WFQ + PQ)? One of the limits of this system is the difficulty of configuring DiffServ routers. Indeed, there are multiple ways of managing the differentiation of services according to the classes.

The monitoring mechanisms (TB and TRTRM) give the required parameters to define the source traffic envelope in a DiffServ domain. However, other indications are needed to ensure the whole network functionalities. These indications should enable edge routers to ensure traffics classification and to specify the desired QoS constraints of each traffic. It is thus, necessary to define a contract between the customer and the service provider. Section 2 presents the required elements for the DiffServ domain configuration. In section 3 we define our Bandwidth Broker model and the allocation strategy. Section 4 presents the implementation of our model using Web-Services and our future work. The conclusion is in section 5.

2. RESOURCES MANAGEMENT AND DIFFSERV CONFIGURATION

2.1 Per-Domain Behavior

The informational RFC 3086[7] defines Per-Domain Behavior (PDB) as the expected edge-to-edge treatment that an aggregate flow will receive within a DS Domain. A PDB sonsists of one or more PHBs and traffic conditioning requirements. Contrary to PHBs, a PDB defines a particular combination of DiffServ components that can be used inside a domain to offer a quantifiable QoS. Five PDBs have been defined.

1 A Best Effort (BE) PDB.
2 The Virtual Wire (VW) PDB.
3 The Assured Rate AR PDB.
4 The one-to-any Assured Rate PDB.
5 A Lower Effort (LE).

For the moment, the definitions of PDBs seem to us insufficient to characterize with precision the edge-to-edge behavior within a DS domain. We need a complete description of the mechanisms to be used within the DS domain that handle the policing, queuing and scheduling functions. The PDB definition is done by the operator of the DS domain. It characterizes the behavior and the operation of his network. That's why it seems to us that it is not necessary to have a normalized PDB.

2.2 SLA - SLS

A Service Level Agreement (SLA) is a contract that specifies forwarding services a customer can expect to receive from a provider. The SLA can cover one or more services a provider can offer to a customer. Each service is technically described in a Service Level Specification (SLS). The SLS contains typically a description of the allowed traffic envelope (peak flow, mean flow, etc.). It's on this basis that a provider can check the traffic conformity and decides which policy to apply for the traffic in excess. The Tequila project gives a more detailed description of the SLS parameters [8].

2.3 Admission Control

The admission control allows the acceptance or the refusal of a new traffic. This could be done by using multi-criteria traffic filters. Only conform traffics with the definite filters are authorized on the network. However, in case of a dynamic service (dynamic SLS) the implementation of an admission control supposes that edge router can access a centralized resources database.

The implementation of an admission control based on per-flow signalization raises significant scalability problem. Thus, a best approach consists in using a minimal description of each flow (SLS) coupled with an admissibility condition. The admissibility criterion determines whether to accept or not a new flow. A criterion can be a threshold of available bandwidth. We made the choice to store in the Bandwidth Broker central database the whole reservations on each link of the DS domain.

A Bandwidth Broker (BB) is a central element in an Autonomous System (AS) that manages network resources within its domain. It also cooperates with other BBs in the neighboring domains to manage the In/out inter-domain communications. The BB gathers and monitors the state of QoS resources within its domain. It uses that information to decide whether to accept or not new traffics. The BB follows the client/server model. It can use COPS protocol to communicate with edge routers within its domain or with adjacent BBs from neighboring domains.

2.4 Policy Based Network

A PDP (Policy Decision Point) is a process that makes decisions based on policy rules and the state of the services those policies manage. The PDP is responsible of the policy interpretation and initiating deployment. In certain cases it transforms and/or passes the policy rules and data into a form of syntax that the PEP (Policy Enforcement Point) can accept. PEP is an agent running on a device (edge router) that enforce a policy decision and/or makes a configuration change. In the DiffServ model, the BB is a PDP.

Signaling protocols are commonly used in policy based network. Policy communication protocols (COPS) enable reading/writing data from a policy repository (SLS database) and communication between PEP and PDP. COPS stands for Common Open Policy Service, is a client/server model that support policy control over QoS signaling protocol[9]. It uses TCP protocol for messages exchange. Many COPS extensions exist like the outsourcing model, the provisioning model etc. In the outsourcing model, PEP can send request, update and delete messages to remote PDP and the PDP returns back its decision to the PEP. The provisioning model (COPS-PR) is used to "push" decisions from PDP to PEP, policy data is described by Policy Information Base (PIB).

According to COPS-PR protocol, the network devices identify their capabilities to PDP using PIB model. PDP can then take into account characteristics of the device, when handling request and/or translating policy rules into PIB parameters. Other extensions like COPS-ODRA for DiffServ, COPS-SLS[10] that supports dynamic SLS were proposed. We thus, believe that COPS is not yet a steady protocol.

3. OUR BANDWIDTH BROKER APPROACH

This section introduces our proposal of an intelligent Bandwidth Broker, its architecture and how it operates. Our goal is to build a tool that handles admission control function and support dynamic SLA. Contrary to other approaches[11] we don't use RAR (Resource Allocation Request) messages because it leads to a two-level resource negotiation. Indeed, the use of RAR can be done only after a previous negotiation of the corresponding SLA.

3.1 DS Domain strategy for resources allocation

The major difficulty that faces a DS domain operator is its capacity to offer the required QoS by supplying in an optimal way the needed resources. Moreover, it is of high importance that a provider can manage easily its domain. We propose a resource allocation strategy of bandwidth based on traffic classes. Each provider has to define the set of traffic classes (DSCP) to use within its domain. For example, he can define four classes: voice, critical, normal and best effort. In this strategy we assume that:

1 All the classes have higher priority compared to the best effort class. In addition, the packets belonging to these priority classes can not be rejected in the core of the domain. This constraint implies that the bandwidth allocated to these classes should be limited and less than the network capacity.

2 The BE traffics are always admitted into the network because they share the remaining bandwidth. The BE class do not offer any guarantees.

3 The remaining bandwidth of a class can be reused by another class.

4 The network operator specifies the DSCP values for each class. In our example we affect EF DSCP to voice class, the AF to critical, the CS to normal and the BE to best effort class.

The provider should specify the maximal allocated bandwidth threshold of each class, allowing the reuse of unused resource by the other classes. We propose to allow resource overbooking of certain classes. This overall strategy for resources allocation is depicted in table 1.

Table 1. Global bandwidth allocation strategy (classes table)

Traffic	Max Allocated BW	Overbooking
Voice (VO)	15%	1.0
Critical (CR)	30%	1.5
Normal (NO)	10%	2.0
Best Effort (BE)	45%	8.0

The provider must limit the EF traffic to guarantee the performance of his DiffServ domain. We thus decided not to authorize EF class overbooking. The other overbooking coefficients are chosen by the operator according to his marketing strategy.

The only constraint imposed by our model is that even with overbooking, the allocated resources of the different priority classes (VO, CR and NO) must remain lower than 100% (15×1 + 30× 1.5 + 10× 2 = 80%). Consequently, the operator guarantees a very low packet loss rate even if all the traffics are active at the same time obviously by degrading the available resources for best effort. However, it is clear that the main priority of an operator is to sell at a higher price the available bandwidth. We presented a simple model for global resource allocation strategy. This approach does not inhibit the operator from adding new classes of traffic.

3.2 Managing resources with Bandwidth Broker

To ensure the admission control function, our Bandwidth Broker requires the reservations statistics within its DS domain. The reservation statistics are grouped by class on each link of the network. The tender of a new SLA implies two important tasks to be done by the Bandwidth Broker. The first one consists of a mapping between the required QoS constraints and a per-domain behavior PDB. This mapping leads to the PHB (or DSCP) assigned to this traffic. At the end of this task the Bandwidth Broker knows the required bandwidth (traffic_throughput in bits/s) and the class of traffic (traffic_class).

The second one consists in determining if the new traffic can be accepted or not in the network. For this reason our BB has the forwarding information of its domain and it knows the whole possible routes between the Ingress and Egress routers. Each route is defined by a set of links ($route_1 = r_{1,1}, r_{1,2} \ldots r_{1,n}$). On each link we have the reservations that were carried out corresponding to the demanded traffic_class. It is then possible to deduce the available bandwidth on this link by the given formula:

$$BW_{avalaible} = BW_{Link} \times (BW_{class\%} \times BW_{overbooking}) - BW_{aLink,class} \quad (1)$$

1 BW_{Link}: BW of the link
2 $BW_{class\%}$: Max allocated BW of this class
3 $BW_{overbooking}$: Overbooking parameter of this class
4 $BW_{aLink,class}$: Sum of the allocated BW on this link for this class.

Example: Supposing that we accepted on a 10Mbit/s link two AF traffics (1.2 Mbits/s and 2.5 Mbits/s). Then the available bandwidth for a new AF traffic is equal to: 10Mbits/s $\times$ (30% $\times$ 1.5) - (1.2 + 2.5) = 0.8Mbits/s.

This computation must be done for each link. We can then deduce the available bandwidth on the route by taking the lowest available bandwidth value on the route's links. If we have more than a possible route, the BB selects the one having the highest available bandwidth. This solution allows a good use of the network resources by selecting the less loaded routes. In case of non point-to-point traffic the BB has to take into account the state of the reservations on the whole set of routes (broadcast). If the allocation of the demanded resources fails on one of these routes, the BB rejects the client's SLA.

3.3 Intelligent Bandwidth Broker Services

Our approach allows the proposition of intelligent services at the same time for customers and the DiffServ operator. Our BB implementation relies on agent technology and Web-Services. An agent is an autonomous program having a goal to attend. In our Bandwidth Broker architecture, agents are Java programs offering services that can be reached using Web-Services. Some of the main functions are the following:

1 A bandwidth allocation request can be rejected because of insufficient resources availability. In this case, an agent can reach the available information in the database and proposes a degraded service by indicating the maximum available bandwidth.
2 Our BB is able to compute the available bandwidth per class of service. If there are no sufficient resource for the required traffic. It can thus, determine if this traffic can be assured by another class. Knowing that clients, do not need to have any information about the operator classes, a software agent can calculate the possible QoS values of this class and propose a new SLA based on new degraded QoS criteria.
3 It is also possible that agents carry out periodic analyses that aim to inform the operator about the overall state of resources within its domain. Thus, a per-class analysis gives the resources allocation rate (i.e. 5% of the overall 30% of AF) and can alert the operator about the less used classes. This alarm means that the operator has to change its business model, either by offering these less used classes at lower prices or simply to stop offering them.

4 After a certain time, the network resources will be consumed. An agent can analyze the levels of available bandwidth on each link, which makes it able to report periodically the load ratio per link (value of BW higher than a threshold). At the issue of this analysis the agent can suggest a network upgrading strategy (to add a new link between two routers).

4. WEB-SERVICES APPROACH

The dynamic SLA management and the interactions between adjacent BB require important exchanges of information. Currently, it seems to us that COPS still unstable and unnecessarily complicates the implementations of Bandwidth Broker. The BB follows the client/server model and that's why we preferred to implement it by using Web-Services.

A Web-Service[12–13] is a software system designed to support interoperable machine-to-machine interaction over a network. It has an interface described in a machine-processable format (specifically WSDL). Other systems interact with the Web Service in a manner prescribed by its description using SOAP messages, typically conveyed using HTTP with an XML serialization in conjunction with other Web-related standards. This section presents the Web-Services technology and our actual Bandwidth Broker architecture.

4.1 Our Bandwidth Broker architecture

The required data for the BB are stored in a relational database management system (RDBMS). From the BB specification presented in section 3, we deduce the tables of the MySQL database.

1. SLA table: contains information about active SLA within the DiffServ domain.
2. Classes table: hold the list of classes that characterize the provider resource allocation policy within a given domain.
3. Routers table: is the set of routers within the DiffServ domain.
4. Links table: is the set of links within the DiffServ domain.
5. Reservations table: lists per link and per class reservations within the DiffServ domain.
6. Route table: can be deduced from the routers, links and SLA tables. However for optimization issues, we decided to pre-calculate all possible routes within the DiffServ domain. This is possible from a technical point of view and corresponds to the Cisco approach (there routers can memorize 600 000 possible routes).

The SQL language allows the management of the database tables and data filtering (i.e. SQL queries). We use Java programs (BASIC Java Service Layer) and the JDBC package to access the data and to implement the basic services. These services allow adding a bandwidth reservation on a link, for all the links of a route, to calculate the available resources on a route for a given class, etc. The Intelligent Agents Layer consists of autonomous agents written in Java. Currently, Madkit is our agents' platform. Several multi-agent platforms have been proposed allowing the development of complex system with the help of agents. The main insufficiency of these approaches is the lack of an organizational structure for the agents. Some researchers[14] have proposed a multi-agent platform named MadKit based on three concepts: agent, group and role. The generic development of a multi-agent system and the agents' organization constitute the central proposal of this platform[15]. Two structure levels are proposed: the group and the role. An agent belongs to one or several groups, and inside a group an agent can play one or several roles. A role can be seen as a particular function of an agent. From the agents' cooperation point of view, these organization concepts allow to structure dialogues between agents. An agent can communicate directly with other agent identified by its address or can broadcast the same message to each agent with a given role in a group.

Finally, the last layer (Web-Services Layer) gathers the whole services accessible from outside and allows the communication with the edges routers and other Bandwidth Broker.

4.2 Previous Bandwidth Broker models

There have been numerous undertakings to propose a Bandwidth Broker model for use within a DiffServ environment, the most notable being the following[16]:

1 CANARIE ANA: Implementation of a basic BB that handles differentiated services. This model uses the BBTP (Bandwidth Broker Transfer Protocol) for the Client/BB communications.
2 University of Kansas Research Group: Implementation of a BB that can handle internal and external differentiated services. This model uses the RAR messages and BBTP for message exchange protocol.
3 Merit: Proposition of a multidomain Bandwidth Broker that support the VLL (Virtual Leased Line). This model focuses on the role of authorizing and establishing one type of service (i.e. VLL).
4 Novel: This model separates the QoS control from core routers. It relies on virtual time reference system for QoS abstraction from the data plane.

4.3 Futures works

Our proposal avoids the use of signaling protocol between the BB and the core routers when establishing a new flow. Thus, the core routers do not have the responsibility to store the customers' traffics information and therefore, we respect the DiffServ model philosophy.

At this stage we have considered that the reservations are active starting from the SLA acceptance date and until its expiry date. Thus, our SLA table contains only the active flows. Therefore, it is possible to enhance this model by adding another table to store non-active flows.

Also it is necessary to consider scheduled SLA with several active/passive phases. Thus, the reservation requests have a start/end time. Consequently, it complicates the computation of the available resources and it is necessary to build the exact reservations state within the time interval of the demanded service. It is thus, necessary to identify all resources requests and releases intervals of this new traffic, and to re-evaluate the whole allocation status per interval. We are evaluating this approach.

For the inter-domain traffics, the BB must preliminary contact the other BB in the adjacent domain to propose an end to end QoS service.

5. CONCLUSION

In this article, we presented the needed functionalities for the operation of a DiffServ network. Additional research on Per-Domain Behavior and signaling protocols seems important to specify the overall behaviors of the devices in an IP network offering QoS. The role of centralized equipment was emphasized in order to provide the admission control function. Our approach allows the exchange of information between the edge-routers and the Bandwidth Broker or between the Bandwidth Broker of different adjacent domains without using signalization between the BB and the core-routers.

The management of an IP network that support QoS and integrate heterogeneous approaches and protocols (IntServ, DiffServ, MPLS, RSVP, COPS, etc) is complex. Many researches have proposed the use of a traditional IP network on the management plan in order to ensure the administrative functionalities. From this point of view, we think that the use of signaling protocols like COPS or RSVP does not benefit from the progress made in the distributed applications domain. We thus built our Bandwidth Broker architecture by using Web-Services concepts.

REFERENCES

1. S. Blake, D. Black, M. Carlson, E. Davies, Z. Wang, and W. Weiss. An architecture for differentiated services. *IETF, RFC 2475*, December 1998.

2. K. Nichols, S. Blake, F. Baker, and D. Black. Definition of the differentiated services field (ds field) in the ipv4 and ipv6 headers. *IETF, RFC 2474*, December 1999.
3. V. Jacobson, K. Nichols, and K. Poduri. An expedited forwarding phb. *IETF, RFC 2598*, June 1999.
4. J. Heinanen, F. Baker, W. Weiss, and J. Wroclawski. Assured forwarding phb group. *IETF, RFC 2597*, June 1999.
5. S. Shenker and J. Wroclawski. General characterization parameters for integrated service network elements. *IETF, RCF 2215*, September 1997.
6. J. Heinanen and R. Guerin. A two rate three color marker. *IETF, RFC 2698*, September 1999.
7. D. Goderis, Y. T'Joens, C. Jacquenet, G. Memenios, G. Pavlou, R. Egan, D. Griffin, P. Georgatsos, L. Georgiadis, and P.V. Heuven. Service level specification semantics and parameters. *draft-tequila-sls-01.txt*, June 2001. Work in progress.
8. K. Nichols and B. Carpenter. Definition of differentiated services per domain behaviors and rules for their specification. *IETF, RFC 3086*, April 2001.
9. D. Durham, J. Boyle, R. Cohen, S. Herzog, R. Rajan, and A. Sastry. The cops (common open policy service) protocol. *IETF, RFC 2748*, January 2000.
10. T.M.T. Nguyen, N. Boukhatem, Y.G. Doudane, and G.Pujolle. Cops-sls: A service level negotiation protocol for internet. *IEEE Communications Magazine*, 40(5):158–165, May 2002.
11. P Chimento and al. Qbone signaling design team. *Final Report*, July 2002. http:qos.internet2.eduwgdocuments-informational20020709-chimento-etal-qbone-signaling.
12. J. Ferber. Multiagent systems for telecommunications: from objects to societies of agents. *networking 2000*, 2000. Paris.
13. Madkit. official web site. *last visited*, June 2004. http:www.madkit.org.
14. S. Sohail and S. Jha. The survey of bandwidth broker. *Technical report UNSW CSE TR 0206*, May 2002. School of Computer Science and Engineering, University of New South Wales.
15. J. McGovern, S. Tyagi, M. Stevens, and S. Mathew. Java web-services architecture. *Morgan Kaufmann*, May 2003.
16. W3C. Web-services architecture. *W3C website*, February 2004. http:www.w3.orgTRws-arch.

A LEARNING AND INTENTIONAL LOCAL POLICY DECISION POINT FOR DYNAMIC QOS PROVISIONING

Francine Krief and Dominique Bouthinon
LIPN Laboratory, UMR CNRS 7030, University of Paris XIII, Avenue Jean-Baptiste Clément 99, 93430 Villetaneuse, France. {krief,db}@lipn.univ-paris13.fr

Abstract: In the policy-based network management, the local policy decision point (LPDP), is used to reach a local decision. This partial decision and the original policy request are next sent to the PDP which renders a final decision. In this paper, we propose to give a real autonomy to the LPDP in term of internal decision and configuration. The LPDP is considered as a learning BDI agent that autonomously adapts the router's behavior to environment changes

Keywords: Policy-Based Management, Self-aware Management, Multi-Agent System, BDI architecture, Learning, Quality of Service, DiffServ mechanisms

1. INTRODUCTION

Today, service providers must provide the quality required by the users. In the context of fierce competition, this quality is negotiated with the customers. A contract, called SLA (Service Level Agreement), is signed between the service provider and the customer[1]. The SLA specifies the service that must be delivered. This implies differentiated treatments and software infrastructures adapted to implement them.

DiffServ is the model accepted by the network providers to allow them this services differentiation. In this model[2], the traffic is separated in traffic classes which are identified by a value coded in the IP header. DiffServ is well adapted to wide networks because the complex operations (e.g., classification, marking) are realized at the network's entry by the edge routers. The core routers only treat packets according to the class coded in the IP header[3], and an adapted behavior, the PHB (Per Hop Behavior).

The IETF proposed a general framework called PBM (Policy-Based Management)[4] for the control and management of these IP networks. This infrastructure provides a certain level of abstraction and allows the network a flexible behavior according to the various events which can occur during its management by using the policy concept.

We proposed an architecture for the self-aware management of IP networks offering quality of service guarantees by using policy-based management and multi-agent systems[5]. The level of autonomy required is reached by introducing the operational objectives and the parameters to be followed in the infrastructure, as well as by providing respective monitoring and adaptation means. The operator does not need to apply corrections and adaptations himself so much anymore. Thus, the management system is simplified and even more oriented towards the definition of policies and operational parameters.

In this paper, we propose to give a real autonomy to the network components in term of internal decision and configuration by introducing a learning and intentional agent in each network element to autonomously adapt the router's behavior to environment changes. This agent can be seen as a learning and intentional Local Policy Decision Point (LPDP).

First, we present policies, intentional agents and the global architecture proposed for the self-aware management of IP networks offering quality of service guarantees. Then we describe the architecture of a LPDP. Finally we present the future work.

2. POLICY APPROACH

The policies can be defined like sets of rules which are applied to the management and control of the access to the network resources[4]. They also allow network administrators or service providers to influence the network element behavior according to certain criteria such as the user's identity or the application type, the traffic required, etc.

In general, the policy rules are in the following form "IF policy_activation_condition THEN policy_action" where the condition describes when policies can be activated.

The IETF introduces the role concept. A role is a type of property that it used to select one or more policies for a set of entities and/or components from among a much larger set of available policies[6].

The policies are centralized in a data base. A Policy Decision Point (PDP) has the responsibility of dispatching the policy rules onto the network elements concerned. The Policy Enforcement Point (PEP), situated in each element, constitutes the application point of the policies.

A policy can be defined at different levels. The highest level corresponds to the business level. Then, this policy must be translated into a network level policy and, then, into a low-level which is understandable by the network element.

The Foundation for Intelligent Physical Agents (FIPA) also defines the policy concept[7]. A policy is a constraint or a set of constraints on the behavior of agents and services.

A policy rule is a conjunction of implications: when a condition holds then an action is permitted, prohibited or whatever.

Policy domains are introduced to efficiently apply policies and simplify policies mechanisms. A policy domain is a set of agents to which a given set of policies apply.

A policy library contains the policies and a distribution mechanism is used to distribute policy rules from originating authorities to mechanisms that have the ability and responsibility of applying policies.

The concept of higher-level policy is introduced to simplify the task of generating specific policy rules for agents and services.

3. INTENTIONAL AGENTS

An agent is a temporal persistent computational system, able to act autonomously to meet its objectives or goals when it is situated in some environment. In order to be perceived as intelligent, a software agent must exhibit a particular kind of behavior, identified by Michael Wooldridge and Nick Jennings[8] as flexible autonomous behavior and characterized by:

- reactivity: intelligent agents must be able to perceive their environment and respond at time to changes on it though its actions;
- pro-activeness: intelligent agents exhibit goal oriented behavior by taking the initiative to satisfy its design objectives;
- social ability : intelligent agents must be able to interact with other agents or humans in order to satisfy their objectives.

The study of intelligent agents has received a great deal of attention in recent years. This paper explores a particular type of intelligent agent, a BDI (Belief-Desire-Intention) agent. BDI agents have been widely used in relatively complex and dynamically changing environments[9]. They are based on the following core data structures: beliefs, desires, intentions, and plans[10]. These data structures represent respectively, information gathered from the environment, a set of tasks or goals contextual to the environment, a set of sub-goals that the agent is currently committed, and specification of how sub-goals may be achieved via primitive actions. The BDI architecture comes with the specification of how these four entities interact, and provides a powerful basis for modeling, specifying, implementing, and verifying agent-based systems.

4. GLOBAL ARCHITECTURE

We proposed an architecture for the self-aware management of IP networks by using policy-based management and multi-agent systems[5]. Using this architecture, the QoS management within the framework of the DiffServ model is dynamic. It includes three levels corresponding to the three mediation components recommended by the architecture of the IST CADENUS project[11]. Moreover, monitoring functions are introduced to allow each level to adapt its behavior to the environment which it is controlling.

Each level implement their own tools of monitoring and have a meta-control level which allows it to adapt its behavior to the dynamicity of the environment it is managing (see figure 1).

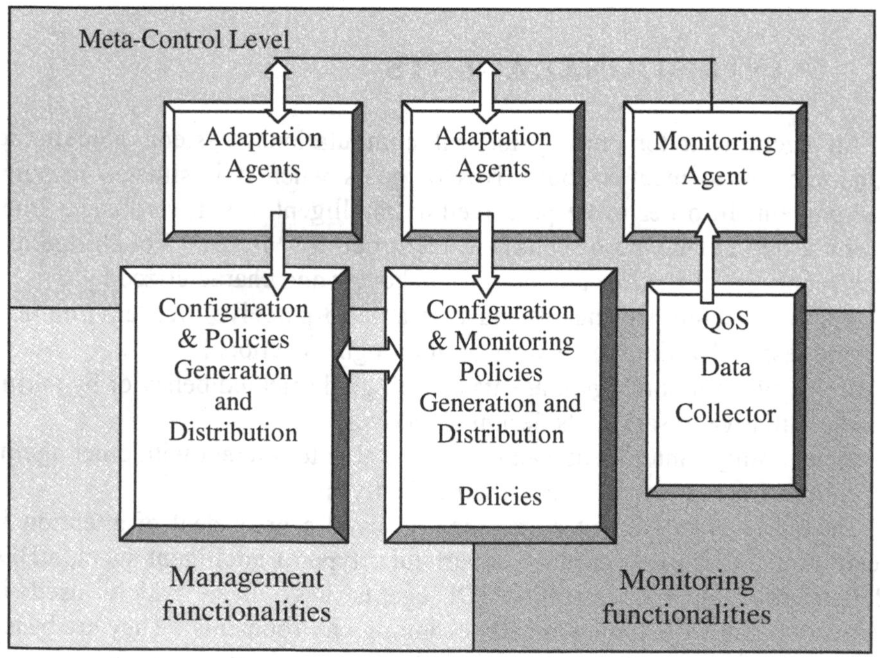

Figure 1. A Service/Resource Mediator

The meta-control level contains two categories of agents:

- The monitoring agent. It controls the coherence of network/network element behavior with the policies which were applied. It makes the decision to inform the others agents or the higher level before a SLA violation;

- The adaptation agent. It modifies the Mediator/Network Element behavior in order to improve its operation and to optimize the service configuration.

5. THE LOCAL PDP

In the architecture proposed, PDPs (i.e. Resource Mediators) send network-level policies which are not directly executable by the network elements in order to give them more autonomy. The Provisioning Agent, situated in each network element, receives the decisions and the policy rules from the PDP (see figure 2). Then, it must translate these policy rules into policy rules/commands understandable by the PEP. Therefore, it can be seen as a local PDP.

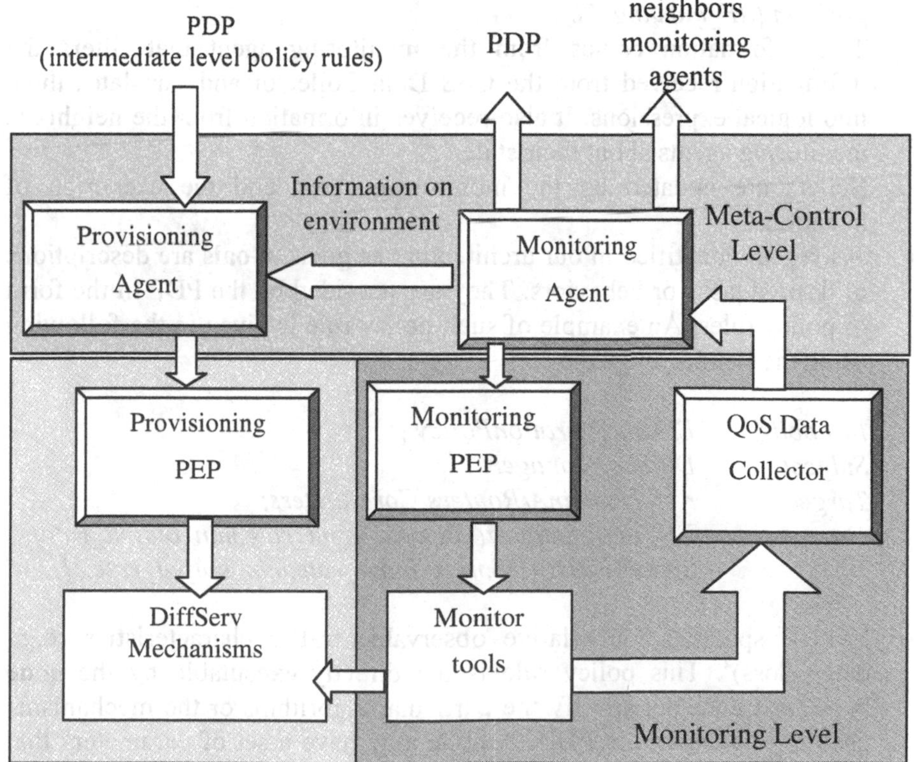

Figure 2. a network element

5.1 bdi architecture

The LPDP has a rational agent's architecture. Its rationality is turned towards the execution of a set of plans to maintain a certain QoS. Depending on the network state and the policy rules sent by the PDP, it pushes new configuration rules to the PEP. Using this architecture, the reallocation and management of network resources is based on current network state and applications QoS requirements.

The LPDP architecture is based on BDI (Beliefs Desires Intention) model and the system implemented by A. Guerra Hernandez[12]. This system is composed of four key data structures : beliefs, desires, intentions and a plan library. In addition, a queue is used to store temporarily the events perceived by the LPDP (see figure 3). Theses structures are presented in the following:

- **Beliefs** can be viewed as the informative component of the system and environment state. This component can be implemented as a set of logical expressions[12, 13]. For example, the fact that the AF queuing size reached a threshold of 70% can be represented by the statement *position (AF_queuing, 70).*
  This information comes from the monitoring agent that filters the information received from the QoS Data Collector and translates them into logical expressions. It also receives information from the neighbors monitoring agents about their state.
  Beliefs are updated by the monitoring agent and the execution of intentions.
- **Desires** are identified in our architecture as goals. **Goals** are descriptions of desired tasks or behaviors. They are provided by the PDP in the form of policy rules. An example of such policy rule is given in the following using the Ponder language[14]:

```
Inst oblig    EFConfigurationPolicy {
Subject       DiffServManager;
Target        r = /DomainA/Routers/CoreRouters;
On            EFConfigRequest(DS,max_input_rate,min_output_rate);
Do            applyEFPHB(DS,max_input_rate,min_output_rate);}
```

EFPHB specifies the relative observable traffic characteristics (e.g., delay, loss)[2]. This policy rule is not directly executable by the node because it does not specify the particular algorithms or the mechanisms used to implement the PHB. A node may have a set of parameters that can be used to control how the packets are scheduled onto an output interface (e.g., N separate queues with settable priorities, queue lengths, round-robin weights, drop algorithm, drop preference weights and

threshold, etc): for example weighted round-robin (WRR) queue servicing or drop-preference queue management[3].
The policy rules are stored in a policy repository and analyzed by a policy conflict detection and resolution module[15]. To be understandable by the intentional LPDP, each policy rule is then translated into a logic expression such as
Achieve (efphb_ds, ds, max_input_rate, min_output_rate)
expressing the desire of the LPDP to associate a certain QoS corresponding to PHB type with a certain DSCP value. The LPDP interacts with its environment through its database and through the basic actions that it performs when its intentions are carried out.
The perceptions of the LPDP are mapped to events stored in a **queue**. Events can be the acquisition or removal of a belief, e.g., the reception of a message coming from the monitoring agent or the acquisition of a new goal coming from the distant PDP.

- **Intentions** are the plans that the LPDP has been chosen for execution.
- **Plan Library** is the set of predefined plans. Plans describe how certain sequences of actions may be performed to achieve given goals or react to particular situations. Each plan has an identifier, a trigger or an invocation condition, which specifies in what situations the plan is useful and applicable, a context or precondition, and a body, which describes the steps of the procedure. An example of plan is given where a Weighted Round Robin scheduling algorithm is used to satisfy the goal presented above.

Plan-id: p012
Trigger:
achieve(efphb_ds, ds, max_input_rate, min_output_rate)
Context:
max_input_rate <= min_output_rate
Body:

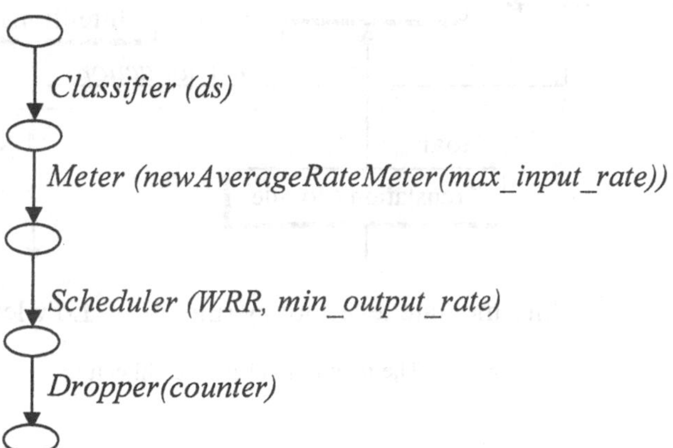

Plan body can be represented as a tree which nodes are considered as states and branches are actions or sub-goals of the LPDP. The executable plans are ordered by utility before selecting the first one. Therefore, a specific queuing or scheduling algorithm can be privileged for example.
Actions are of two kinds, internal and external ones. External actions are low-level policy rules directly executable by the PEP. They affect the environment where the LPDP is situated. Internal actions affect only the beliefs of the LPDP. Once a plan instance is executed, the LPDP executes a sequence of internal actions, i.e. add and delete beliefs. These internal actions are predefined for each plan in the plan library.

An **interpreter** manipulates these components by selecting appropriate plans based on the system's beliefs and goals, placing those selected on the intention structure, and executing them. It is responsible for the behavior of the LPDP. The interpreter verifies that the terms associated with an action are grounded before executing an external action. Different algorithms are proposed in literature to execute intentions. Some of them are well adapted to dynamic environment[16].

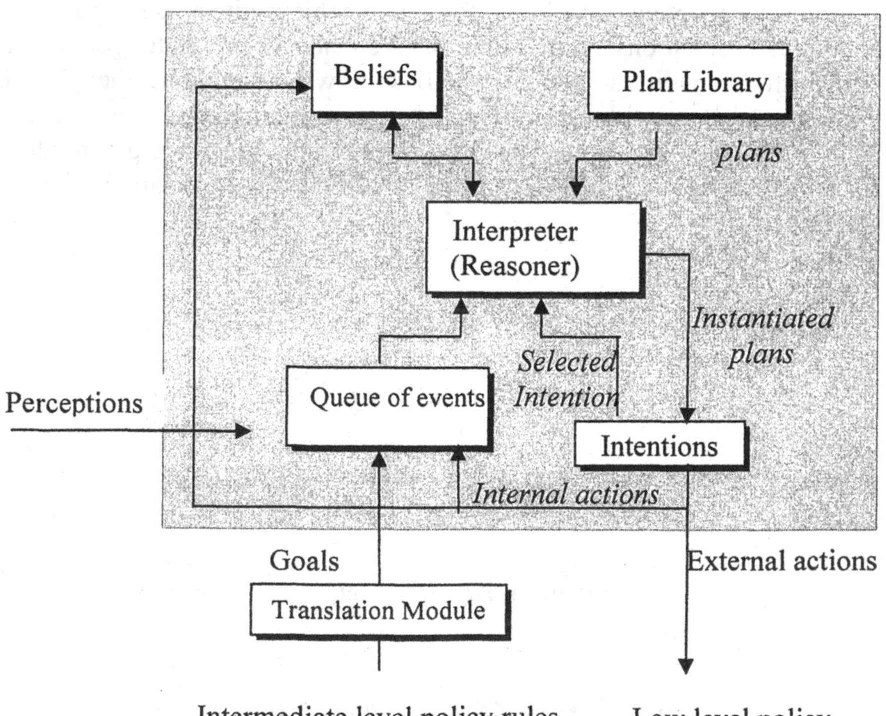

Figure 3. The intentional LPDP architecture

5.2 Learning

The success or failure in the execution of the LPDP's plans depends on many factors connected with the environment such as the input traffic type or the network load. All theses factors are difficult to well identify. Moreover, the implantation of PHBs, which is the basis for DiffServ operation, involves a hard task of choosing among a set of buffer management and scheduling techniques. This is a crucial issue to an effective QoS management. Adaptation by learning are the most suitable policy configuration strategies[14]. Therefore, the BDI architecture presented is extended with the introduction of a learning module. This module allows the LPDP to use its passed experience (i.e. plans instances that have succeed or failed in a particular context and environment) to refine the context of its plans. This context represents the reasons a LPDP has to act in a particular way. By learning the context of plans execution, it selects plans, and consequently policies which are the most suitable depending on the environment.

5.2.1 Learning and intentional LPDP architecture

The learning BDI architecture we propose is an extension of the intentional LPDP architecture based on supervised concept-learning (see figure 4).

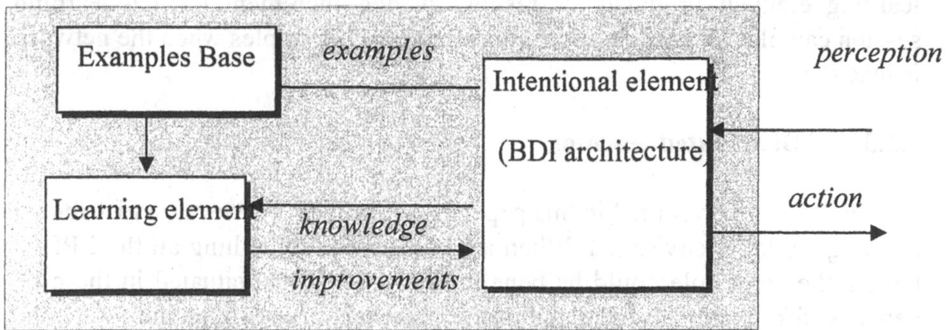

Figure 4. The learning and intentional LPDP architecture

The **examples base** contains labeled examples provided by the intentional element, i.e. the BDI architecture. Each example refers a plan and a set of beliefs that represents a given state of the environment, where the plan has succeeded or failed.

The **learning element** is responsible for the learning task. It acts from examples base and background knowledge provided by the intentional element. In result the learning element proposes improvements, precisely refined contexts of the plans, to the intentional element.

5.2.2 Learning Process

The building of the examples base precedes the learning phase. The intentional element uses the evaluation of its actions from the environment to fill the examples base with labeled examples. Each labeled example characterizes a success or a failure in the execution of a plan in a given state of the environment (i.e. a given set of beliefs). A positive example *(plan-id, e, success)* is built each time the intentional element selects and applies the plan *plan-id* in an environment *e* represented by a set of beliefs. A negative example *(plan-id, e', failure)* is built when the intentional element detects that a previously applied plan *plan-id* is no longer available in the current environment *e'*.

The learning task starts at the end of an outputs sequence of the intentional element, or ideally when the intentional element is idle and the examples base contains a sufficient number of examples. Tilde (Top-Down Induction of Logical Decision Trees) is used as learning method[12]. Tilde represents learnt concepts as decisions trees which suit the disjunctive form of the plans contexts.

At the end of the learning process the learning element gives the intentional element modified contexts of applied plans. These new learnt contexts match the environments in which these plans must be applied.

Cyclic incremental learning sessions can be activated by emptying the examples base and restarting the process described above each time the learning element produces its results. A non incremental batch learning session can also be carried out from all collected examples when the network is inactive.

5.2.3 Distributed learning

The learning presented in this paper is centralized. However, a distributed learning could be envisaged. When a LPDP learns something all the LPDPs having the same role could be beneficiary. All LPDPs, situated in the core network for example, have the same internal structure including goals, background knowledge and possibly competence. They also have the same procedure to select their actions. The only difference among them is their experience, i.e., their perception since they are situated differently in the environment. Thus the performance of a LPDPs group (i.e. core router LPDPs) could be improved by direct interactions, as exchanging the learnt contexts of their plans, among neighbors LPDPs having the same role.

6. CONCLUSION AND FUTURE WORK

We have proposed an architecture for the Self-aware management of IP networks by using policy-based management and multi-agent systems. In this architecture, the role of the administrator is limited to the guidance of these processes in their laying down operational objectives and parameters.

In this paper we propose to give a real autonomy to the LPDP in term of internal decision and configuration. The LPDP is considered as a learning BDI agent that autonomously adapts the router's behavior to environment changes.

This approach presents many advantages : The performance of the LPDP is improved because the introduction of BDI concept allows it to adapt its behavior in an autonomous manner. It chooses the policy rules according to the policy rules sent by the distant PDP. Therefore, the network element becomes relatively autonomous. Policies are distributed to all the routers that are concerned and every one of them according to its context will seek the good parameters to configure its part of the service.

By introducing learning methods, the LPDP is able to use its past actions to improve its future actions. It learns more accurately the environment in which a plan must be applied.

Our future work concerns the distributed learning and the simulation of the LPDP by adapting A. Guerra Hernandez's system[12].

REFERENCES

1. E. Marilly, O. Martinot, S. Betgé-Brezetz, "Service Level Agreement Management: Business Requirements and Research Issues", DNAC'01, December 2001
2. S. Blake, D. Black, M. Carlson, E. Davies, Z. Wang, W. Weiss,"An Architecture for Differentiated Services", RFC 2475, December 1998
3. K. Nichols, S. Blake, F. Baker, D. Black, "Definition of the Differentiated Services Field in the IPv4 and IPv6 Headers", RFC 2474, December 1998.
4. R. Yavatkar, D. Pendarakis, R. Guerin, "A Framework for Policy Based Admission Control", RFC 2753, January 2000.
5. F. Krief, "Self-aware management of IP networks with QoS guarantees", International Journal of Network Management, Vol.14, Issue 5, September-October 2004
6. B. Moore and al. : "Policy Core Information Model (PCIM)"- IETF- RFC 3060.
7. FIPA, "Fipa Policies and Domains specification", Document number PC00089D, August 2001.
8. M. Wooldridge and N.R. Jennings. "Intelligent agents: Theory and practice. The Knowledge Engineering Review, 10(2): 115-152, 1995.
9. C.Olivia, C.-F. Chang, C. F. Enguix, A. K. Ghose, "Case-Based BDI Agents: an Effective Approach for Intelligent Search on the World Wide Web", AAAI Symposium on Intelligent Agents, Stanford, CA., USA, 1999.
10. Rao A., Georgeff M., "BDI Agents:From Theory to Practice", Tech. Note 56, 1995
11. CADENUS Project. QoS Control in SLA Networks. IST-1999-11017, March 2001.

12 Alejandro Guerra Hernandez, "Learning in intentional BDI Multi-Agent Systems", PHD thesis, University of Paris 13, December 2003
13 M. Ljungberg, A. Lucas, "The OASIS Air Traffic Management System", PRICAI'92, Seoul, Koea, 1992
14 J.C. Strassner, " Policy-based network management: Solutions for the next generation", Morgan Kaufmann, 2003
15 L. Lymberopoulos, E. Lupu and M. Sloman, "An Adaptative Policy Based Management Framework for Differentiated Services Networks", Policy 2002
16 M. Wooldridge. "Reasoning about rational agents". MIT Press, Cambridge, MA., USA, 2000.

GENERIC IP SIGNALING SERVICE PROTOCOL

Thanh Tra Luu, Nadia Boukhatem
GET-Télécom ENST Paris ; LTCI-UMR 5141 CNRS. 46 rue Barrault, 75 013 Paris. Email : {luu, boukhatem} @enst.fr

Abstract: The Next Steps In Signaling (NSIS) working group has been recently created to design a generic IP signaling protocol supporting various signaling applications. This paper presents the Generic In Signaling Service Protocol (GISP) we designed considering the current outputs of the NSIS working group. In particular, we focus on the state management and message fragmentation using a simple mechanism to detect the path MTU.

Key words: Signaling, transport, protocol design and implementation

1. INTRODUCTION

During the last few years, several IP signaling protocols have been defined to support different signaling applications such as resource reservation, label distribution and middlebox configuration. Recently, the IETF has created the NSIS (Next Steps in Signaling) Working Group to design and standardize a generic signaling protocol supporting a large variety of signaling applications and managing general-purpose states.

The NSIS WG is considering protocols for signaling information about a data flow along its path in the network. The NSIS signaling problem is very similar to that addressed by RSVP[1,2]. Thus, the NSIS WG explicitly intends to re-use, where appropriate, the RSVP protocol and generalize it to support various signaling applications rather than the single case of reservation resource application.

To achieve generalization, the NSIS protocol stack is decomposed into two layers: a generic lower layer responsible for transporting the signaling messages, and an upper layer, which is specific to each signaling application. The lower layer is called NTLP (NSIS Transport Layer Protocol) and the upper layer is called NSLP (NSIS Signaling Layer Protocol).

In this paper, we propose a specification of the NTLP layer, called GISP (Generic sIgnaling Service Protocol), considering the requirements[3], the guidelines[4] of the NSIS working group and the analysis of the existing RSVP implementations.

2. NSIS CONSIDERATIONS

The NSIS protocol is envisioned to support various signaling applications that need to install and/or manipulate state related to a data flow on the NSIS Entities (NEs) along the data flow path through the network.

Each signaling application has its own objective. For example, the QoS signaling application[5] is only used to reserve resources on the data path, the NAT/FW signaling application[6] is used to configure NAT/Firewall devices, etc. To achieve its objectives, the signaling application requires support to exchange signaling messages between the signaling entities to install/manipulate state in these entities.

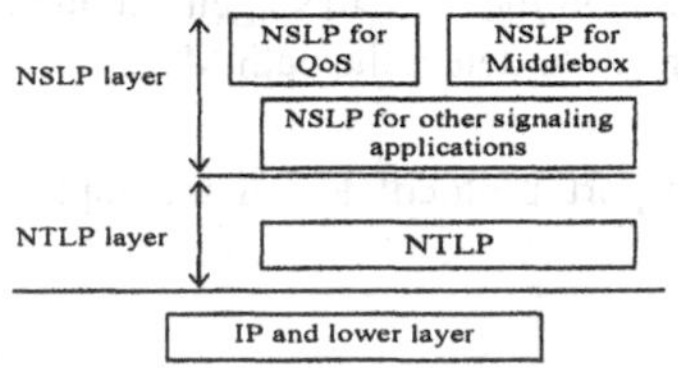

Figure 1. Protocol Signaling Structure

While the objective is tied to the signaling application itself, transporting the signaling messages and managing the state are the common functions of all or a large number of signaling applications. In order to achieve genericity, extensibility and flexibility, the NSIS framework proposed to split the protocol architecture in two layers to be able to support various signaling applications (figure 1).

The NSIS **Transport Layer** Protocol (NTLP) is responsible for transporting the signaling messages and supporting the state management. These functions are independent of any particular signaling application. Note that the transport layer has a peer-to-peer scope (or local scope). This means that NTLP only transports the signaling messages to its adjacent NSIS aware peers.

The NSIS **Signaling Layer** Protocol (NSLP) contains functionalities such as message formats and sequences, specific to a particular application. One NSLP is associated with only one signaling application. There can be an NSLP for QoS[5], an NSLP for middlebox configuration[6], etc. However, different NSLPs will use the common NTLP protocol to transport the signaling messages for their own signaling purposes. The NSLP has a scope, which is larger than the NTLP local scope. The NSLP signaling messages can be transported by the NTLP messages through many NTLP hops.

The main advantage of this layer model is its extensibility. Indeed, this facilitates the development of new signaling applications by using the transport and state management functions supported by the NTLP layer. However, this requires that the NTLP must be flexible and efficient enough to support existing and future signaling applications.

A NSIS signaling-aware entity (NE) can support many NSLPs. However, an NSLP can be installed on few NEs along the data path. For example, the NAT/Firewall signaling application[6] is only installed on the edge router of a private domain, whereas the QoS resource reservation signaling application[5] can be installed on all routers on the data path.

Actually, it is not reasonable to assume that all the equipments of an administrative domain will support the NSIS protocol. The NSIS protocol must therefore provide correct protocol operation even when there are non-NSIS-aware entities between two NSIS-aware entities. Furthermore, the NSIS protocol should support multiple signaling applications; it is very likely that a particular NSLP will only be implemented on a subset of the NSIS-aware nodes on a path. Therefore, in a heterogeneous environment, different kinds of nodes can be defined: non-NSIS-aware nodes, NTLP-only-aware nodes and NSIS-aware nodes supporting one or more NSLP.

2.1 NSIS transport layer functionalities

An overall NSIS signaling is the joint result of the NSLP and NTLP operations. As mentioned above, the NTLP has a peer-to-peer scope and operates only between adjacent signaling entities. Larger scope issues including end-to-end aspects are supported by the NSLP layer.

The functionalities of the NSIS protocol layer have been identified in the NSIS framework[4]. The NTLP is responsible for transporting the signaling messages between the peer NE nodes (upstream and downstream peer). The transport functions are the reliable message delivery, congestion control, bundling of small message, message fragmentation, and security protection.

Internet signaling requires the existence and management of state within the network for several reasons. Therefore, the NSLP should maintain the state for signaling sessions. However, the NTLP manages its own state to support the signaling application and it is unaware of the difference in state semantics between different types of signaling application.

2.2 Transport functions

Reliable and non-reliable message delivery: the NTLP should support mechanisms to confirm the reception of a signaling message if this is required.

Overload control: the NTLP should support overload control mechanisms. This allows the signaling protocol to coexist with other traffic flows and holds the performance of the signaling to degrade gracefully rather than catastrophically under overload conditions.

Messages bundling: the NTLP should support means to bundle signaling messages.

Fragmentation and assembling: the NTLP should be capable of fragmenting/assembling the signaling messages if the total length of message exceeds the link MTU (Maximum Transmission Unit).

Security protection: The NTLP only has a local scope. Thus, the NTLP only supports the security protection for the message transport between the adjacent NEs. The security protection that the NTLP should support includes integrity, anti-replay and confidentiality.

2.3 Signaling functions

Soft-state management: the NTLP should be capable of supporting soft-state management to support the signaling applications. The soft-state means that after being established on an entity, a state will be deleted on that entity if it is not periodically refreshed during a specific time period. The soft state is used to avoid the faults and the cases in which using explicit commands to delete an established state cannot be done (e.g. an intermediate router is shut down).

2.4 Other functions

Besides the transport and the signaling functions, the NTLP should also support other functions such as follows.

Multiplexing and demultiplexing signaling messages of different NSLPs.

The NTLP should be capable of sending notification reflecting the changes occurring in the network status (e.g. routing change, congestions) to the NSLP applications and other NSIS-aware entities.

3. GISP PROTOCOL

The GISP protocol is designed to run directly on the IP layer or UDP protocol. It is easier to create a signaling protocol without constraints by using IP or UDP as an underlying layer. However, all functions of the signaling protocol must be designed and implemented within the protocol to satisfy the requirements of signaling applications and the requirements of NSIS working group.

A GISP message consists of a common header, followed by a body consisting of a variable number of variable-length objects. The description of GISP objects can be found in our draft[7] with some minor changes. In this paper, we only focus the analysis on state management and message fragmentation.

The GISP is responsible for transporting signaling messages and supporting generic signaling functions. When the GISP is required to support a signaling session by its own NSLP upper layer or by receiving a signaling message, the GISP layer will establish a GISP state for that session. Once the GISP state of the signaling session is established, the GISP layer can filter, (de)multiplex the signaling messages, manage the state and notify NSLP applications about changes in the network (e.g. routing change, congestion).

3.1 Signaling service functions - State management

In order to support NSLP signaling applications, the GISP must install and maintain its own states on the data path. This is the prime purpose of the NSIS protocol. In the NSIS framework, the NTLP state can be installed on all NEs or only some NEs on the data path. In the following part of the paper, we will present how the GISP establishes and manages the state on the data path. We will also describe how the GISP messages are defined to efficiently support the state management tasks for signaling applications.

The GISP protocol can operate in the stateful or stateless mode. In the stateful mode, the GISP must register and maintain the state concerning a particular data flow. In the stateless mode, the GISP processes the message without installing any state concerning that message. Note that, in the same signaling session, some signaling-aware nodes on the data path run in the stateful mode (i.e. establish the state) while the others run in the stateless mode (i.e. do not establish the state for this session)[8].

3.2 State management modes

As we discuss above, the GISP supports both stateful and stateless mode for signaling applications. The GISP defines five state management modes

(SM) to indicate how the state is establish on the signaling path. The SM value is specified by the SM field in the GISP message header (see our draft[7] for more details).

3.2.1.1 SM=0.

The SM=0 can be used for directly sending signaling messages between two signaling-aware entities (SE) without interference of intermediate nodes. In other terms, the message is forwarded by the intermediate nodes without establishing state.

In figure 2, the signaling entity which initiates the session (signaling initiator) sends the message with the SM field = 0. Thus, the state is only established on the signaling initiator (SI) and the signaling responder (SR) which terminates the signaling path.

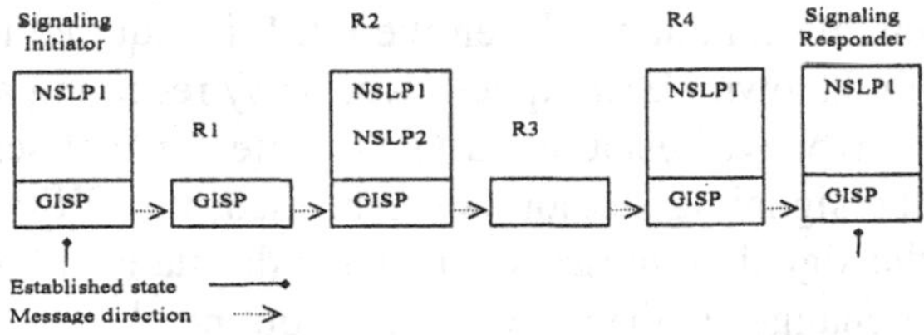

Figure 2. Establishing states in case SM=0

A message sent with SM=0 may have a router alert option (RAO) in its IP header to assure the hop by hop security or other purposes. The IP source and destination address are the address of the entity that sends and receives the message respectively.

3.2.1.2 SM=1.

In case of SM=1, the GISP state is established on all signaling-aware nodes along the data path regardless of signaling applications. The SM=1 is therefore only used for the signaling applications requiring the signaling path must be seriously tied to the data path (e.g. resource reservation). A message sent with SM=1 has a RAO in its IP header to allow the next signaling nodes to examine the message.

In case a signaling-aware node does not support the NSLP specified in the message, this node only registers the content of message without analyzing it. If this node detects the changes in the network, it can send the registered message to establish rapidly the session on the new path. This provides a fast adaptation to route changes. Note that, the signaling application message content is always opaque to the GISP.

3.2.1.3 SM=2.

In case of SM=2, the GISP state is established only on the SEs that support the NSLP specified in the message. The NSLP signaling application type is indicated by the NSLP-ID field in the GISP message. A message sent with SM=2 also has a RAO in its IP header to allow the next signaling nodes to examine the message.

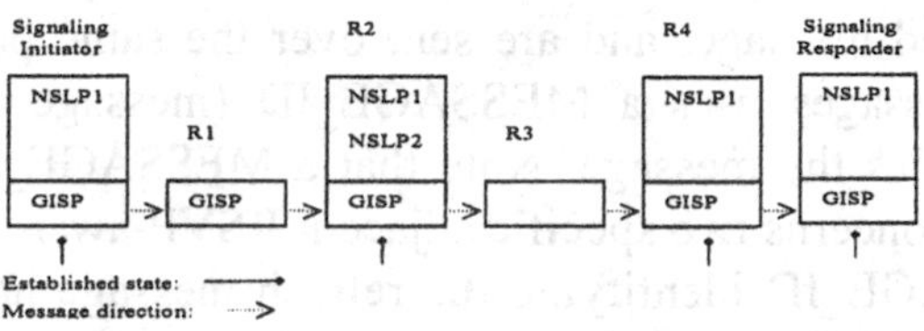

Figure 3. Establishing states in case SM=2

The SM=2 can be used for signaling applications which are not sensitive to routing changes as NAT/FW_NSLP[6]. In figure 3, the SI sends the signaling message with the SM=2 to the SR to establish a session for the NSLP1 signaling application. The intermediate nodes R2 and R4 support the NSLP1, and the state is only established on these nodes. The routing changes between the node R2 and the node R4 do not heavily influence the NAT/FW_NSLP application. Normally, the NAT/FW_NSLP is only installed on the edge nodes of private domain. Thus, the routing changes between two domains do not affect the signaling application.

3.2.1.4 SM=3.

The case of SM=3 looks like to SM=2. However, if the state is not yet established, the GISP must wait the decision of the NSLP signaling application to know whether the state will be established. In some cases, the GISP cannot decide to establish the state by itself (e.g. lack of information).The mode SM=3 allows the NSLP signaling application decide to establish the state or not.

3.2.1.5 SM=4.

In case of SM=4, the message is not involved in state establishment. It is used by the GISP to exchange information about the states that have been established (refresh, delete states) or other information (congestion control, security control...).

3.2.1 State refresh mechanism

In the original RSVP version[1], an established state is refreshed by resending the Path and Resv message on the data path. Every Path (or Resv) must be totally examined even though the session is established. This increases the cost of message processing.

The RSVP extension[2] has improved the refresh mechanism. The RSVP messages are categorized into two types: trigger and refresh message. Trigger messages are the messages that advertise state or any other information not previously transmitted. Refresh messages transmit previously advertised state and contain exactly the same objects and same information as a previously transmitted message, and are sent over the same path. Every trigger and refresh messages uses a MESSAGE_ID (message identifier), which uniquely identifies the message. Note that a MESSAGE_ID has a hop-by-hop scope and concerns two specific adjacent RSVP-aware routers.

The MESSAGE_ID identifying the refresh message has the same value as the MESSAGE_ID of the trigger message that was previously sent.

When an established state needs refreshing, the RSVP only sends the MESSAGE_ID of that state. This avoids sending the whole message to refresh the state. If the trigger message content needs to be modified, the RSVP sends a new trigger message with new incrementally changed MESSAGE_ID. As a result, the MESSAGE_IDs values of established sessions on two adjacent nodes are not necessarily consecutive. Therefore, the RSVP must send each MESSAGE_ID to refresh a specific session.

In the GISP, we define a new mechanism to reduce the refresh cost. This mechanism is also applicable to RSVP to reduce the refresh cost. Each signaling-aware entity (SE) has a list of established sessions with a particular adjacent SE. We call this list Socket_ID list. The index numbers of each entry in the list are called Socket_ID. Each Socket_ID is associated with a Session_ID value to identify a specific established session between two adjacent SEs. The GISP will choose the Socket_ID for each new state and this value is not reused until the state is deleted. When establishing a new state, the GISP chooses the first unused Socket_ID (i.e. there is no established session corresponding to this Socket_ID) in the list. As a result, the Socket_ID of established states are consecutive. Therefore, the GISP only sends the first and the last Socket_ID values to refresh all the sessions having the Socket_ID values in this range. The figure 4 illustrates this mechanism.

Suppose that the node A sends a trigger message for Session_ID=K for establishing a session on the node B (see figure 4). When receiving the trigger message, the node B looks for the first empty Socket_ID (Socket_ID=2) and sends this value back to the node A. When A wants to refresh the states that are established on the node B (Session_ID I, K and A), it sends the socket blocks (Socket_ID=1, Socket_ID=3). If an established session is deleted, the socket entry value is set empty (null). This null socket entry will be filled with a new session.

Note that, the A's socket list is a mirror of B's socket list. The node A can deduce the first null socket in the socket list of the node B. However, the value of the first null Socket_ID must be sent to A to make the GISP more robust. In some cases, a signaling-aware can loose the information about the established states (e.g when the node reboots). The node A can use the Socket_ID sent back from the node B to synchronize the two lists and detects the errors.

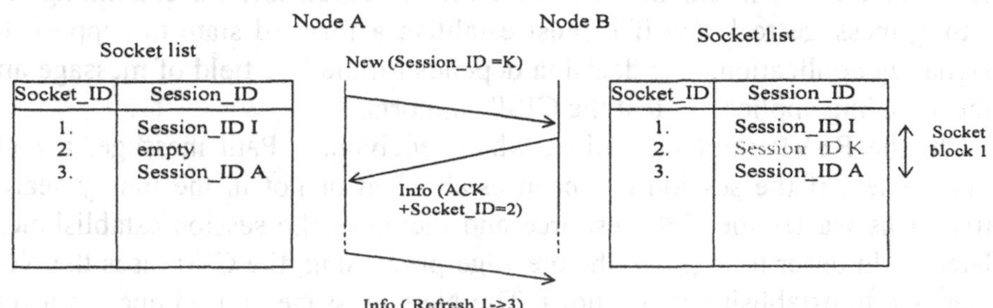

Figure 4. States refreshing in the NTLP

This mechanism aims to reduce the payload used to refresh the state between two nodes. Note that this mechanism allows refreshing a list of session rather than refreshing each individual session. All the sessions in a list are refreshed at the same time. It saves the CPU resources to manage the timeout for each session.

In addition, the GISP proposal allows two adjacent NSIS-aware nodes have more than one Socket_ID list. Each list has a different timeout value. If a particular session is more stable, we can place it in a list having a longer timeout. This allows the signaling application change the refresh period for a specific state in the same manner as the staged refresh timer mechanism[11] but still avoid managing the soft state for each individual state.

3.2.2 The GISP message

In the GISP protocol, two states are associated with a session: forward state and backward state. The forward state concerns the information sent by the SI along the downstream path. The backward state concerns the information sent by the SR along the reverse data path. The forward state is usually created before the backward. The backward state is not always created. If the forward state of a session is deleted, the backward state of that state is also deleted. However, if the backward state is deleted, the forward state of a session still exists.

The GISP supports three message types that are used to establish and remove the states (forward and backward) on the data path. They are also used to exchange signaling information between the SEs.

3.2.3.1 New message.

A New message is used to establish the forward state for the downstream path between two signaling entities. The SI will send a New message along the data downstream path. The New messages can be sent with the Router Alert IP (RAO) option in their IP headers to allow the GISP on intermediates nodes to examine signaling messages.

When the New message is sent with a RAO option in the IP header, intermediate SEs will examine it. The GISP is responsible for examining signaling message to know if it must establish a forward state to support the signaling application. The decision depends on the SM field of message and the signaling application that the GISP supports.

In the RSVP implementation, when receiving a Path message, a node must detect if the session has been established or not in the binary search tree. This wastes the CPU resource and increases the session establishment latency. In order to improve the message processing, the GISP uses the New message to establish new sessions. The New message is used once in a session. If the state on the path changes, the New message cannot be used to re-establish the session. In this case, the GISP sends a Mod message to re-establish the session.

3.2.3.2 Mod message.

The Mod message has the same structure as the New message. It is used to modify a forward state of a session. If the sender wants to modify the signaling session, he sends a Mod message along the data path toward the receiver. The GISP on the data path will examine the message and decide to deliver it to the NSLP layer or not.

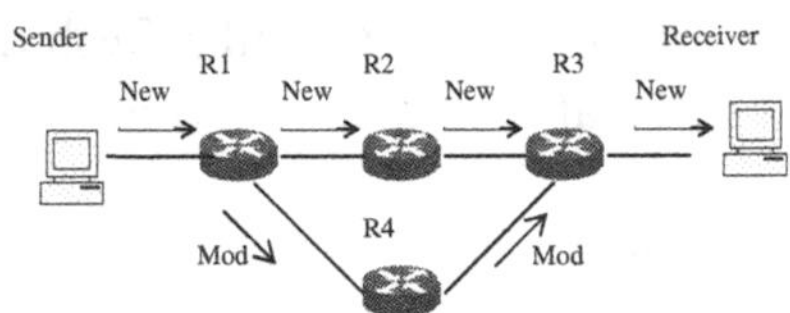

Figure 5. Establishing forward states by using New and Mod message

The sender or the intermediate nodes can periodically send the Mod message to refresh and detect a route change in the network. When a SE receives a Mod message that is was previously received, it considers this message as a refresh notification from the previous node.

In case a SE node detects a route change (ex. receive a route change notification from the routing table), it sends a Mod message to establish the state on the new path. This is illustrated in figure 5. When the router R1 detects a route change, it sends a Mod message on the new data path until the message reaches the node R3.

The Mod message is also used to establish the backward state if the receiver requires this. In this case, the Mod message will be sent hop by hop along the reverse data path. Note that the backward state is established if and only if the forward state is established. Thus, the SE always verifies the existence of the forward state before establishing the backward state. That is the reason why the New message cannot be used to establish backward states.

3.2.3.3 Info message.

The Info message can be sent hop by hop along the data path or the reverse data path. It is responsible for exchanging information between adjacent SEs nodes. The information can concern state management (refresh, delete an established state), reliable message delivery (acknowledge a message) and other information (congestion control, error notification, etc).

Moreover, the Info message allows to GISP to detect route changes (as the Mod message) and discover the path MTU (Maximum Transmission Unit).

The Info message does not carry the NSLP message content to refresh an established state; it only carries the Session_ID of the established session. This reduces the signaling payload in the network. When receiving the message, the GISP detects if the Session_ID corresponds to one of the established states. If there is a corresponding state, the GISP considers this message as a refresh notification for that state. Otherwise, the GISP will send back a notification to require sending a Mod message to establish the state.

In the following section, we present how the Info message is used to detect the path MTU.

3.3 Transport service functions

The GISP directly runs on the IP or on UDP, which do not meet the requirements in the NSIS WG. Thus, the GISP must implement all the generic transport service functions specified in the NSIS requirements[3,4]. The GISP has its own transport functions such as follows: reliable message delivery, message bundling, message fragmentation and the path MTU discovery, congestion control, security transport

3.3.1 Reliable message delivery

The reliable message delivery is considered as a significant service for the signaling applications in the NSIS framework. This allows the signaling applications to receive the confirmation of signaling message reception. A

NSLP signaling application can invoke the reliable delivery service for its own purpose. This is always optional to use.

The GISP supports mechanisms to confirm of the reception of a signaling message if it is required. If a message is required to transport reliably, the GISP sets the ACK flag in the MSG_SEQ object, which contains the sequential number of the message. When a SE receives this message, it will send back a MSG_ACK that contains the sequential number of the MSG_SEQ object.

3.3.2 Message bundling.

The GISP supports the message bundling for the signaling applications. The GISP can send more than one NSLP message in a GISP message. The GISP puts each NSLP messages in a GISP object and transports these objects in an only one GISP message. When receiving a bundle message, the GISP decapsulates it and process each NSLP message as it was received individually.

3.3.3 The message fragmentation and the path MTU discovery

For NSLPs that generate large messages, it will be necessary to fragment them efficiently within the network. The GISP is required to support the message fragmentation. However, using the IP fragmentation can be incompatible.

Note that in the RSVP, the IP source address and the IP destination address of an RSVP downstream trigger are the address of the sender and receiver respectively. If an intermediate router does not fragment the message (e.g. Don't Fragment bit is set or IPv6 packet) a "Packet Too Big" ICMP message will be sent to the sender address rather than to the previous node that sent the message. Thus, the previous node cannot know why the message is lost.

The GISP protocol uses a simple mechanism to detect MTU between signaling aware nodes. The GISP only uses this mechanism to send a trigger message (New or Mod) if the message length is larger than 576 bytes (IPv4) or 1280 bytes (IPv6). After sending the trigger message, the GISP will send an Info called Info Detect message. The Info Detect message only carries the information about the SE that sent this message (e.g. IP address) and Session_ID of the session. Because the Session_ID length is 16 bytes, the Info Detect message length is always smaller than 576 bytes (IPv4) or 1280 bytes (IPv6).

In case an intermediate node drops the trigger message, the next SE node only receives the Info Detect message. This node waits for a short time interval before sending back an Info message, called Info Failure message. This Info Failure message includes the next SE's address and the link MTU value of the interface through which the Info Detect message was received.

When the SE that sent the trigger message receives the Info Failure message, it knows the address of the SE and the link MTU of the interface through the Info Detect message was received. Thus, the GISP can use the path MTU discovery mechanisms for IP to detect the path MTU which are defined in RFC1191 and RFC1981. With this path MTU value, the GISP will fragment the trigger message and send individual fragments towards the next SE. Theses fragments will be assembled at the next SE and sent to the signaling application layer.

3.3.4 The congestion control

The RSVP does not implement any kind of congestion control algorithm. In case the RSVP traffic is low, the co-existence between RSVP and other TCP-like protocol is acceptable. Nevertheless, when the RSVP traffic is quite high, its flows are very aggressive and starves other TCP and TCP-like flows.

In this paper, we only show the two main difficulties to support the congestion control in the GISP. The solution for congestion control in the GISP is still under investigation. Firstly, the GISP has two ways to address a signaling message as the RSVP:

Peer-to-peer: the message is addressed to an adjacent signaling-aware. In this case, the message is sent directly between two signaling nodes without the interference of the intermediate nodes.

End-to-end: the message is addressed to the flow destination directly, and intercepted by intervening signaling-aware nodes.

In the first way, the signaling messages are sent between two explicit nodes, the GISP can reuse the congestion control mechanism for TCP as in the RFC2581[9]. However, in the second way, the GISP does not know which SE on the data path will intercept the message. Thus, the congestion control for the TCP cannot be used.

Secondly, the GISP supports the reliable and unreliable message delivery for the signaling applications. It is not always required to send back the acknowledgement of a message. In the congestion control mechanisms[9,12], the acknowledgement of packet is used to analyze the congestion in the network. If a packet sender does not receive the acknowledgement of a specified packet, it considers the packet is lost and there is congestion in the network. However, the GISP cannot consider that this message is lost if it does not receive the acknowledgement of that message since the acknowledgement is not always required.

3.3.5 Security protection

The GISP only supports the security protection of the message transport between two adjacent NEs. Each SE must establish security associations (SA) with the other adjacent SEs before sending the GISP message. The GISP uses existing protocols to establish the SAs (e.g IKE). To support the message authentication, integrity and anti-replay security service, the GISP reuses the security mechanism proposed in the RSVP[10].

4. CONCLUSION

The NSIS WG has proposed a generic framework for the design of an IP signaling protocol. The NSIS protocols architecture is decomposed in the NSLP and NTLP layers. We have designed GISP protocol to be the generic NTLP layer taken into account the requirements of the NSIS WG and based on the analysis of the existing RSVP implementations.

In this paper, we presented the main characteristics of GISP protocol and proposed a new refresh mechanism to reduce the refresh payload and the CPU resource used for the soft-state management. We also proposed a mechanism to discover the path MTU between two signaling-aware nodes. This allows GISP to determine the appropriate length of the signaling messages and decreasing the load of the header message processing.

REFERENCES

1. R. Braden et al., "Resource ReSerVation Protocol (RSVP) -- Version 1 Functional Specifica tion", RFC 2205, September 1997.
2. L. Berger et al., "RSVP Refresh Overhead Reduction Extensions", RFC2961, April 2001
3. M. Brunner et al., "Requirements for Signaling Protocols", RFC 3726, April 2004.
4. R. Hancock, Next Steps in Signaling: Framework, draft-ietf-nsis-fw-05, Internet Draft, Work in progress, October 2003
5. S. Van den Bosch et al., NSLP for Quality-of-Service signaling, draft-ietf-nsis-qos-nslp-02, Internet Draft, work in progress, February 2004.
6. M. Stiemerling et al., "A NAT/Firewall NSIS Signaling Layer Protocol (NSLP)", Internet draft, work in progress, draft-ietf-nsis-nslp-natfw-01, February 2004
7. Thanh Tra Luu et al., "NTLP Considerations and Implementation", Internet Draft, work in progress, draft-luu-ntlp-con-imp-01, May 2004
8. A. Bader et al.,"RMD (Resource Management in Diffserv) QoS-NSLP model", Internet Draft, Work in progress, draft-bader-RMD-QoS-model-00, February 2004
9. M. Allman et al., TCP Congestion Control, RFC 2581, April 1999
10. F. Barker et al., RSVP Cryptographic Authentication, RFC 2747, January 2000
11. Ping Pan et al., "Staged refresh timers for RSVP", Global Telecommunications Conference, 1997. GLOBECOM '97., IEEE , Volume: 3 , 3-8 Nov. 1997 Pages:1909 - 1913 vol.3
12. Datagram Congestion Control Protocol working group http://www.ietf.org/html.charters/dccp-charter.html

ON DISTRIBUTED SYSTEM SUPERVISION - A MODERN APPROACH: GENESYS

Jean-Eric Bohdanowicz[1], Laszlo Kovacs[2], Balazs Pataki[2], Andrey Sadovykh[3] and Stefan Wesner[4]

[1]EADS SPACE Transportation, 66, route de Verneuil, BP 3002, 78133 Les Mureaux, France; [2]MTA SZTAKI, Computer and Automation Research Institute of the Hungarian Academy of Sciences, Budapest, Hungary; [3]LIP6, Laboratoire d'Informatique, Université Paris 6, 8, rue du Capitaine Scott, 75015 Paris, France; [4]HLRS - Stuttgart University, High Performance Computing Centre Stuttgart, Allmandring 30, 70550 Stuttgart, Germany.

Abstract: This article presents limitations of current network management standards in the context of comprehensive distributed system supervision. As a proposed solution, the article describes the GeneSyS project achievements, a modern approach allowing straightforward integration of all monitoring/control means, as well as providing basic intelligence capabilities. These issues are illustrated on several industrial examples.

Key words: SNMP, GeneSyS, supervision, distributed management, intelligent agent, Web-Services.

1. INTRODUCTION

For more than 30 years, the information systems have moved forward from a single computer to distributed computer societies.

These new information systems involve different heterogeneous components working together and include groupware, collaborative engineering, distributed simulation and distributed computation resources management (GRID) systems

The maintenance activities of such systems include:

- **Application management** including deployment, set-up, start, stop, hold/resume, configuration management (for instance, for redundancy management purpose) and resource management;

- **Operating systems management** comprising resource usage monitoring and control;
- **Time synchronisation**;
- **Network management** including parameterisation and performances (e.g. dynamic control of bandwidth allocation) **and monitoring**;
- **Security management**, like authentication/authorisation control;
- **Archiving** and etc.

The difficulty of these tasks depends on the network deployed, the components spread, the number of components, the availability of compatible management tools and the infrastructure. Working with thousands of parameters simultaneously becomes uneasy without intelligent solutions categorising, synthesising and filtering them, as well as situation pattern recognition and prediction mechanisms.

Historically, the principles of network supervision are older than those governing the frameworks and mainly based upon the SNMP protocol (and its extensions). There exist many commercial frameworks, like Unicenter TNG, HP OpenView, etc, using SNMP not only for network management, but also for application management. These frameworks inherit SNMP advantages: performance, maturity for network management, multiple compatible devices; as well as well known disadvantages: lack of security (UDP based), lack of complex data types support that makes it difficult to build intelligent solutions.

There exists a JMX (Java Management Extension) specification that solves earlier mentioned issues for Java platform systems. Besides, there exist frameworks (IBM Tivoli, etc.) using middleware standards, like the OMG CORBA.

However, several common constraints can be identified for available supervision technologies and frameworks:

- **Interoperability issues:** Components written on different languages using different toolkits, which are supposed to use the same architecture specification, may not be capable to co-operate on a full scale.
- **Components portability:** Often components are built to work only under their native operating systems like MS-Windows or UNIX. They are very sensible to transport mechanism and to low level communication protocols, in general.
- **Development/deployment complexity:** Many commercial applications have proprietary APIs that makes it difficult to create new agents and to plug them to the existing supervision systems.
- **Non-flexible architecture:** When agent and visualisation tools are realised in the same component, upgrades of the console impact agent functionality and vice versa.

- **Dedication to a particular monitoring layer, lack of comprehensive solutions:** For instance, there exist various application layer tools to supervise Oracle database. It would be very useful to get simultaneously the system information and network statistics to better control the system.
- **Lack of intelligence support:** Dealing with thousands of relevant parameters simultaneously is a laborious task
- **Lack of integration capability:** Often application management can't be supplemented with existing network management solutions due to the lack of integration capability.

The next section introduces the GeneSyS project intended to overcome these limitations.

2. GENESYS

GeneSyS (Generic System Supervision) is a European Union project (IST-2001-34162) co-funded by the Commission of the European Communities (5th Framework). EADS SPACE Transportation (France) is the project Co-ordinator, with University of Stuttgart (Germany), MTA SZTAKI (Hungary), NAVUS GmbH and D-3-Group GmbH (both of Germany) as participants. GeneSyS started in March 2002 and is due to be completed in October 2004[1]. The project is aimed at developing a new, open, generic and modular middleware for distributed systems supervision. Besides, the consortium intends to make GeneSyS an open standard in the distributed system supervision domain.

2.1 Proposed Solution

The protocol based supervision architectures (ICMP, SNMP) have the most remarkable interoperability characteristics due to the fact that, their message format is strictly fixed and they do not impose any limitations on a component implementation, requiring only the protocol support. This makes their usage independent from operating systems and programming languages.

Their force is also their weakness. The strict message format makes it difficult and often impossible to operate with a custom data required for modern supervision systems. The network management protocols are inseparable from their transport protocols.

Meantime, Web technologies provide with flexible means to build custom, XML based protocols and portable transport mechanisms independent from network protocols (Web Services).

Our proposal is to combine Web technologies to build a supervision middleware, which shares advantages of protocol based architectures: operating system and programming language independency; and provides flexible and customisable messaging protocol, as well as network portability.

2.2 Web Technologies as a Platform for a Supervision Framework

With the advancement of Web technologies, more and more works appeared to introduce these technologies in the world of supervision (DMTF WBEM, OASIS WSDM, etc.). GeneSyS was one of the firsts to bring the Web Services to this domain.

As a result of our research, an agent based approach was implemented which separates the monitoring/controlling and visualisation of monitoring data. Web Services technologies were chosen as the base for GeneSyS messaging protocol.

Basing the supervision infrastructure on agents seems logical, because the monitoring of IT entities requires properties that are available with software agents. A software agent is a program that is authorised to act for another program or human (see[2]). Agents possess the characteristics of delegacy, competency and amenability that are the exact properties needed for a monitoring software component.

Delegacy for software agents centres on persistence. Delegacy provides the base for an agent, which makes it an autonomous software component. By taking decisions and acting on their environment independently, software agents reduce human workload by interacting with their end-clients when it is time to deliver results. In case of GeneSyS, the agents reside either on the computer hosting the monitored entity or on a computer that is able to communicate with the monitored entity.

Competency within a software environment requires knowledge of the specific communication protocols of the domain (SQL, HTTP, API calls). A monitoring agent competency is to have knowledge about the monitored entity to be able to collect runtime information from it or to control it with commands.

Amenability in intelligent software agents can include self-monitoring of achievement toward client goals combined with continuous, online learning to improve performance. GeneSyS makes no restriction on its agents or on their intelligence or autonomous operations, but provides the ability to include it as found necessary by agent writers and also provides some middleware components (like monitoring data repository) that can be used to implement amenability.

Openness and standards based solution is one of the key requirements of GeneSyS especially in the light of the Consortium's intention to turn GeneSyS itself into an industry standard. After a number of iterations, we had two candidates for the realisation of the communication protocol:

- InterAgent Communication Model (ICM - FIPA based) (cf.[3])
- Web Services technologies (see[4])

The ICM framework has not been designed for monitoring or supervision needs but is a general communication framework for inter-agent communication. The Web Services framework standardised by the W3C is a generic framework for the interaction of Services over the Internet and is designed to exploit as much as possible existing protocol frameworks such as SOAP and HTTP. The Web Services framework is in contrast to ICM more a hierarchical or client-sever communication model.

Web Services has a major problem with respect to performance. The use of an XML based protocol cannot be as efficient as a binary protocol due to the text processing, which is highly performance consuming. Additionally, the most common transport protocol used for SOAP messages, the Hypertext Transfer Protocol (HTTP), is not very efficient as it lacks stateful connections. However we are convinced that these problems can be solved as Web Services potentially can use different protocols. The feature of alternative protocol bindings is already used for example in the .NET framework using Remoting, which uses different (proprietary) protocols. As this problem is not solely part of GeneSyS but the whole community including the major software vendors that are committed to Web Services will face this problem, the assumption that this limitation will disappear seems reasonable.

After a detailed comparison of these two technologies, we selected Web Services, including the SOAP XML based communication protocol as a base for GeneSyS. Going on the Web Services path, we have a strong industry backing with tools available for many languages. With this decision, we also defined the first instance of a Web Services based supervision system that has recently been followed by other companies and standards organisations (OASIS WSDM, DataPower Technology[5])

On top of SOAP and Web Services, a new layer of the GeneSyS protocol has been established called the GeneSyS Messaging Protocol (GMP). GeneSyS Messaging Protocol is a lightweight messaging protocol for exchanging structured supervision information in a decentralised, distributed environment. It is an XML protocol based on XML 1.0, XML Schema and XML Namespaces. GMP is intended to be used in the Web Services Architecture, thus, SOAP is considered as a default underlying protocol. However, other protocol bindings can be equally applied. Using XML to represent monitoring data was a natural choice. XML is a widely accepted industry standard that supports structured representation of complex data types, structures (enumerations, arrays, lists, hash maps, choices, and sequences) and it can be easily processed by both humans and computers. With the wide acceptance of XML, integration with supervised application and 3d party monitoring solutions can be smoothly achieved, since XML toolkits are available for every platform.

2.3 Basic Components and Communication Model

This section provides implementation details, illustrating common supervision framework architecture.

Fig.1 depicts the basic GeneSyS functionality.

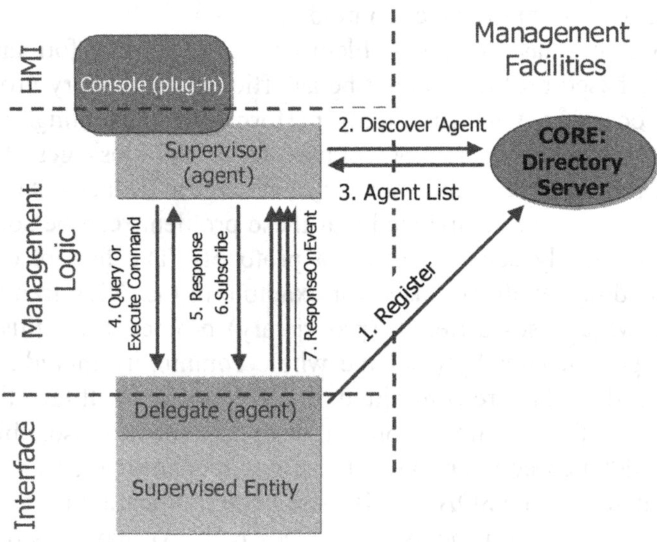

Figure 1. GeneSyS Communication Model

As showed above, supervision process involves several generic components. The Delegate implements an interface to the Supervised Entity (Operating System, Network, Applications, etc.), retrieves and evaluates monitoring information and generates monitoring events. The Supervisor is a remote controller entity that communicates with one or more Delegates. It may encapsulate management automation functionality (intelligence), recognising state patterns and making recovery actions. The Console is connected to one or many Supervisors to visualise the monitoring information in a synthetic way, and to allow for efficient controlling of Supervised Entity. The Core implements Directory Server, a location storage being updated dynamically.

The agents are registered in the Core with the purpose to be discoverable by other agents. Hereafter, the "agent" is a generic term comprising the Supervisor and the Delegate.

Both "pull" and "push" interaction models are available. The pull model is realised by the Query/Response mechanism, while the Event Subscribe mechanism secures the push model. All interactions between agents are provided for by the SOAP-RPC. The flexibility of XML standard is used to encode communication messages (GeneSyS Messaging Protocol) supporting complex data structures and custom data types.

2.4 Integration Capability

As a result, transport mechanism (SOAP-RPC) and the GeneSyS architecture itself are very flexible. That allows smooth integration with existing management frameworks.

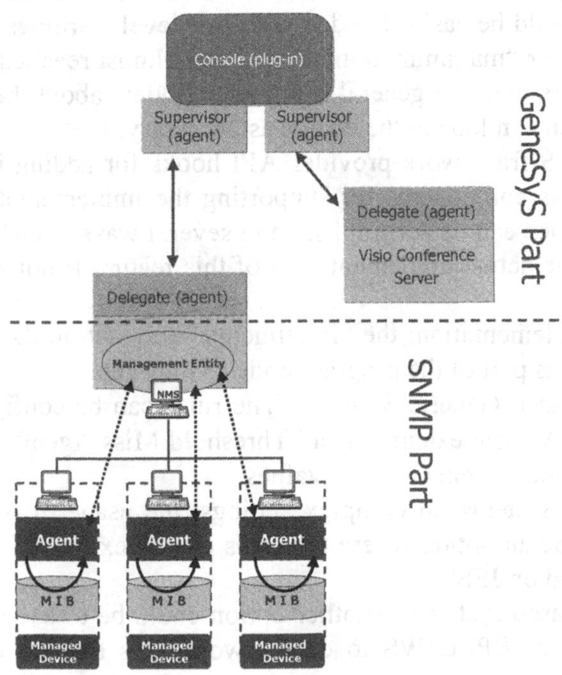

Figure 2. GeneSyS - SNMP Collaboration

Fig. 2 gives an example of collaboration between GeneSyS and SNMP. An SNMP Management Entity, a Network Management System, is plugged to a GeneSyS delegate, which makes network management information available at the administrator console. Thus, this information can be processed together with information of other agents (operating system, middleware, applications, like Visio Conference Server on the Fig. 2) in order to synthesise all the metrics in a single global view.

Hence, this SNMP bridge concept gives opportunity to benefit from both network and application level management. In that way, it is used in some of GeneSyS network agents mentioned lately in the following applicability sections.

2.5 Intelligence

An inherent property of software agents is autonomy, that is, the ability to work without the intervention of other programs or humans. Autonomous work requires some level of intelligence so that the agent can react on changes in its environment or can make decision based on its internal logic driven by rules or other means. Intelligence in agents is also required because in a complex environment with some 10 or 100 monitored entities, an administrator could be easily flooded with low level warnings like "memory is running low" or "maximum number of users almost reached". Instead, the administrator first needs a general, summarised view about the health of the systems and then can look at the details as necessary.

The GeneSyS framework provides API hooks for adding intelligence to agents as well as components for supporting the implementation of intelligence. Intelligence can be accomplished in several ways, which are only outlined here, as the actual implementation of this feature is not a main goal of GeneSyS:

- Specific Implementation: the "intelligence" to react on the system status can be done as part of the program code of the agent.
- Parameter based Generic Solution. The rules can be configured through parameters. A basic example is a "Threshold Miss Agent" where the parameters would be min and max values.
- Rule Based Systems. In complex settings, the usage of rule based systems could be an option where the rules can be expressed in an external file e.g. based on JESS.
- Workflow based systems. Another option could be to use workflow languages such as BPEL4WS to define workflows that act depending on events receive.

GeneSyS provides a data Repository that is connected to the middleware bus via the same API as any other agents, which means its functionality is available to all other agents connected to a given CORE. The Repository provides a generic XML data storage facility. Agents can store monitoring or control messages in the Repository, which can later be queried. With the use of the Repository, an agent can base its decisions on archived data, for example, by analysing past messages for detecting trends in the operation of the monitored entity. More over, the Repository is also capable for storing control messages – or a list of control messages – which can be "replayed" any number of times at any time it is necessary.

The Agent Dependency Framework (ADF) is another aid for adding intelligence to monitoring. ADF allows defining dependencies of monitored entities. To be more precise, not directly the dependencies of monitored entities but the dependencies of the agents monitoring the entity can be described. Each delegate agent can describe in its component description (which is stored in the Core) what agents it depends on. The dependency forms a directed graph that should never cause a circular reference. Once the dependency of each delegate is described, the dependency graph can be queried from the Core. Based on the dependency graph, a special supervisor console view can be created that draws a tree view of the dependent entities and gives a quick overview of the health of the system with green, yellow and red light depicting a healthy, questionable or erroneous state of the dependant systems. This way of visualising the monitored system with all its dependent components provides a way for tracking root cause of problems. For example, an administrator seeing a red light in the top of the dependency hierarchy can expand the tree until he finds the subsystem that generates the red light and which has been "propagated" up in the dependency tree. In the same way, an autonomous intelligent agent can walk this tree and find the root cause of the problem and can work only with that subsystem that was the source of the problem.

3. APPLICABILITY RESULTS

This section illustrates flexibility, integration capability and basic intelligence features with several real-life examples of the GeneSyS framework in use. The scenario was intended to prove the viability of the GeneSyS concept. Common system and network agents were developed to reflect system administrator needs. Custom application agents were used to monitor the system functional status (application load, resources used by applications, etc), user activities (documentation in use, on-line meetings, access violation, etc).

The main goal of these scenarios was to prove the capability of Web Services based distributed system to work in a heterogeneous environment. It includes support of different operating systems (Windows, Linux), programming languages and toolkits (C/C++/gSOAP, Java/Axis, .Net). Besides, while developing custom application agents (Oracle, EDB, GTI6-DSE, Mbone, Tomcat), the integration capability was ensured.

3.1 Distributed Training Scenario

This scenario was brought by EADS SPACE Transportation, the European aerospace industry leader. It concerns HLA-based simulations. HLA (see[6]) is a DoD standard for real-time interactive simulations. This standard is widely used in military, aerospace and automotive industries. The Distributed Training Scenario involves 4 real-time simulators playing different roles in joint training sessions of astronauts and ground controllers in order to prepare them in advance for contingency situation during the ATV to International Space Station (ISS) approach manoeuvre. The trainee teams are located in different places all over the world (Toulouse, Houston, Moscow), which imposes performance constraints on a supervision solution.

The flexible GeneSyS information allowed customising of System and Network agents and development of scenario specific Middleware and Application agents (RTI middleware, DIS-RVM application).

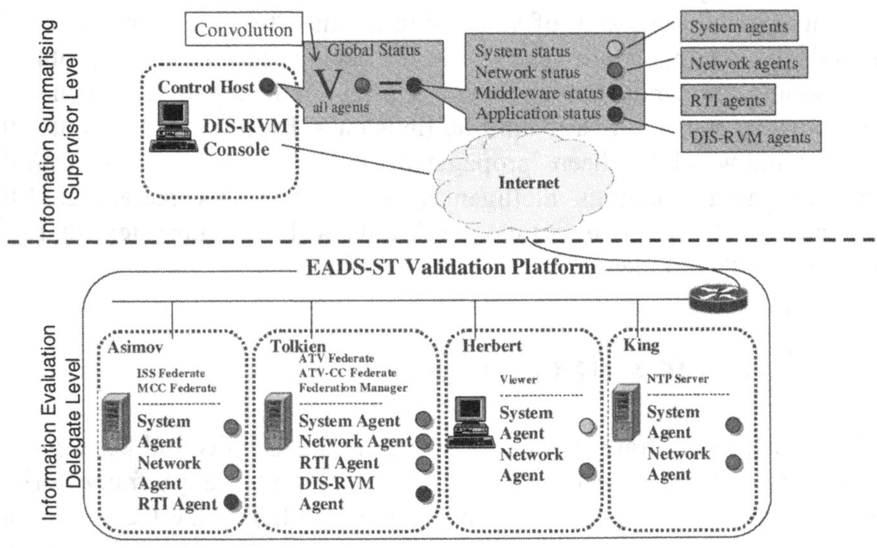

Figure 3. Intelligence in Distributed Training Supervision

Figure 7 depicts the deployment schema and gives intelligence implementation hints.

The "synthetic view" and "agent dependencies framework" approaches were used to provide administrators with a run-time system operation status summary and to allow a fast problem location.

Thus an administrator could browse down the agents to find a problem origin and then maintain the system.

3.2 Web Servers Scenario

The Web Servers Monitoring validation scenario aims at using GeneSyS for monitoring and controlling web servers and web based on-line services.

A Web Server is typically more than just an HTTP daemon: it may invoke external programs and those programs may use other programs for their execution, and so on. A typical Web Server can include, for example, an Apache server with a PHP interpreter and a MySQL database used by a number of PHP application. The Web Server is considered "healthy" only if all of these components are in good condition. Because these components may be dependent of each other it is not enough to have separate agents for all entities but these agents must be connected in a way to reflect the dependencies of the monitored entities.

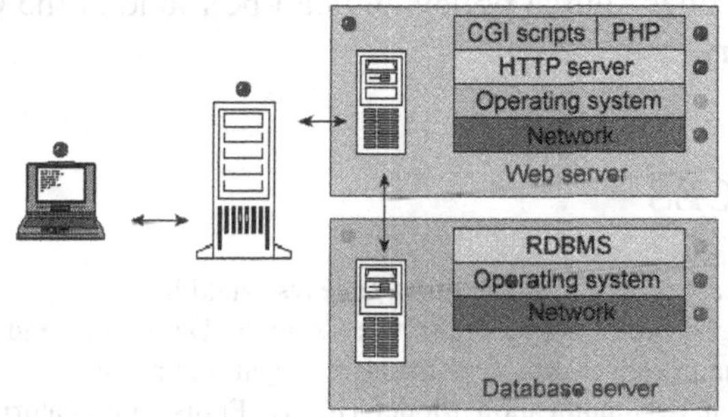

Figure 4. Web Application - A Common Deployment

Going on with the previous example, a Web Server could be considered healthy if the Apache daemon is up and running, the PHP applications it hosts respond in an acceptable time interval and the MySQL server has enough space for new records. If any of these conditions are not met the system should notify the administrator. More over, the unresponsiveness of Apache may be the result of a number of other dependent subsystems, like the operating or network system. So the "monitoring entity" could be divided into some more elements, namely the Apache server itself, the underlying operating system and the network connecting the server machine to the outer world. In this case even if Apache is found to be alive the operating system agent may report that the CPU load is too high and this could cause in a short time the Apache server being unable to respond to requests.

The Web Servers Monitoring scenario extensively uses the Agent Dependency Framework of GeneSyS, which provides the ability to describe the dependencies of system components and use this dependency graph to detect and find root cause of an erroneous system state.

4. CONCLUSION

This article presents an innovative supervision middleware intended to supplement classical distributed management approaches based on SNMP. The proposed framework has great integration capability illustrated on different real-life applicability examples. That permits using it in conjunction with existing network management infrastructure.

In comparison with other solutions, among other advantages, the authors would like to emphasise that the GeneSyS architecture is open to be extended with custom agents for all kind of applications.

The validation showed that, besides some ergonomics and performance issues, the solution is ready for the large community of the Internet users. That is why, generic components for system and network monitoring, as well as, visualisation tools, service components and development toolkits were released under open source policy and can be found at the GeneSyS SourceForge repository (see[7])

REFERENCES

1. GeneSyS project official web-site: http://genesys.sztaki.hu
2. Wallace Croft, David, "Intelligent Software Agents: Definitions and Applications", 1997, http://www.alumni.caltech.edu/~croft/research/agent/definition
3. The Inter-Agent Communication Model (ICM), Fujitsu Laboratories of America, Inc., http://www.nar.fujitsulabs.com/icm/about.html
4. Web Service Activity of W3C, http://www.w3.org/2002/ws/
5. DataPower Offering Web Services-Based Network Device Management, http://www.ebizq.net/news/2534.html
6. HLA, Institute of Electrical and Electronic Engineers - IEEE 1516.1, IEEE 1516.2, IEEE 1516.3
7. GeneSyS project SourceForge file repository, http://www.sourceforge.net/projects/genesys-mw

MULTIGROUP COMMUNICATION

Using active networks technology

Agnieszka Chodorek[1] and Robert R. Chodorek[2]

[1]*Department of Telecommunications and Photonics, The Kielce University of Technology, al. Tysiąclecia Państwa Polskiego 7, Kielce, Poland;* [2]*Department of Telecommunications, The AGH University of Science and Technology, al. Mickiewicza 30, Kraków, Poland*

Abstract: Common multicast tree shared by all layers/streams belonging to one session, is essential to provide multicast-based congestion avoidance. It enables both synchronization of layered data (in the case of layered multicast) or streams (in the case of multicast stream replication) and stable congestion control. Although many authors have been addressed the problem of multicast transmission, the methodology of building common delivery trees still remains an unresolved issue. In the paper, a new solution of that problem is proposed – a multigroup communication based on active network technology.

Key words: multicast communication, congestion control, active networks

1. INTRODUCTION

There are two main architectures of multicast-based congestion avoidance, able to preserve real-time characteristics of multimedia transmission: receiver-driven layered multicast and stream replication multicast (Kim and Ammar, 2001). In the both cases, the multimedia stream is divided into several complementary layers (layered multicast) or independent streams (stream replication) with different QoS requirements. All layers/streams are synchronised and simultaneously transmitted trough the network as separate multicast groups. Receivers can individually subscribe or unsubscribe to the appropriate multicast groups to achieve the best quality signal that the network can deliver.

The great advantage of multicast-based congestion avoidance is that it is a good solution of network heterogeneity problem. However, such a transmission assumes, that all multicast groups (layers or streams) will follow the same multicast tree even if they are sent separately (Matrawy et al., 2002). Above assumption is especially necessary in the case of layered multicast, where identical propagation parameters for each layer allow to avoid loss of synchronism of transmitted layers.

Moreover, common multicast tree, shared by all layers/streams belonging to one session, is needed to provide stable congestion control. Receiver-driven congestion avoidance forms a close loop control from point of congestion to the receiver and to the point of congestion again. The receiver acts as a controller, which motivates the control device (network node at the point of congestion) to perform rate adaptation. Stable control process will be possible only if group membership stay in one-to-one relationship with effective transmission rate measured at the point of congestion.

Unfortunately, common multicast tree cannot be guaranteed in, connectionless in nature, IP network. Although many authors have been addressed the problem of layered multicast transmission management, also using active network technology (Yamamoto and Leduc, 2000), the methodology of building common delivery trees still remains an unresolved issue. In the paper, a one of possible solution of that problem is proposed – a multigroup communication based on active network technology. Instead of the, mentioned above, previous propositions of active networks based layered multicast, the proposed management scheme allows full utilisation of existing multicast infrastructure.

The rest of the paper is organized as follows. Section 2 proposes multigroup communication management. Section 3 describes examples of multigroup management while Section 4 addresses the evaluation of proposed management scheme. Section 5 concludes the paper.

2. MULTIGROUP COMMUNICATION USING ACTIVE NETWORKS TECHNOLOGY

2.1 A concept of multigroup communication

W define multigroup as *a set of multicast addresses, belonging to one multicast sessions, which follow the same multicast tree, while multicast forwarding is adapted individually to particular group membership.* Multigroup comprises of a set of multicast groups, tied together by common delivery tree. It was assumed, that multicast delivery tree will be constructed on the basis of multicast distribution of the lowest-quality stream (base layer or base stream). This assumption arose directly from the concept of layered multicast, where receiver have to connect to the multicast group, which transmits the base stream and then connect to one or more supplementary groups (in order of relevance of carried substreams). In result, all receivers are always connected to the multicast group of the base layer and the base layer's delivery tree is, in fact, the base delivery tree, which always connects possible session participants. In the case of multicast stream replication, the base (lowest-quality) stream also is the only stream to which each receiver must connect, at least once (at the beginning of transmission).

The proposed multigroup management, able to construct multigroup from a separate layers, is implemented using active network (AN) technology. Code, which defines the behaviour of the multigroup, is moving along delivery tree. Instead of typical AN-based solutions, the code isn't distributed from the sender to receiver(s), but it is distributed hop by hop, from receiver to the sender (exactly: the router closest to the source) – as the other multicast management messages are moved.

For multigroup management purposes, the session description transmitted by Session Description Protocol (SDP) must be extended by code, which defines the behaviour of the multigroup. In particular, the code conveys information about multicast group behaviour within the multigroup and packet marking policy. Receiver obtains the code together with other session data transmitted by SDP. During the connection's establishment, a copy of a code is installed at network nodes within a delivery tree. Thus, each network node obtains definition of a multigroup behaviour (in particular, a method of joining and leaving multigroup as well as a single layer).

2.2 Joining and leaving multigroup

A receiver connects to the multigroup sending the `join_multigroup` message. The `join_multigroup` message is sent to the default Designated Router (DR), as the IGMP (or MLD) reports are. The DR router is usually the nearest router supporting IP multicast. If DR router belongs to the shared tree (there is a receiver in link-local network connected to the base layer), DR router will activate data delivery on interface, from which it receives the `join_multigroup` message. Otherwise, the router will have to connect to the delivery tree.

A router (or DR router) which want to connect to common delivery tree install the code and sends the `join_multigroup` message to the NH router. NH router is the next hop router in delivery tree, toward the root. The address of the NH router can be obtained from the routing protocol. Delivery tree is stored in local cache for fast lookup purposes. If, from any reason, routing protocol changes delivery tree, cache will be automatically updated.

Code is forwarded along delivery path to the next hop toward the root. NH router checks the uniqueness of the code, and (if needed) install the code and then sends code to the successive hop on delivery path. The sequence is repeated until the root or the first router served the base layer is achieved. In result, each router in the path obtains code. To rebuild distribution tree standard multicast routing protocols are used. Its worth remarking, that (instead of IP multicast) `join_multigroup` message is used for communication both between receiver and DR and between routers.

Receiver leaves multigroup by sending `leave_multigroup` message to the DR router, which checks the global number of receivers connected to the layer 1 in router's directly attached networks. This number is stored into the global variable *rcv_all*[1]. If the *rcv_all*[1] is equal to zero, the router will send the `leave_multigroup` message to the NH router and will activate prune mechanism of routing protocol. In result, the router leaves multicast delivery tree, then terminates execution of the code and removes it.

2.3 Joining and leaving layers

If the receiver connects to the upper layer, sends the `join_layer` message. The message propagates along the delivery tree, until the first router served the upper layer is achieved. The `join_layer` message activates the transmission of upper layer (activates proper interfaces along the delivery tree). The upper layer transmission follows the same delivery path from the sender to the receiver as the base layer transmission. A simplified algorithm for joining layer is as follows:

```
receive join_layer(layer) message on interface
if (rcv_all[layer] == 0) {
  lookup route cache for NH router
  send join_layer(layer) to NH router
  update forwarding table }
if (rcv[layer, interface] == 0) {
  enable forwarding for interface }
increment rcv[layer, interface]
increment rcv_all[layer]
```

The above pseudo-code defines two counters. The *rcv*[*layer*, *interface*] describes the number of receivers at given interface, counted for given layer. The *rcv_all*[*layer*] counter describes global number of receivers connected to the layer *layer* in router's directly attached networks.

Receiver, which decides to leave layer *layer*, sends the `leave_layer` message to its DR router, which increments both *rcv*[*layer*, *interface*], and *rcv_all*[*layer*]. If the number of receivers at the given interface reaches 0, the router won't propagate data via this interface. If the global number of receivers reaches 0, the router will send the `leave_layer` message to the upper (the next) router. When a layer is leaved, the remaining datagrams of the layer are useless and should be removing from the queue rather than be carried to the receiver. The last operation carried out during the leaving layer is `flush` procedure, which discards these datagrams. A simplified algorithm for leaving layer is as follows:

```
receive leave_layer(layer) message on interface
decrement rcv[layer, interface]
if (rcv[layer, interface] == 0) {
  disable forwarding for interface }
decrement rcv_all[layer]
if (rcv_all[layer] ==0) {
  lookup route cache for NH router
  send leave_layer(layer) message to NH router
  update forwarding table
  flush(layer)}
```

Receiver joins and leaves, successively, layer by layer, in order of their relevance. In the case of emergency (e.g. if the receiver will be switch off without prior `leave_layer` message), layers will be leaving using the time-out signalization from IGMP (or MLD) protocol.

2.4 Maintenance of multicast connections

DR routers provides typical, defined by IGMP (or MLD) standard, maintenance procedures to determine if there is any group member on their directly attached networks. Such a group checking is performed using standard mechanisms based on IGMP (or MLD) Query message.

This maintenance procedure actualize information about hosts belonging to each multicast group and allows verifying status of particular group membership. Differences between real and counted status of group membership may be caused by e.g. connection failures or other emergency situations, while receivers leave groups without sending typical `leave_multigroup` or `leave_layer(`*layer*`)` messages.

3. EXAMPLES AND SCENARIOS

3.1 Joining the multigroup

Assume that receiver R1 decide to join multigroup, which serve the MPEG-4 layered video transmission. Initially no one receiver is connected to the multigroup. In this example, message with code is distributed along delivery tree from the DR router to the root. The process of joining multigroup is performed in 7 steps (Fig. 1):

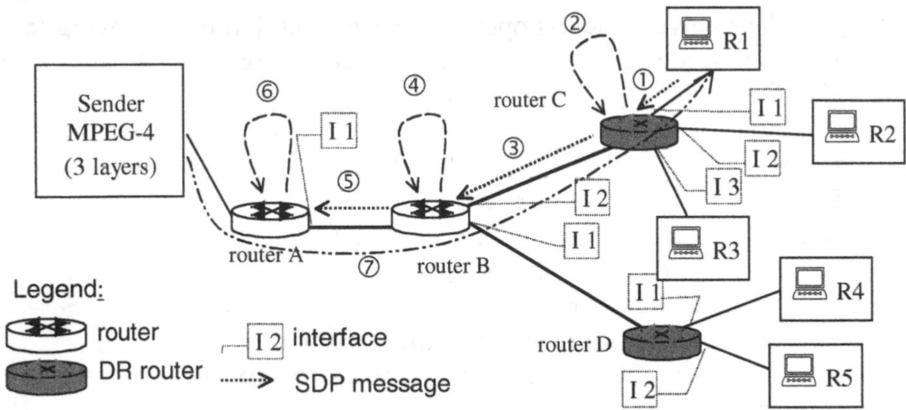

Figure 1. Receiver R1 joins the multigroup

1. Receiver R1 sends the `join_multigroup` message to the router C (a DR router of the receiver's link-local network).
2. Router C gets the `join_multigroup` message via the interface I 1. Check the uniqueness of the code. The code hasn't been yet installed, so the router installs it and sets parameters. In particular, the global number of receivers connected to the base stream *rcv_all*[1] and the number of receivers at given interface *rcv*[1,1], are set to one. Router creates delivery tree using multicast routing protocol (updates forwarding table).
3. Router C lookup for the address of the NH router in local cache and sends the `join_multigroup` message to the NH router (router B).
4. Router B gets the `join_multigroup` message via the interface I 2 and repeats the processing of the obtained code (see 2).
5. Router B lookup for the address of the NH router in local cache and sends the `join_multigroup` message to the NH router (router A).
6. Router A gets the `join_multigroup` message via the interface I 1.
7. Router A activates interface I 1. Router B transmits the multicast packets which convey the base layer to router C via the interface I 2. Base layer is transmitted from the router C via the interface I 1 to the receiver R1.

Router A is a root of delivery tree and receives multimedia data strictly from the sender.

3.2 Joining upper layers (streams)

Assume that receivers R1 and R5 decide to join the upper (namely, 2nd) layer. Initially all receivers are connected to the multigroup, but only R2 and R3 are connected to the layer 2. In the first example, message with code is transmitted only from the receiver to the DR router. The process of joining the upper layer by receiver R1 is performed in 3 steps (Fig. 2, scenario A):

1. Receiver R1 sends the `join_layer`(2) message to the DR router (C).
2. Router C gets the `join_layer`(2) message via the interface I 1. Check the status of counter *rcv_all*[*layer*]. Receivers R2 and R3 are connected to layer 2, so *rcv_all*[2] = = 2. Thus, router C activates forwarding of multicast packets to interface I 1 and increments variables *rcv*[2,1] and *rcv_all*[2]. In result, *rcv*[2,1]=1 and *rcv_all*[2]=3.
3. Layer 2 is transmitted from the router C (via I 1) to the receiver R1.

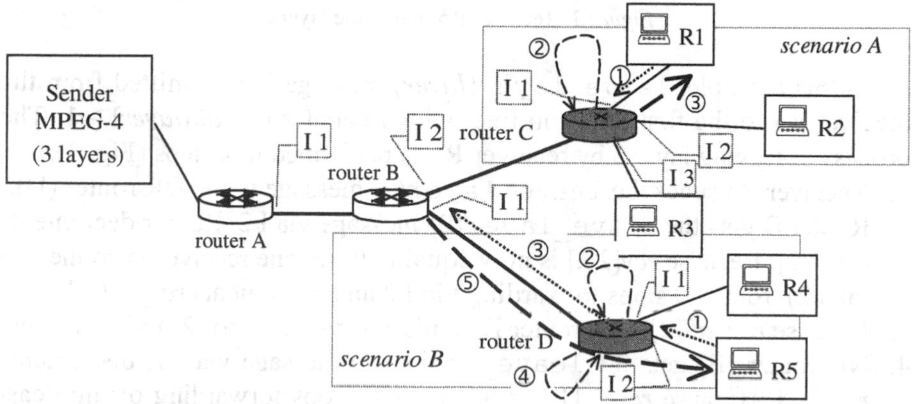

Figure 2. Receivers R1 and R5 joins the layer 2

The second example shows transmission of the `join_layer`(layer) message from the receiver to the router B. The process of joining the upper layer by receiver R5 is performed in 5 steps (Fig. 2, scenario B):

1. Receiver R5 sends the `join_layer`(2) message to the DR router (D).
2. Router D gets the `join_layer`(2) message via I 2, checks the status of counter *rcv_all*[*layer*]. No one receiver is connected to layer 2 in router's directly attached networks (*rcv_all*[2] = = 0). Thus, router C activates forwarding of multicast packets to interface I 2 and increments variables *rcv*[2,1] and *rcv_all*[2]. In result *rcv*[2,1] = 1 and *rcv_all*[2]=1.
3. Router D sends the `join_layer`(2) message to the NH router (B).
4. Router B gets the `join_layer`(2) message via I 1. Because *rcv_all*[2] = = 3, router B activates interface I 1 and increments variables *rcv*[2,1] and *rcv_all*[2]. In result *rcv*[2,1] = 1 and *rcv_all*[2] = 4.
5. Layer 2 is transmitted from the router B through the router D to R1.

3.3 Leaving upper layers (streams)

Assume that receivers R1, R2 and R3, are connected to the layer 3 and receiver R5 to layer 2. Due to possibility of congestion, receiver R5 have to leave layer 2 to reduce effective transmission rate in the access network.

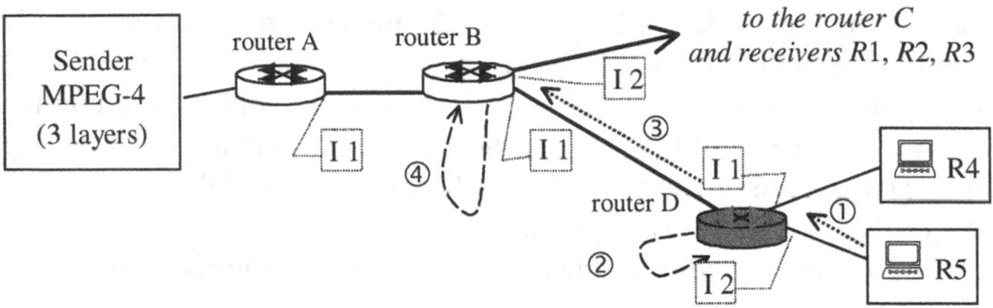

Figure 3. Receiver R5 leaves the layer 2

In this example, `leave_layer`(*layer*) message is transmitted from the receiver R5 to the first router in tree, which counter *rcv_all*[*layer*] > 1. The process of leaving layer 2 by receiver R5 is performed in 4 steps (Fig. 3):

1. Receiver R5 sends the `leave_layer`(2) message to the DR router (D).
2. Router D gets the `leave_layer`(2) message via I 2. Router decrements *rcv*[2,2]. Because *rcv*[2,2] is now equal to 0 (no one receiver is connected via I 2), router D stops forwarding via I 2 and decrements *rcv_all*[2].
3. Because *rcv_all*[2]= =0, router D sends `leave_layer`(2) to NH router.
4. NH router (B) gets the `leave_layer`(2) message via I 1, decrements *rcv*[2,1]. Because *rcv*[2,1] = = 0, router B stops forwarding of multicast packet via I 1 and decrements *rcv_all*[2]. Counter *rcv_all*[2] = = 3, so router B forwards multicast packets and don't propagate `leave_layer`(2) to NH router (router A).

4. RESULTS

Described multigroup management was successfully implemented in Berkeley's *ns*-2 network simulator environment. As the source of elastic traffic (ST), FTP over TCP (SACK version) was used. TCP packets have a size of 1000 bytes. As a video application, the ECN-capable layered multicast was used – see (Chodorek, 2003) for details. Video traffic sources (SV) were modeled as three-layer VBR streams, generated from 13 publicly available video traces, encoded in high, medium and low picture quality. Properties of traces can be found in (Fitzek and Reisslein, 2001). The video streams use ECN-capable RTP protocol (Chodorek, 2002) and RTP packets have maximum 188 bytes.

The proposed transmission scheme has been simulated in a 3 different topologies (Fig. 4), to expose the performance issues as well as scalability. Senders are connected to the router at 100 Mbps and 1 ms delay. Receivers are connected to the router through 1 ms delay link. Routers are connected with a link at 5 ms delay. All routers are ECN-capable (RED queue marks packets instead of dropping) and utilize 3-level RED queue.

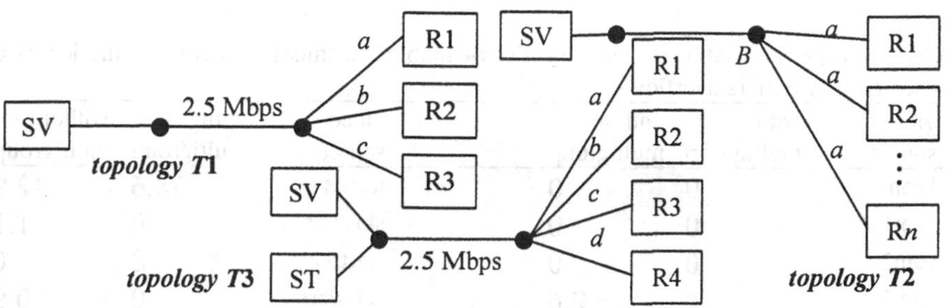

Figure 4. Simulation topologies. Parameters of T1: $a = 0.4$ Mbps, $b = 0.8$ Mbps, $c = 2$ Mbps. Parameters of T2: $a = 100$ Mbps. Parameters of T3: $a = b = c = 2$ Mbps, d = 0.2 Mbps.

Topology *T1* was used for adaptability tests. Receiver R3, connected via non-congested link, always was able to receive full video information (layer 1 to 3). Receiver R2 was connected via lightly congested link and always was able to receive layer 2. If the realistic video source was characterized by low-detail, slowly dynamic content, R2 could connect to layer 3. Receiver R1 was connected through heavy congested link. R1 was able to receive layer 1 and, if the video trace was small enough, could connect to layer 2.

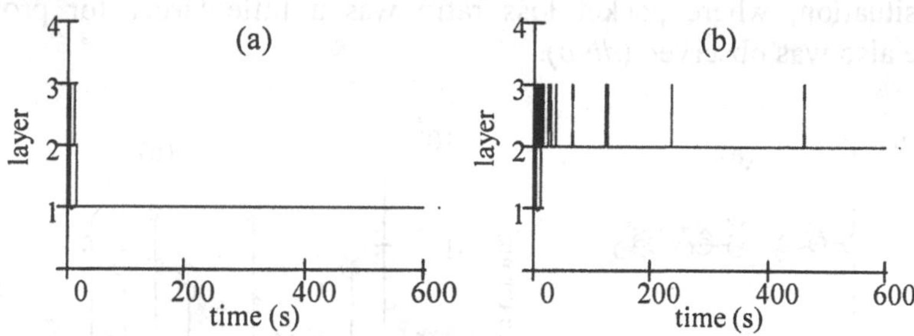

Figure 5. Adaptability (R1, *starwars* video): a) without multigroup, b) with multigroup.

Experiments shows, that adaptability of the receiver-driven layered multicast is the same or better when the multigroup communication is used, because of faster reaction on the variable network conditions (here: VBR traffic). In Figure 5 an example of such a situation is depicted. The slow reaction of the system without multigroup (layer 3 was leaved about 2 times slower without the multigroup) has resulted in longer congestion during the first seconds of transmission, what lead to the saturation of RED's average queue size, so the receiver gets ECNs even if congestion is over. Last, but not least, reason of better behavior of the system implementing multigroup communication is `flush` procedure, which discards packets belonging to unsubscribed layer from an output queue, what increases the bandwidth utilization and decreases delay of the rest of packets.

Table 1. Packet loss ratio. Because the ECN-capable transmission was used, the losses are caused only by buffers overflows.

video source	with multigroup	without multigroup	video source	with multigroup	without multigroup
bean	0	0	lambs	38.6	42.9
cam1	0	0	ski	0	1.1
cam2	0	0	startrek	0	0
cam3	0	2.6	starwars	0	0.9
dieh	0	3.5	troop	0	3.1
dino	5.3	5.2	vclips	0	1.6
form1	0	4.4	-	-	-

Adaptability can be measured as the function of packet loss – if the system is fine-tuned to the network conditions, packet loss ratio will be small. As shown in Table 1, packet loss ratio equals 0 (the best adaptation) in the case of 11 movies of 13 while the multigroup was used. Only 4 movies of 13 have achieved the best adaptability without the multigroup. If the network is not well-dimensioned (losses larger than 5% – *dino* and *lambs* video sources), the multigroup improves adaptability (*lambs*), although the other situation, where packet loss ratio was a little larger for proposed scheme also was observed (*dino*).

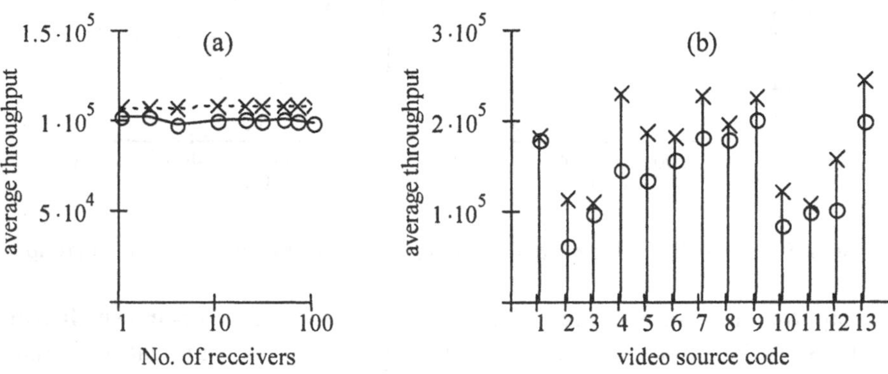

Figure 6. Session-size scalability of a system with (x's) and without multigroup (o's): a) average throughput as a function of number of receivers (*starwars* video source), b) average throughput observed for population of 30 receivers. Video source codes: 1 – *bean*, 2 – *cam1*, 3 – *cam2*, 4 – *cam3*, 5 – *dieh*, 6 – *dino*, 7 – *form1*, 8 – *lambs*, 9 – *ski*, 10 – *startrek*, 11 – *starwars*, 12 – *troop*, 13 – *vclips*.

Session-size scalability was investigated using topology *T2*. Bandwidth *B* was large enough to assure that receivers can connect to layer 2. Results of experiments are depicted in Figure 6. Although both systems (with and without multigroup communication) achieve very good scalability (Fig. 6a), achievable throughput is usually larger for system with multigroup communication (Fig. 6b). This is caused both by faster reaction of the system and discarding IP datagrams belonging to unused layers.

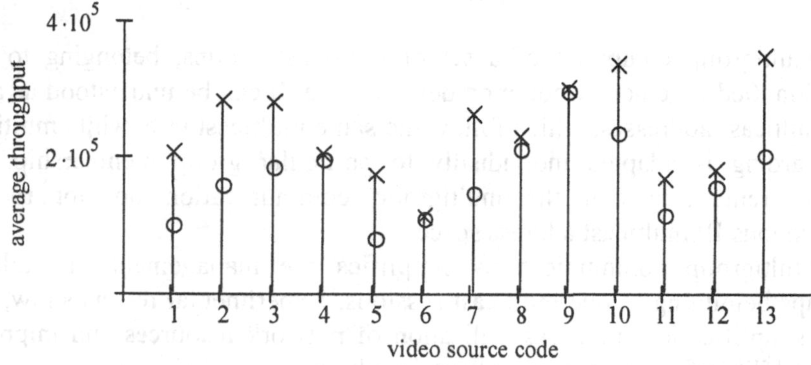

Figure 7. RTP competing with TCP flow. Video source codes: as in Fig. 6.

Other investigations were carried out using topology *T3*, where R1...R3 receives VBR stream and R4 receives TCP packets. During the experiment, ECN-capable layered multicast share link with ECN-capable TCP flow. Results shows, that TCP flow always (with and without multigroup, for all tested video sequences) achieves throughput close to the nominal (0.2 Mb/s), what confirms effects described in (Chodorek, 2003). Thus multigroup communication doesn't influence on TCP-friendliness of layered multicast. However, multigroup communication increases the average throughput of competing video from 1% (*dino*) to more than 110% (*bean*) and in more than 60% of movies (8 of 13) this growth was significant (Fig. 7).

Some elements of the management scheme, essential from the transmission point of view, were tested in the Linux active routers. The remarks from field trials are as follows:

- the multigroup communication together with IGMPv2 gives faster layer switching than usage of IGMP alone – receiver leaves layer about 2 seconds faster than without multigroup management, while duration of layer joining remains unchanged,
- in the case of dense mode multicast routing protocols, which doesn't require rebuilding of delivery tree during layer switching, multigroup communication doesn't give significant advantage,

- in the case of sparse-mode multicast routing protocols, where the layer switching rebuilds delivery tree, the usage of multigroup communication results in faster layer switching.

Above results do not cover additional mechanisms, which accelerate join/leave operations, as for example `flush` procedure.

5. CONCLUSIONS

Multigroup comprises of a set of multicast groups, belonging to one session, tied together by common delivery tree. It can be understood as a set of multicast addresses, which follow the same multicast tree, while multicast forwarding is adapted individually to particular group membership. It's worth remarking that the multigroup communication do not require continuous IP multicast address space.

Multigroup communications simplifies the management of multiple groups belonging to one multicast sessions. Experimental results show, that this simplification improves utilization of network resources and improves adaptability of receiver driven layered multicast.

ACKNOWLEDGEMENTS

This research was supported by State Committee for Scientific Research (KBN) under grant No. 4 T11D 015 24 (years 2003-2005)

REFERENCES

Chodorek, R. R., 2002, A simulation of ECN-capable multicast multimedia delivery in ns-2 environment, *Proc. of 14th European Simulation, ESS'*2002, pp. 233-237.

Chodorek, R. R., 2003, ECN-capable multicast multimedia delivery, *Springer LNCS* 2720, pp. 62-72.

Fitzek, F.H.P., and Reisslein, M., 2001, MPEG-4 and H.263 video traces for network performance evaluation, *IEEE Network* 15(6): 40-54.

Kim, T., and Ammar, M.H., 2001, A comparison of layering and stream replication video multicast schemes, *Proc. of NOSSDAV* 2001.

Matrawy, A., Lambadaris, I., and Huang, Ch., 2002, Comparison of the use of different ECN techniques for IP multicast congestion control, *Proc. ECUMN'*02, pp. 74-81.

Yamamoto, L., and Leduc, G., 2000, An active layered multicast adaptation protocol, *Springer LNCS* 1942, pp. 180-194.

POLICY USAGE IN GMPLS OPTICAL NETWORKS

Belkacem Daheb[1,2], Guy Pujolle[1]
[1] University of Paris 6, LIP6 Lab, 8, rue du Capitaine Scott, 75015 Paris ; [2] Institut Supérieur d'Electronique de Paris (ISEP) ; {Belkacem.Daheb,Guy.Pujolle}@lip6.fr

Abstract: Recently, a great consortium on Generalized Multi Protocol Label Switching (GMPLS) is emerging as the control plane for next generation optical backbone networks. This article proposes guidelines for managing optical networks controlled by GMPLS in a policy based fashion. A flexible management solution is presented especially for optical network issues, where the service management system efficiently impacts the control plane offering the possibility to dynamically change network functionality to enhance the controllability of optical networks.

Key words: GMPLS, Optical Networks, Policy based Management, Service Management.

1. INTRODUCTION

In recent years, there has been a dramatic increase in data traffic, driven primarily by the explosive growth of the Internet as well as the proliferation of virtual private networks (VPN). On the other hand the huge amount of bandwidth, offered by optical links, render optical networks the ideal candidate for next generation backbone networks.

The initial use for optical fiber communication and its prevalent use today, is to provide high-bandwidth point-to-point pipes. At the ends of these pipes, data is converted from the optical to the electrical domain, and all the switching, routing, and intelligent control functions are handled by higher-layer equipment, such as SONET or IP boxes[1].

With IP routers emerging as among the dominant clients of the optical layer, there has been a great deal of interest recently in trying to obtain a closer interaction between the IP layer and the optical layer from a control and management perspective. This is done using distributed control protocols, which are widely used in data communications networks such as IP and ATM. These protocols set up and take down connections, maintain topology databases, and perform route computations. In any event, there is emerging consensus on basing these optical layer protocols on a modified version of Multi-Protocol Label Switching (MPLS), now called Generalized MPLS[2] (GMPLS). One of the key aspects of MPLS is the addition of a new connectivity abstraction: explicitly routed point-to-point path. This is accomplished by the concept of explicitly routed label switched paths (LSP).

Nonetheless, maturing distributed optical intelligence solutions lead to a new control-management interaction scheme. In addition, since the control plane affects pivotal functions in classical network management, it is instrumented to fit only a question-answer relation with that centralized management plane[3]. To accomplish an efficient control-management inter-operation, i.e. to reduce management complexity and delay and to enhance automation, one of the most commonly used approaches is the policy-based management (PBM). An effort of standardization has been made on the field of PBM at the IETF[4,5]. The PBM approach provides an overall, network-wide regulatory infrastructure and allows network administrators and Service Providers to simplify end-to-end configuration and regulation of the network behavior with enhanced Quality of Service (QoS) and Traffic Engineering (TE) features rather than configure individual devices. For instance, network administrators may find a way to preprogram their networks through rules or policies that define how the network should automatically respond to various networking situations.

To wrap up things, Figure 1 shows a layered architecture of optical networks, independently of the transmission technology. In the middle stems a Data Plane, which forms the data transmission mechanisms based on wavelength paths. This plane is over hanged by a Control Plane which in turn processes generic provisioning mechanisms, that are independent from transmission ones. On the bottom of the figure, we have a management plane which performs management tasks that are less real-time than the previous.

This paper provides a brainstorming on the integration of the policy management approach to immerging optical networks controlled by a GMPLS control plane. Next section addresses the possible policy based interaction between the control and management plane. It is followed by a Section that details the different mechanisms involved in controlling such optical networks. While the last section details every component of the control plane by showing where would policies impact.

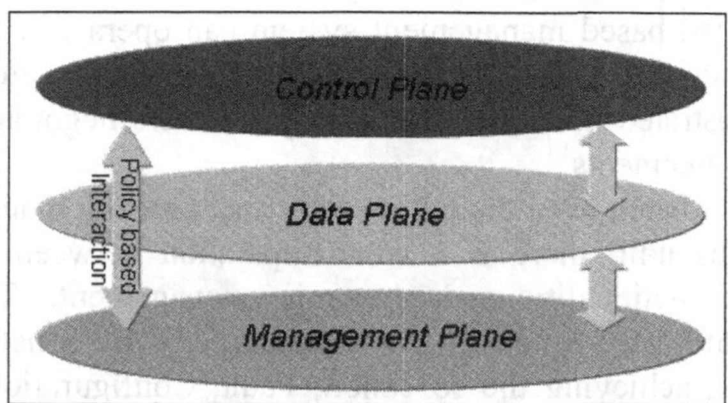

Figure 1. Optical network functional planes

2. CONTROL AND MANAGEMENT PLANE INTERACTION

Emerging requirements brought by the growing amount data traffic and need of automation lead to the introduction of GMPLS as a control plane for automating the provisioning process. GMPLS allows the control of optical Label Switched Paths (LSP) closer to the data plane than the management plane. Because GMPLS is an extension from MPLS to non packet switched networks, and further difficulties come from the optical layer. For example, optical network may contain entirely photonic, hybrid, and opaque network nodes.

Consequently, basic control plane functionality – carrying out autonomous topology management, routing, signaling, and required QoS enforcement for traffic engineering – raises some issues of consistency when applied without appropriate inter-operations with the management plane, and constitute a challenge to efficient network management. On the other hand, in regard to configuration management and especially in provisioning, the manual establishment of explicit LSPs with associated QoS parameters is slow, error prone, and laborious to network administrators.

Previous studies tend to the introduction of a policy-based management system on MPLS networks[6,7,8,9]. This proves that even though MPLS efficiently controls a network but higher abstraction layer mechanisms are needed for the controllability of LSP life cycle. Since GMPLS brings the opportunity to establish network wide paths within the optical network, and therefore implicitly offer a rationale and also a background for end to end service deployment, a reinforced management system including a Policy-based Management (PBM) infrastructure is adjoined to the previous functional planes structuring, see Figure 1. Globally PBM is viewed as part of the management plane. It deals specifically with node intelligence configuration, decision at network level, and with service related actions. A policy controlled based management system can operate at a higher, more service focused level where control for admission and security as well as QoS are orchestrated in accordance with parameters negotiated in different service level agreements[10].

On the other hand, even though it seems that network management forms a homogeneous unit, there is a clear separation between the traditional management functionality and dynamic management. The traditional management of GMPLS networks is to manage GMPLS network elements the TMN way, achieving the so called, Fault, Configuration, Accounting, Performance, and Security (FCAPS) management processes. The dynamic management tasks are performed essentially via the policy-based management system. The former can be seen as unaware to the optical nature while the later is devoted, amongst other, to handle different constraints, especially optical ones.

3. GMPLS CONTROL PLANE

Generalized multi-protocol label switching, also referred to as GMPLS, supports not only devices that perform packet switching, but also those that perform switching in the time, wavelength, and space domains. This part briefly presents the traffic engineering (TE) features introduced by GMPLS to extend the MPLS control plane to non packet networks.

The GMPLS protocol suite could be devised into two blocks: a routing block for choosing the route that must take the traffic and a signaling one that allocate resources for the path chosen in the routing step, Figure 2. Since GMPLS has two building blocks, the TE enhancements affect both of them[11,12].

Some of these enhancements concern the routing block[11], and are cited in what follows. Among the routing enhancements, the LSP hierarchy is the notion that LSP of different types (FSC, LSC, TDMC, PSC) can be nested inside others. At the top of this hierarchy are nodes that have fiber-switch-capable (FSC) interfaces, followed by nodes that have lambda-switch-capable (LSC) interfaces, followed by nodes that have TDM-capable interfaces, followed by nodes with packet-switch-capable (PSC) interfaces. One of the most important enhancements is Link Bundling, which means aggregating several links of similar characteristics, and assigning these aggregated attributes to a single "bundled" link. In so doing, the size of the link state database (maintained by the routing protocol, OSPF-TE for instance) is reduced by a large factor. Especially in optical networks, with a large number of links and ports on each router, an Unumbered Link feature allows to locally number some links and ports to prevent the lake of addresses.

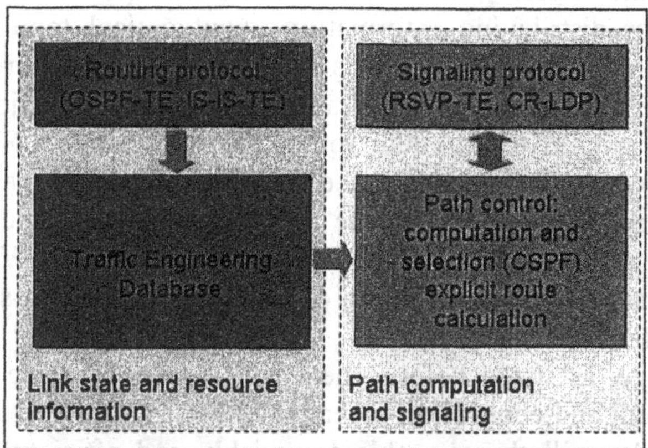

Figure 2. GMPLS building blocks

While in the signaling part a new feature named the Hierarchical LSP Setup, has been introduced. It means that an LSP could encompass several regions and take advantage of the link bundling respecting the LSP Hierarchy defined. In optical networks, the signaling protocol is responsible of choosing the adequate wavelength on each link maintaining as much as possible the wavelength continuity constraint. The set of wavelengths chosen define the different labels of an optical LSP. GMPLS signaling allows a label to be suggested by an upstream node, thanks to the Suggested Label. In the basic MPLS architecture, LSP are unidirectional; but Bidirectional optical LSP (or lightpaths) are a requirement for many optical networking service providers. This is possible in GMPLS thanks to the Bidirectional LSP setup. The last enhancement to signaling is the Notify Messages that are sent to notify the node(s) responsible for restoring the connections when failures occur.

It is expected that in very large networks, such that controlled by a GMPLS control plane, a network operator could face many failures. To facilitate the management of the occurred failures a Link Management Protocol (LMP) was added to the GMPLS protocol suit. A key service provided by LMP is to set-up and verify associations between neighboring nodes.

One of the merits of GMPLS stems from its ability to automate circuit provisioning in optical networks. Connection management complexity related to LSP setup/modification/teardown is thus reduced. This simplification is realized through a suite of protocol extensions currently under standardization in the IETF[2].

Figure 2 presents the functional GMPLS building blocks that would be distributed along the different network nodes. The link state Internet Gateway Protocol (IGP), which can be either OSPF or Intermediate System to Intermediate System (IS-IS) with optical-specific extensions, is responsible for distributing information about optical topology, resource availability, and network status. This information is then stored in a traffic engineering (TE) database. A constraint-based routing function acting as a path selector is used to compute routes for the desired LSP. This route calculation accounts for the information collected in the TE database as well as the traffic requirements. Once the route has been computed, a signaling protocol such as Resource Reservation Protocol with TE (RSVP-TE) is used for path activation (i.e., instantiation of the label forwarding state along the computed path).

For instance, as stated before, the GMPLS control plane is capable of performing fault and connection management in a fast distributed way. The service provider will thus be able to quickly and efficiently build high-capacity optical infrastructures supporting fast connection provisioning. Hence, new types of services requiring stringent connection setup times such as bandwidth-on-demand services would have the desired fast deployment time.

4. USING POLICIES WITH GMPLS

The GMPLS technology enables the setup of Label Switched Paths (LSP) through an optical network. Initially, the idea of GMPLS was to automate the service provisioning and apply traffic engineering. However, in the meantime traffic engineering and QoS in optical networks became the dominant driving force behind GMPLS. But GMPLS can't deal with all aspects of QoS or TE expected, since it is not guided by a higher level abstraction system to assist it reaching the goal of gracefully managing and controlling a network. To apply policy-based management concepts to manage a GMPLS network is an appropriate way of dealing with large sets of managed elements instead of manually managing each network element[3].

With regard to the previous section it is clear that the GMPLS control plane must be guided with some directives expressed under the form of policy rules. But what would policies impact in the GMPLS control plane?

To answer this question, let's go back to Figure 2 that represents the building blocks of an optical node controller. Policies would mainly impact these boxes. Requirements and conditions of policy usage are exemplified for each block in what follows through provisioning cases with respect to optical networks.

4.1 Routing Policies

This set of policies translates various types of constraints provided by the network operator and the client expectations such as constraints related to performance and business model.

The client expectations would introduce constraints on the path selection. A client may require a certain boundary to the time taken to establish his connection. He may also require a specified protection for his connection or even a restoration instead. Knowing that the operator applies rerouting of some connections from time to time, in order to enhance the performance of his network, a client can forbid to the operator rerouting his connection. Or at least he can limit the number of rerouting of his connection for a period of time. A client may also ask for certain confidentiality, in order to be sure that his traffic cannot be read by an intruder. Finally, the client can ask for a quality connection respecting some quality of signal features of the wavelength (BER for example). These constraints can be inferred through Service Level agreements[10].

It is clear that these user constraints participate in the choice of the path taken by a LSP. To handle these constraints lets go more in detail with a look at the GMPLS path computation process. The GMPLS path computation is performed by a Constraint Shortest Path First (CSPF) algorithm which uses the link state database maintained by the routing algorithm and some other constraints applied to the Shortest Path First algorithm. These constraints are directly derived from the user constraints described above and some other constraints defined by the administrator. The path computation is based on a Constraint Based Routing which computes an explicit route on the originating node and passes this information to the signaling protocol which is responsible of setting up the light-path.

4.2 Signaling Policies

After choosing the path for a connection, a signaling phase consists in choosing along the path the adequate wavelength used on each link. The main goal of this step is to reduce the wavelength conversion and minimize the set of wavelength needed in an optical network.

There are different schemes to perform wavelength selection named Spectral Routing in Papadimitriou[13]. The spectral routing is subject to constraints that may be a combination of both external and internal (or intrinsic) constraints due to the physical transmission medium for instance. Notice that external spectral routing constraints are generally dictated by Bit Error Rate (BER) or inferred through Service Levels. As a matter of fact, external constraints must be injected into the node performing the spectral routing; this is the role of the signaling policies.

More over, a spectral route can be computed by the ingress node, or each hop of the path or even by the egress node, depending on the scheme used. In each scheme, a node must choose and propose a set of possible wavelengths to use. This set is computed based on the set of available wavelengths and the set of wavelengths proposed by the ingress node. However, the choice is not trivial and can be guided via policies. For instance, the network operator can forbid the use of a wavelength and reserve it for signaling purposes.

4.3 Dimensioning policies

For optimization reasons, it may be desirable for a photonic cross-connect to optically switch multiple wavelengths as a unit. This unit is a waveband that represents a set of contiguous wavelengths, which can be switched together to a new waveband[2]. Hence, another stage is added in the LSP hierarchy, defined above. In Noirie[14], it is shown that the waveband stage reduces the complexity and cost of the optical cross-connects. Waveband coverage of the network is given first before routing the wavelength LSP. Every waveband setup is declared as a Forwarding Adjacency by the routing protocol. By definition, a Forwarding Adjacency (FA) is a TE link between two GMPLS nodes whose path transits one or more other (G)MPLS nodes in the same instance of the (G)MPLS control plane[2]. In other words, the wavebands are seen as links by the nodes; hence, they constitute the virtual topology over which the wavelength LSP will be setup.

This way of constructing the topology is static. The waveband coverage is based on some statistical traffic analysis but cannot predict and fit all the future connections demand. Therefore, a more dynamic way of constructing the topology is needed. For instance, it would be interesting to add some wavebands during the network life cycle.

Dimensioning policies are a mean to manage the virtual topology. Their role is to setup, modify or even tear down some wavebands depending on the network state and the connections demand. For example, when the whole wavebands are over loaded, a policy is needed to add a new waveband in order to permit future wavelength LSP establishment.

Another issue concerns the bundling of existing wavelength into wavebands during the waveband coverage step, which is not trivial. Two complementary nesting strategies have long been studied in Noirie[14]. The first grooming strategy consists in nesting lambda-LSP in a single waveband-LSP from end to end. The second grooming strategy consists in nesting a common sub-path of wavelength LSP into a waveband. However, the choice of the bundling strategy is not made at the node level but at higher levels. The dimensioning policies can be used to guide the grooming strategy from a higher level.

Finally, in order to enhance the flexibility of network configuration, the waveband construction must be policy-guided.

4.4 Logical Policy Levels

Having presented some directives that can be applicable for a policy management of GMPLS optical networks, here a logical view of these policies is inferred, Figure 3. The lowest level is concerned with the optical device-level configuration, the mid-level deals with the network-wide (GMPLS) configurations, and the upper layer handles services.

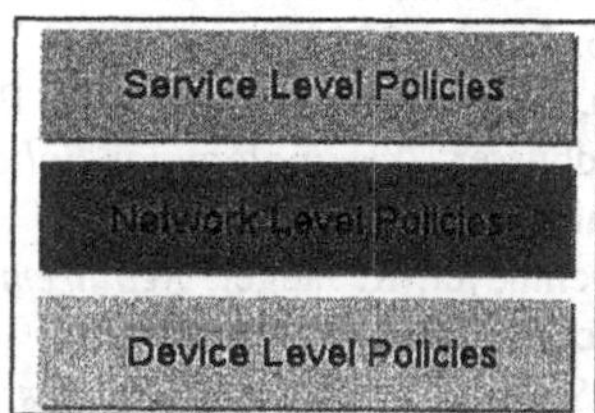

Figure 3. Logical Policy Levels

The device level policies deal essentially with optical network element configuration. It consist mainly in configuring the switching matrix of the optical cross-connects. The upper level policies are defined more network wide and include routing, signaling and dimensioning policies. These kinds of policies could eventually affect the GMPLS protocol stack, configuring both routing and signaling protocols the way described before. So, the control plane policies discussed above are a subset of these policies. The upper level defines policy rules depending on service exigencies that can emanate from service level agreements.

Notre that above the service oriented policies, we could have a higher level comprising some policies defined by the administrator itself, named Business Level Policies. These policies would respond to the manner that the administrator want to use his network and could be totally independent from service or network objectives.

5. CONCLUSION

This article dealt with policy based management of GMPLS optical networks. The need for policy rules was highlighted in optical networks controlled by GMPLS. The paper showed where and for what purpose do policies impact the GMPLS tool box. The policies discussed enhance the flexibility of provisioning and dimensioning of an optical network. We expect that this work will contribute in defining a policy based system for GMPLS optical networks.

REFERENCES

1. R. Ramaswami, *Optical Fiber Communication: From Transmission To Networking*, Invited Article, IEEE Communications Magazine, May 2002.
2. Eric Mannie, *Generalized Multi-Protocol Label Switching Architecture*, IETF draft, draft-ietf-ccamp-gmpls-architecture-07.txt, May 2003.
3. B. Berde, E. Dotaro, M. Vigoureux, *Evolution of Policy-enabled GMPLS Optical Networking*, Alcatel-CIT, Research and Innovation, Technical Report.
4. R. Yavatkar, D. Pendarakis, R. Guerin, *A Framework for Policy-based Admission Control*, Request for Comments: 2753, January 2000.
5. A. Westerinen, J. Schnizlein, J. Strassner, M. Scherling, B. Quinn, S. Herzog, A. Huynh, M. Carlson, J. Perry, S. Waldbusser, *Terminology for Policy-Based Management*, Request for Comments: 3198, November 2001.
6. S. Wright, S. Herzog, F. Reichmeyer, R. Jaeger, *Requirements for Policy Enabled MPLS*, IETF draft, draft-wright-policy-mpls-00.txt, March 2000.
7. M. Brunner, J. Quittek, *MPLS Management using Policies*, IEEE/IFIP, 2001.
8. K. Isoyama, M. Brunner, M. Yoshida, J. Quittek, R. Chadha, G. Mykoniatis, A. Poylisher, R. Vaidyanathan, A. Kind, F. Reichmeyer, *Policy Framework MPLS Information Model for QoS and TE*, IETF draft, draft-chadha-policy-mpls-te-01.txt, December 2000.
9. T. Hamada, P. Czezowski, T. Chujo, *A Policy-enabled GMPLS-based Control Plane for Bandwidth Brokering*, NOMS 2002 (IEEE2002).
10. B. Daheb, F. Wissam, M. Du-Pond, A. Olivier, B. Berde, M. Vigoureux, G. Pujolle, *Service Level Agreement in Optical Networks*, IEEE-IFIP Net-Con 2003, Kluwer ed.
11. A. Banerjee, J. Drake, J. Lang, B. Turner, K. Kompella, Y. Rekhter, *Generalized Multiprotocol Label Switching: An Overview of Routing and Management Enhancements*, IEEE Communications Magazine, January 2001.
12. A. Banerjee, J. Drake, J. Lang, B. Turner, D. Awduche, L. Berger, K. Kompella, Y. Rekhter, *Generalized Multiprotocol Label Switching: An Overview of Signaling Enhancements and Recovery Techniques*, IEEE Communications Magazine, July 2001.
13. D. Papadimitriou, M. Vigoureux, E. Dotaro, K. Shiomoto, E. Oki, N. Matsuura, *Generalized MPLS Architecture for Multi-Region Networks*, IETF draft, draft-vigoureux-shiomoto-ccamp-gmpls-mrn-00.txt, October 2002.
14. L. Noirie, M. Vigoureux, E. Dotaro, *Impact of Intermediate Traffic Grouping on the Dimensioning of Multi-Granularity Optical Networks*, Proc. OFC 2001.

BEYOND TCP/IP: A CONTEXT-AWARE ARCHITECTURE

Guy Pujolle[1], Hakima Chaouchi[1], and Dominique Gaïti[2]
[1]LIP6-University of Paris 6, 8 rue du Capitaine Scott, 75015 Paris, France; [2]ISTIT - M2S team- UTT, 12 rue Marie Curie - 10000 Troyes, France

Abstract: To configure current networks a large number of parameters have to be taken into account. Indeed, we can optimize a network in different ways: Optimize battery capacity, optimize reliability of the network, optimize QoS, optimize the security, optimize the mobility management, and so on. In this paper we begin by proving that the TCP/IP architecture is not the best protocol in a specific wireless experiment: the energy consumption on a Wi-Fi network. Then, we propose a new architecture, Goal-Based Networking (GBN) architecture, using adaptable protocols named STP/SP protocols (Smart Transport Protocol/Smart Protocol) able to optimize locally the communications through the networks. Finally we discuss the pros and cons of this new architecture.

Key words: TCP/IP, energy consumption, wireless networks, protocol, goal-based management.

1. INTRODUCTION

Wireless communications are more and more involved in different applications, and are already part of our daily lives. These new wireless technologies are growing faster than the networking technologies. In fact, the basic networking technology is TCP/IP communication model. This model has been designed in wired networks. Wireless networks requirements and applications were not considered during the design of this model. TCP/IP model is facing a limitation in wireless and mobile environment[1-3]. This limitation could become important in future wireless environments[4-6]. In this paper, we illustrate via Wi-Fi network measurements on energy consumption the non efficiency of TCP/IP. This proves the need for a new architecture that includes a smart mechanism able to decide which communication protocol is suitable for a certain situation (Goal) in the network, and also adapt the parameters of the selected protocols to better react to the present and future state of the network.

The rest of the paper is organized as follows. First we illustrate via a Wi-Fi measurement the problems of the TCP/IP stack. Then, we introduce the new STP/SP model, followed by the description of the smart architecture (Goal Based Networking architecture) to support the deployment of the new STP/SP model. Finally, we present an analysis of this architecture and we conclude this work.

2. TCP/IP LIMITATION IN WI-FI NETWORKS

We are looking in this section at a Wi-Fi network and we are interested in measuring energy consumption in different experiences. We measured a file transmission on a Wi-Fi network. The transmission was at a 100 mW level and the receiver was at 2 meters in a direct view so that the quality of the signal is sufficient to avoid retransmission at the MAC layer. We transmit a 100 Mbytes file fragmented into 100 bytes TCP/IP packets. To send one useful bit of the payload on a Wi-Fi wireless network of a TCP/IP packet, we got the energy of approximately 700 nJ.

As a cycle of the processor asks for approximately 0,07 nJ, the transmission for one bit of the payload is approximately equal to 10 000 cycles of the processor.

When measuring the consumption of just one bit out of the 100 Mbytes, we obtained 70 nJ for the transmission of one bit. Therefore, we can deduce that the TCP/IP environment is asking on the average 10 times more energy to transmit one useful bit than for the transmission of one bit. It has to be noted that the transmission of small IP packets asks for a large number of small signalling packets. This explains in part the high energy consumption. Another part of the energy consumption comes from the number of overhead bits produced by the TCP/IP architecture.

Several other experiments on sensor networks lead to the same outcomes.

Two significant conclusions can be provided from these measurements:

- The TCP/IP protocol over Wi-Fi and more generally over wireless systems is very energy consuming when the segmentation provides small packets.
- As it is necessary to send ten bits for one efficient bit, the question is: is it possible to find another protocol able to improve the energy consumption?

It turns out that TCP/IP is not an efficient protocol for wireless networks as soon as small packets have to be sent. This is almost expected since TCP/IP was designed to cope with wired networks.

On this example, we have shown that the TCP/IP is not very efficient for wireless networks. We can find other examples where the TCP/IP architecture is not optimal at all on QoS, reliability or security issues. The sequel of this paper is a proposal for a new architecture able to optimize not only the energy but also different performance parameters.

3. A NEW SMART ARCHITECTURE STP/SP

TCP/IP architecture was created for the interconnection of networks running with different architectures. Then, the TCP/IP architecture was chosen as the unique architecture for all communications. The advantage is clearly to permit a universal interconnection scheme of any kind of machines. However, TCP/IP is only a tradeoff and we wonder if specific architectures but, may be, IP compatible could not be a better solution to optimize the communications. The idea is to propose a Smart Protocol (SP) that can adapt to the environment, for optimizing battery or optimizing reliability or optimizing QoS or any other interesting functionality. The design of a Smart Protocol at the network layer that is aware of the upper and the lower layers and adapts their communication to a set of parameters is obviously the ultimate communication architecture that can support current and emerging wireless networks. This new context-aware architecture that we named STP/SP Smart Transport Protocol/Smart Protocol could be compatible with IP.

Indeed, the SP protocol is a set of protocols SP1, SP2, ….SPn that could be either derived from the IP protocol or could be adapted to specific environments. In the same way the STP protocol is a set of protocol that could be derived from the TCP protocol or from independent protocols. In this paper, we are interested in the compatibility of STP/SP with the TCP/IP architecture. Indeed, the TCP/IP functionalities are rich enough to cope with the different situations.

All the different architectures are easily interconnected through a classical TCP/IP protocol. For instance, a sensor network will deploy its STP/SP protocol stack that support the requirements of the application set up over the sensor network. This sensor network will be interconnected through a classical TCP/IP gateway to another network that deploys another STP/SP protocol stack which supports the requirements of this other network. This might sound as going back to the period where the networks deploy their proprietary protocols. Then, IP was designed to interconnect these networks. Next IP was generalized and today reached the point where this protocol cannot cope with all types of environment such as wireless environments. The difference between the STP/SP approach and the former proprietary solutions is that STP/SP will basically use the TCP/IP concepts and functionalities, but in a smart way. In fact, rather than deploying TCP/IP in the same way in any environment without being aware of the requirements of this environment, STP/SP will offer a smart TCP/IP like environment. This will keep the simplicity and efficiency of TCP/IP, but will add a smart process that is totally absent in TCP/IP. This smart process will be deployed using a new architecture in the network guided by a set of objectives named Goals.

We describe this global architecture in Figure 1. The objective of this architecture is to implement the smart process of selecting the sub-protocol of the STP/SP protocol that fulfils the requirements of the concerned network. This is a goal-based networking architecture and the control is a goal-based control.

4. A GOAL-BASED NETWORKING ARCHITECTURE

The goal-based architecture is composed of mainly two mechanisms: The smart mechanism to select the STP/SP protocol and its parameters, and the enforcement mechanism to enforce the decisions of the smart mechanism. We propose to use agent-based mechanism to implement the smart mechanism, and to use some concepts of the policy based networking[7] such as the enforcement procedures to implement the enforcement mechanism.

An agent-based platform permits a meta-control structure such as the platform described in[8]. Assuming that for each network node we associate one or several agents, the network can be seen as a multi-agent system in which the main goal is to decide about the control to use for optimizing a given functionality described in the goal distributed by the meta-agent.

Intelligent agents are able to acquire and to process information about situations that are "not here and not now", i.e., spatially and temporally remote. By doing so, an agent may have a chance to avoid future problems or at least to reduce the effects. These capabilities allow agents to adapt their behavior according to the traffic flows going through the node.

Finally, interruptible agents may be used. They are able to act intelligently and to respond in real time. This capability allows implementing complex mechanisms able to take care of real time decisions.

It is important to note that other works has proposed a decision mechanism in the network to enforce decision or policies in the network. This typical architecture named Policy-based Networking (PBN) enforces high level decisions without unfortunately considering the problem optimization of parameters related to lower levels of the network. It's only a top down approach. In our proposed architecture, we intend to use the enforcement procedure of policy-based networking architecture that is an interesting concept for automating the enforcement of the smart mechanism decisions. The Goal-based architecture considers the optimizing problem related to the higher but also the lower layers of the network, and enforces the most suitable STP/SP protocols and parameters for the given network and application.

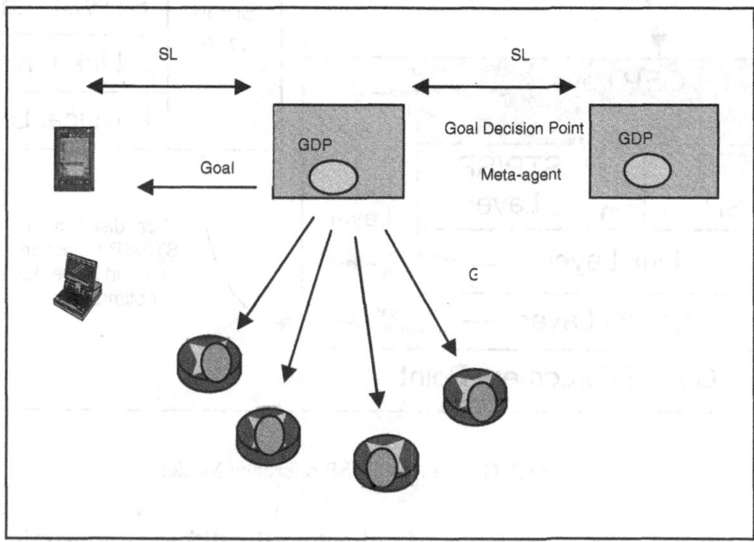

Figure 1. The global architecture.

5. THE PROPOSED GOAL-BASED ARCHITECTURE

Figure 1 depicts the global Goal-based architecture (GBN) and Figure 2 depicts the GBN and STP/SP reference model.

First, users can enter their SLA through a Web service scheme for example. The manager of the network can also enter the network configurations corresponding to the goals of the network. A meta-agent in the smart Layer is able to decide about the global goal of the network. This meta-agent is supported by any kind of centralized servers if any. As soon as defined, the goal is distributed to the different network nodes or Goal Enforcement Point (GEP). Knowing the goal, the different nodes have to apply policies and then configuration to reach the goal.

The agents in the GEP are forming a multi-agent system. The problem is to group these agents for providing a global task. The Distributed Artificial Intelligence (DAI) systems may be seen as such a grouping.

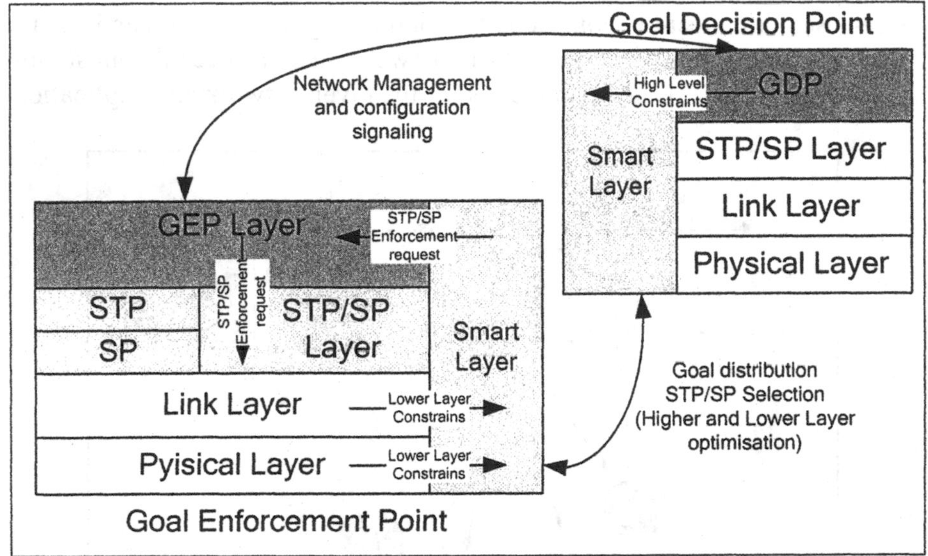

Figure 2. GBN and STP/SP reference Model

The Smart Layer is in charge of collecting the different constraints from the lower layers but also from the higher layer (business level policies), then specify and update the goal of the network which is about *what to optimise in the network and what to be offered by the network.* Note that the classical approaches consider only, what to be offered by the network. After specifying the network goal, the smart layer selects the STP/SP protocols and parameters that will optimize the specified goal. The smart layer will keep updating the goal of the network based on the current state of the network or on a new policies introduced by the Goal Decision Point.

The smart Layer can be implemented by an agent-based architecture. It is very suitable to provide the specification of the network goal and also the selection of the suitable STP/SP protocols.

The choice of the protocol can be seen at two levels: the local and the global level. One agent in each node (Smart Layer) may be defined for deciding the local protocol in cooperation with the other agent of the multi-agent system. Each agent has to perform a specific procedure, which is triggered according to the state of the node, to the QoS required, and to any other reason. This constitutes a local level for the decision. Moreover, agents can periodically interact to exchange their knowledge and ask to other agents if they need information they do not have. This constitutes the global level.

The smart layer interacts with the Goal Enforcement point (GEP) in order to enforce the STP/SP selected protocol that realizes the global goal. This implies also the definition of the algorithms to manage the CPU, the sensor, the radio or any parameter of the traffic conditioner.

Indeed, the traffic conditioner is replaced by an extended traffic conditioner (XTC) where different algorithms can be supported. The GEP is in charge to decide the value of the parameters and to decide about the protocol to be used. Within the entities that can be configured, classical droppers, meters, schedulers, markers, etc. may be found but also resource of the battery, availability, security parameters, radio parameters, etc. This XTC is shown in Figure 3.

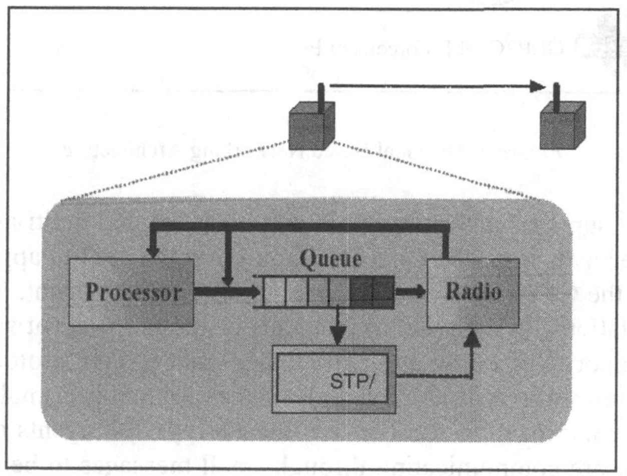

Figure 3. The Extended Traffic Conditioner XTC

6. GENERALISED GOAL-BASED ARCHITECTURE

We described in section 3, the global architecture. The agents managing the Goal Enforcement Point have to be installed in the nodes of the networks. However, an important part of the equipment is the end terminal. This terminal could be either powered or not. So, the routers cannot decide about the best protocol to use without knowing some characteristics of the terminal. If, for example, the terminal is a wireless terminal with low-level batteries, the goal of the network could be to take care of the consumption of the batteries and to decide to optimize the lifetime of the system. So it is necessary to implement a Goal Enforcement Point (GEP) in the terminal equipment. This architecture is presented in Figure 4.

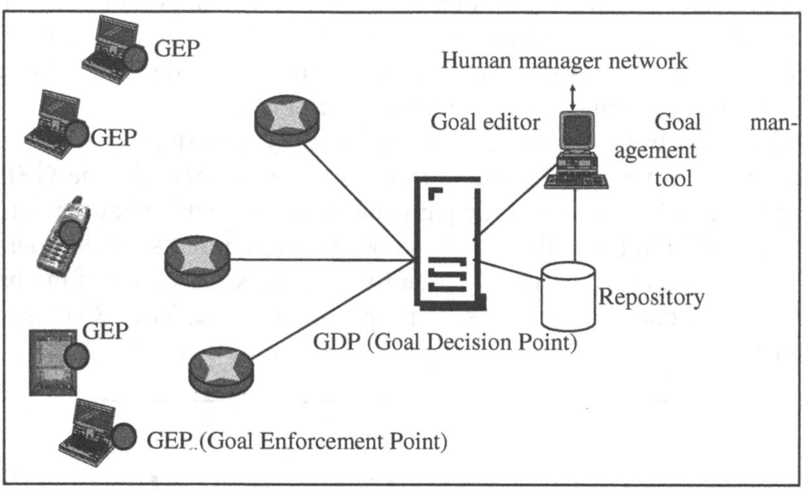

Figure 4. The Goal Based Networking Architecture

The GEP supports an agent that is part of the global multi-agent system. The agent receives from the Goal Decision Point the goal to apply. The goal received by the GEP has to be implemented through the agent. So, the agent using the relationship with the other agents of the multi-agent system has to choose the algorithms to use and particularly what STP/IP protocol is appropriate to optimize the goal. For example, if the terminal is a small portable or a sensor, the optimization can concern the energy. The agents of the multi-agent system are communicating through small messages to be coordinated. The agent is also deciding about the values of the different parameters of the protocol. For security reasons the agent is stocked in a smartcard connected to the terminal as an USB key for example. The user cannot modify the agent and the goal due to the restricted access to the card. The smartcard is also a way to authenticate the user and to apply a goal that could be dependent of the profile of the user memorized in the smartcard. The architecture with the smartcard is shown in Figure 5.

Two kinds of agents are defined in our architecture:

- The meta-agent which monitors the other agents that sends goal and receives alarms;
- Intelligent cooperative agents deciding about protocols to be used and monitoring local parameters of the extended traffic controller.

We can distinguish two levels of decision within a network node. These two levels are the following:

- A control mechanism at the lower level. This level is composed of the different node control mechanisms currently activated. Each control mechanism has its own parameters, conditions and actions, which can be monitored and manipulated by the entity lying at the higher level, the meta-agent. The functioning of a control mechanism is limited to the execution of the loop (goal and condition → actions). This is realized by a reactive agent. There is a causal link between the different stimuli (events) that a control mechanism recognizes and the actions it executes.
- The meta-agent chooses the goals to undertake by consulting the current system state (nodes state, requested SLA, goal of the company) and goals memorized in a goal repository. The meta-agent activates goals.

The nodes, thanks to the two control levels, respond to internal events (loss percentage for a class of traffic, load percentage of a queue, etc.) and external ones (a message sent by a neighbor node, reception of a new goal, etc.).

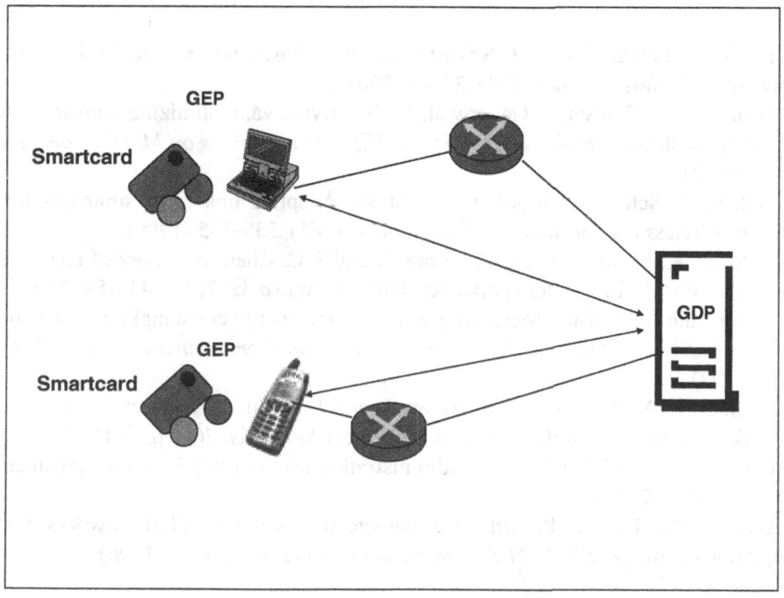

Figure 5. The architecture with the smartcard

7. CONCLUSIONS

This paper introduced a new communication architecture to better support wireless networks and new applications. STP/SP (Smart Transport Protocol/Smart Protocol) is a smart communication model that will use different transport and network protocols for each wireless environment. These protocols consider not only the policies provided by the business plan but also the constraints of the lower layers of the network. A smart architecture is proposed to provide the selection mechanism of the corresponding protocols and parameters. This smart architecture interacts with and enforcement architecture in order to configure the network with the selected protocols and parameters. A first analysis of our architecture, shows that smart selection of the communication protocols to use for a certain network environment and application bring better results than the classical TCP/IP architecture. Further research intend to focus on different wireless networks environments and design the STP/SP corresponding architecture that optimizes all the lower and higher layers parameters. The first considered environment is sensor networks.

REFERENCES

1. A. Boulis, P. Lettieri, Mani B. Srivastava, Active Base Stations and Nodes for Wireless Networks, *Wireless Networks* **9**(1), 37-49 (2003).
2. C. Schurgers, V. Tsiatsis, S. Ganeriwal, M. B. Srivastava, Optimizing Sensor Networks in the Energy-Latency-Density Design Space, *IEEE Transactions on Mobile Computing* **1**(1), 70-80 (2002).
3. P. Lettieri, C. Schurgers, Mani B. Srivastava, Adaptive link layer strategies for energy efficient wireless networking, *Wireless Networks* **5**(5), 339-355 (1999).
4. C. E. Jones, K. M. Sivalingam, P. Agrawal, and J. C. Chen, A survey of energy efficient network protocols for wireless networks, *Wireless Networks*, 7(4), 343-358 (2001).
5. M. Stemm and R. H. Katz, Measuring and reducing energy consumption of network interfaces in hand-held devices, *IEICE Transactions on Communications*, E80-B(8), 1125-1131 (1997).
6. C. E. Jones, K. M. Sivalingam, P. Agrawal, and J. C. Chen, A survey of energy efficient network protocols for wireless networks, *Wireless Networks*, 7(4), pp. 343-358, July 2001.
7. D. C.Verma, Simplifying Network administration using policy-based management, *IEEE Network* **16**(2), (2002).
8. D. Gaïti, and G. Pujolle, Performance management issues in ATM networks: traffic and congestion control, *IEEE/ACM Transactions on Networking*, 4(2), (1996).

Index of Contributors